Student's Solutions Manual to Accompany
APPLIED MATHEMATICS FOR BUSINESS AND THE SOCIAL AND NATURAL SCIENCES

Chester Piascik
Bryant College

Prepared by
Gloria Langer

West Publishing Company
Minneapolis/St. Paul   New York   Los Angeles   San Francisco

WEST'S COMMITMENT TO THE ENVIRONMENT

In 1906, West Publishing Company began recycling materials left over from the production of books. This began a tradition of efficient and responsible use of resources. Today, up to 95% of our legal books and 70% of our college texts are printed on recycled, acid-free stock. West also recycles nearly 22 million pounds of scrap paper annually—the equivalent of 181,717 trees. Since the 1960s, West has devised ways to capture and recycle waste inks, solvents, oils, and vapors created in the printing process. We also recycle plastics of all kinds, wood, glass, corrugated cardboard, and batteries, and have eliminated the use of styrofoam book packaging. We at West are proud of the longevity and the scope of our commitment to our environment.

Production, Prepress, Printing and Binding by West Publishing Company.

Contents

Student's Solution Manual

CHAPTER
R
ALGEBRA REVIEW

R–1 The Real Numbers and Interval Notation

1. True

3. False, since –2 > –5.

5. True

7. True

9. True

11. False, since 8 < 10.

13. True

15. True

17. True

19. False, since the whole number "0" is not a counting number.

21. True

23. True

25. True

27. False, since 4.7065 is a rational number.

29. $-5 \le x \le -1$

31. $-4 < x < -2$

33. $-3 < x \le 2$

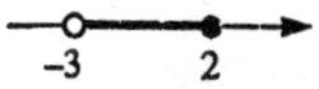

35. $5 \le x$

37. $x \le -3$

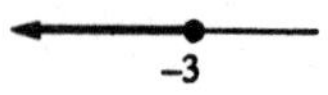

39. $x > -2$

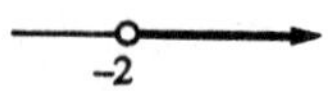

41. $2 < x$

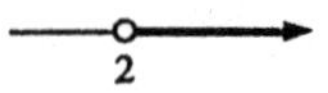

43. $x \ne 2$

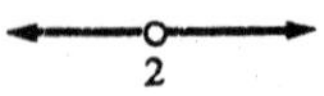

45. Closed

47. Open

49. Closed

51. $[3, 9]$

53. $(-\infty, -4]$

55. $(-\infty, 6)$

57. $(4, 9)$

59. $|0| = 0$

61. $|1| = 1$

63. $|-2| = -(-2) = 2$

65. $|-15| = -(-15) = 15$

67. $|20| = 20$

69. $|9 - 5| = |4| = 4$

71. $|7 - 16| = |-9| = -(-9) = 9$

73. $|9 - 15| = |-6| = -(-6) = 6$

75. $|-4 - 9| = |-13| = -(-13) = 13$

77. Distance $= |5 - 11| = |-6| = -(-6) = 6$

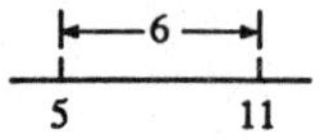

79. Distance $= |-9 - (-4)| = |-9 + 4| = |-5|$
$= -(-5) = 5$

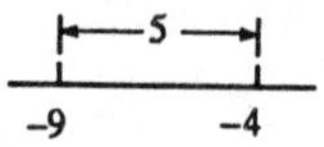

R–2 Linear Equations and Inequalities

1. $4x + 8 = 32$
$4x = 24$
$x = 6$

3. $-2x + 6 = -12$
$-2x = -18$
$x = 9$

5. $4x - 3 = 21 + 2x$
$2x = 24$
$x = 12$

7. $\frac{1}{2}x + 6 = 10$

$\frac{1}{2}x = 4$
$x = 8$

9. $5y + 2 = 7y - 18$
$-2y = -20$
$y = 10$

11. $2x + 4 \leq 15$
$2x \leq 11$

$x \leq \dfrac{11}{2}$ or $\left[-\infty, 5\dfrac{1}{2}\right]$

$5\dfrac{1}{2}$

13. $4x - 5 < 25$
$4x < 30$

$x < \dfrac{15}{2}$ or $\left[-\infty, 7\dfrac{1}{2}\right]$

$7\dfrac{1}{2}$

15. $-3x + 17 \geq -14$
$-3x \geq -31$

$x \leq \dfrac{31}{3}$ or $\left[-\infty, 10\dfrac{1}{3}\right]$

$10\dfrac{1}{3}$

17. $-6x - 5 \geq -23$
$-6x \geq -18$
≤ 3

3

19. $3(x - 5) \geq 18$
$x - 5 \geq 6$
$x \geq 11$

11

R–3 Exponents and Radicals

1. $3^2 = 9$

3. $(-5)^2 = 25$

5. $(4^2)^3 = 4^{2 \cdot 3} = 4^6 = 4096$

7. $2^{-4} = \dfrac{1}{2^4} = \dfrac{1}{16}$

9. $\dfrac{2^7}{2} = 2^{7-1} = 2^6 = 64$

11. $5^3 \cdot 5^4 = 5^7 = 78{,}125$

13. $\left[\dfrac{3}{5}\right]^2 = \dfrac{3^2}{5^2} = \dfrac{9}{25}$

15. $(2^{-3})^2 = 2^{-6} = \dfrac{1}{2^6} = \dfrac{1}{64}$

17. $64^{-\frac{1}{2}} = \dfrac{1}{64^{\frac{1}{2}}} = \dfrac{1}{\sqrt{64}} = \dfrac{1}{8}$

19. $16^{\frac{1}{2}} = 4$

21. $49^{\frac{1}{2}} = 7$

23. $64^{\frac{5}{2}} = (8^2)^{\frac{5}{2}} = 8^5 = 32{,}768$

25. $49^{\frac{3}{2}} = (7^2)^{\frac{3}{2}} = 7^3 = 343$

27. $216^{\frac{2}{3}} = (6^3)^{\frac{2}{3}} = 6^2 = 36$

29. $867^0 = 1$

31. $49^{\frac{5}{2}} = (7^2)^{\frac{5}{2}} = 7^5 = 16{,}807$

33. $\dfrac{1}{5^6} = 5^{-6}$

35. $\dfrac{1}{(-5)^3} = (-5)^{-3}$

37. $\dfrac{1}{x^8} = x^{-8}$

39. $(\sqrt{4})^9 = 4^{\frac{9}{2}}$

41. $\sqrt[5]{2} = 2^{\frac{1}{5}}$

43. $\sqrt[7]{9^4} = 9^{\frac{4}{7}}$

45. $(\sqrt[3]{5})^7 = 5^{\frac{7}{3}}$

47. $\dfrac{1}{\sqrt{5^3}} = \dfrac{1}{5^{\frac{3}{2}}} = 5^{-\frac{3}{2}}$

49. $\dfrac{1}{\sqrt{x^5}} = \dfrac{1}{x^{\frac{5}{2}}} = x^{-\frac{5}{2}}$

51. $\dfrac{1}{\sqrt[3]{x^2}} = \dfrac{1}{x^{\frac{2}{3}}} = x^{-\frac{2}{3}}$

53. $(-3\,x)^2 = (-3)^2 x^2 = 9x^2$

55. $(5xy)^3 = 5^3 x^3 y^3 = 125 x^3 y^3$

57. $(4 \cdot 81)^{\frac{1}{2}} = 4^{\frac{1}{2}} \cdot 81^{\frac{1}{2}}$

59. $\left[\dfrac{5}{6}\right]^3 = \dfrac{5^3}{6^3} = \dfrac{125}{216}$

61. $\left[\dfrac{x}{3}\right]^5 = \dfrac{x^5}{3^5} = \dfrac{x^5}{243}$

63. $\left[\dfrac{x}{5}\right]^2 = \dfrac{x^2}{5^2} = \dfrac{x^2}{25}$

65. $\left[\dfrac{-2}{x}\right]^3 = \dfrac{(-2)^3}{x^3} = \dfrac{-8}{x^3}$

67. $\left[\dfrac{2^{-3} \cdot 2^5}{2^{-2}}\right]^3 = (2^{-3+5-(-2)})^3 = (2^4)^3 = 2^{12} = 4096$

69. $\dfrac{3^{-\frac{7}{2}} \cdot 3^{\frac{5}{2}}}{3^{\frac{1}{2}} \cdot 3^{-\frac{3}{2}}} = \dfrac{3^{-\frac{7}{2}+\frac{5}{2}}}{3^{\frac{1}{2}-\frac{3}{2}}} = \dfrac{3^{-2}}{3^{-1}} = 3^{-2-(-1)} = 3^{-1} = \dfrac{1}{3}$

71. $496 = 4.96 \times 10^2$

73. $8{,}000{,}000{,}000 = 8 \times 10^9$

75. $0.0000008 = 8 \times 10^{-7}$

77. $0.56 = 5.6 \times 10^{-1}$

79. $0.00357 = 3.57 \times 10^{-3}$

R–4 The Distributive Law and Factoring

1. $3(x + 4) = 3x + 3 \cdot 4 = 3x + 12$

3. $-2(x + 6) = -2x - 2 \cdot 6 = -2x - 12$

5. $9(x + 3) = 9x + 9 \cdot 3 = 9x + 27$

7. $-4(x^2 - 3x + 7) = -4x^2 - 4(-3x) + (-4)(7)$
$= -4x^2 + 12x - 28$

9. $5x(x^3 - 4x^2 + 4x + 5)$
$= 5x(x^3) - (5x)(4x^2) + 5x(4x) + 5x(5)$
$= 5x^4 - 20x^3 + 20x^2 + 25x$

11. $3x^2y^3(x^4 - 5y^2 + 6xy)$
$= 3x^2y^3(x^4) - 3x^2y^3(5y^2) + 3x^2y^3(6xy)$
$= 3x^6y^3 - 15x^2y^5 + 18x^3y^4$

13. $5x + 20 = 5x + 5(4) = 5(x + 4)$

15. $3x - 27 = 3x - (3)(9) = 3(x - 9)$

17. $6x - 30 = 6x - (6)(5) = 6(x - 5)$

19. $3x^2 - 27x = 3x(x) - (3x)(9) = 3x(x - 9)$

21. $-6x^2 + 48x = (-6x)(x) - (-6x)(8) = -6x(x - 8)$

23. $7x^2 + 28x = 7x(x) + (7x)(4) = 7x(x + 4)$

25. $3x^2y^4 + 6xy^2 = 3xy^2(xy^2) + 3xy^2(2) = 3xy^2(xy^2 + 2)$

27. $-5x^4y^6 + 20x^3y^7 = (-5x^3y^6)(x) - (-5x^3y^6)(4y)$
$= -5x^3y^6(x - 4y)$

29. $3x^2y^6 + 9x^5y^3 + 6x^3y^4$
$= 3x^2y^3(y^3) + 3x^2y^3(3x^3) + 3x^2y^3(2xy)$
$= 3x^2y^3(y^3 + 3x^3 + 2xy)$

31. $P + Prt = P(1 + rt)$

33. $P + Pi = P(1 + i)$

35. $ax^2 - bx = x(ax - b)$

37. $P(1 + i)^2 + P(1 + i)^2i = P(1 + i)^2(1 + i) = P(1 + i)^3$

39. $3x(x - 7) + 3x(x + 4) = 3x(x - 7 + x + 4)$
$= 3x(2x - 3)$

41. $2x(x + 3) - y(x + 3) = (x + 3)(2x - y)$

43. $5xy(x - 2) + 8x^2(x - 2) = (x - 2)(5xy + 8x^2)$
$= x(x - 2)(5y + 8x)$

R–5 Multiplying Binomials

1. $(x - 2)(x + 3) = x^2 + 3x - 2x - 6$
$= x^2 + x - 6$

3. $(x + 1)(x + 5) = x^2 + 5x + x + 5$
$= x^2 + 6x + 5$

5. $(x - 8)(x - 7) = x^2 - 7x - 8x + 56$
$= x^2 - 15x + 56$

7. $(3x + 5)(x - 1) = 3x^2 - 3x + 5x - 5$
$= 3x^2 + 2x - 5$

9. $(4x - 7)(2x + 3) = 8x^2 + 12x - 14x - 21$
$= 8x^2 - 2x - 21$

11. $(7x - 2)(3x + 1) = 21x^2 + 7x - 6x - 2$
$= 21x^2 + x - 2$

13. $(x - 3)(x + 3) = x^2 - 9$

15. $(x - 9)(x + 9) = x^2 - 81$

17. $(3x - 2)(3x + 2) = 9x^2 - 4$

19. $(x - 2)^2 = x^2 + 2(-2x) + 4$
$= x^2 - 4x + 4$

21. $(x + 5)^2 = x^2 + 2(5x) + 25$
$= x^2 + 10x + 25$

23. $(2x - 3)^2 = 4x^2 + 2(-6x) + 9$
$= 4x^2 - 12x + 9$

R–6 Factoring

1. $x^2 + 7x - 18 = (x + 9)(x - 2)$

3. $x^2 + 2x - 15 = (x + 5)(x - 3)$

5. $x^2 - 3x + 2 = (x - 2)(x - 1)$

7. $x^2 - 13x + 40 = (x - 8)(x - 5)$

9. $x^2 - 81 = (x - 9)(x + 9)$

11. $x^2 - 49 = (x - 7)(x + 7)$

13. $x^2 + 6x + 9 = x^2 + 2(3x) + 3^2$
$= (x + 3)^2$

15. $x^2 - 10x + 25 = (x - 5)^2$

17. $x^2 + 18x + 81 = (x + 9)^2$

19. $x^2 - 6x - 27 = (x - 9)(x + 3)$

21. $x^2 - 4x - 45 = (x - 9)(x + 5)$

23. $x^2 + 7x + 6 = (x + 6)(x + 1)$

R–7 More Factoring

1. $2x^2 - x - 28 = (2x + 7)(x - 4)$

3. $5x^2 + 18x - 8 = (5x - 2)(x + 4)$

5. $6x^2 - 13x - 5 = (2x - 5)(3x + 1)$

7. $9x^2 - 36 = 9(x^2 - 4)$
$= 9(x - 2)(x + 2)$

9. $4x^2 + 20x + 25 = (2x + 5)^2$

11. $25x^2 - 70x + 49 = (5x - 7)^2$

R–8 Rational Expressions

1. $\dfrac{4x - 28}{2} = \dfrac{4(x - 7)}{2} = 2(x - 7) = 2x - 14$

3. $\dfrac{-(2x + 14)}{2} = \dfrac{-2(x + 7)}{2} = -(x + 7) = -x - 7$

5. $\dfrac{x^2 - 6x}{x - 6} = \dfrac{x(x - 6)}{x - 6} = x$

7. $\dfrac{x^2 + 2x - 15}{x^2 - 25} = \dfrac{(x + 5)(x - 3)}{(x + 5)(x - 5)} = \dfrac{x - 3}{x - 5}$

9. $\dfrac{x^2 + 3x - 4}{x^2 - 1} = \dfrac{(x + 4)(x - 1)}{(x + 1)(x - 1)} = \dfrac{x + 4}{x + 1}$

11. $\dfrac{x^2 + 8x + 15}{x^2 + 7x + 10} = \dfrac{(x + 5)(x + 3)}{(x + 5)(x + 2)} = \dfrac{x + 3}{x + 2}$

13. $\dfrac{9}{6x^2} \cdot \dfrac{2x}{3} = \dfrac{18x}{18x^2} = \dfrac{1}{x}$

15. $\dfrac{x^2 - 4x - 5}{x^2 + 3x + 2} \cdot \dfrac{x^2 + 5x + 6}{x^2 - 7x + 10}$

$= \dfrac{(x-5)(x+1)}{(x+2)(x+1)} \cdot \dfrac{(x+3)(x+2)}{(x-5)(x-2)} = \dfrac{x+3}{x-2}$

17. $\dfrac{x^2 - 81}{x^2 + 10x + 9} \div \dfrac{x^2 - 7x - 18}{x^2 + 4x + 3}$

$= \dfrac{x^2 - 81}{x^2 + 10x + 9} \cdot \dfrac{x^2 + 4x + 3}{x^2 - 7x - 18}$

$= \dfrac{(x-9)(x+9)}{(x+9)(x+1)} \cdot \dfrac{(x+1)(x+3)}{(x-9)(x+2)}$

$= \dfrac{x+3}{x+2}$

19. $\dfrac{x+5}{4} + \dfrac{x+7}{4} = \dfrac{x+5+x+7}{4} = \dfrac{2x+12}{4}$

$= \dfrac{2(x+6)}{4} = \dfrac{x+6}{2}$

21. $\dfrac{5}{x} + \dfrac{8}{2x} = \dfrac{10}{2x} + \dfrac{8}{2x} = \dfrac{18}{2x} = \dfrac{9}{x}$

23. $\dfrac{5}{x+6} - \dfrac{8}{x} = \dfrac{5x}{x(x+6)} - \dfrac{8(x+6)}{x(x+6)}$

$= \dfrac{5x - 8x - 48}{x(x+6)}$

$= \dfrac{-3x - 48}{x(x+6)}$ or $\dfrac{-3(x+16)}{x(x+6)}$

25. $\dfrac{5}{x^2 - 36} + \dfrac{9}{x+6} = \dfrac{5}{(x-6)(x+6)} + \dfrac{9(x-6)}{(x+6)(x-6)}$

$= \dfrac{5 + 9x - 54}{(x-6)(x+6)}$

$= \dfrac{9x - 49}{(x-6)(x+6)}$ or $\dfrac{9x - 49}{x^2 - 36}$

27. $\dfrac{7}{x-7} - \dfrac{5}{x(x-9)} = \dfrac{7x}{x(x-9)} - \dfrac{5}{x(x-9)}$

$= \dfrac{7x - 5}{x(x-9)}$

29. $\dfrac{2x+7}{x} = \dfrac{2x}{x} + \dfrac{7}{x} = 2 + \dfrac{7}{x}$

31. $\dfrac{3x^2 - 36x}{x} = \dfrac{3x^2}{x} - \dfrac{36x}{x} = 3x - 36$

33. $\dfrac{4x^3 - 8x^2 + 6x}{x} = \dfrac{4x^3}{x} - \dfrac{8x^2}{x} + \dfrac{6x}{x} = 4x^2 - 8x + 6$

35. $4x^2 \left[1 + \dfrac{2}{x} - \dfrac{3}{x^2} \right] = 4x^2(1) + 4x^2 \left[\dfrac{2}{x} \right] - 4x^2 \left[\dfrac{3}{x^2} \right]$

$= 4x^2 + 8x - 12$

37. $5x^3 \left[1 - \dfrac{1}{x} + \dfrac{6}{x^2} - \dfrac{2}{x^3} \right]$

$= (5x^3)(1) - (5x^3) \left[\dfrac{1}{x} \right] + (5x^3) \left[\dfrac{6}{x^2} \right] - (5x^3) \left[\dfrac{2}{x^3} \right]$

$= 5x^3 - 5x^2 + 30x - 10$

39. $x^3 - 4x^2 + 7x + 5 = x^3 \dfrac{(x^3 - 4x^2 + 7x + 5)}{x^3}$

$= x^3 \left[\dfrac{x^3}{x^3} - \dfrac{4x^2}{x^3} + \dfrac{7x}{x^3} + \dfrac{5}{x^3} \right]$

$= x^3 \left[1 - \dfrac{4}{x} + \dfrac{7}{x^2} + \dfrac{5}{x^3} \right]$

41. $x^2 + 7x + 9 = x^2 \dfrac{(x^2 + 7x + 9)}{x^2}$

$= x^2 \left[\dfrac{x^2}{x^2} + \dfrac{7x}{x^2} + \dfrac{9}{x^2} \right]$

$= x^2 \left[1 + \dfrac{7}{x} + \dfrac{9}{x^2} \right]$

43. $\dfrac{(1+i)^{20}-1}{i} \div (1+i)^{20} = \dfrac{(1+i)^{20}-1}{i} \cdot \dfrac{1}{(1+i)^{20}}$

$$= \dfrac{(1+i)^{20}-1}{i(1+i)^{20}}$$

$$= \dfrac{\dfrac{(1+i)^{20}}{(1+i)^{20}} - \dfrac{1}{(1+i)^{20}}}{i}$$

$$= \dfrac{1-(1+i)^{-20}}{i}$$

45. $\dfrac{(1+i)^{39}-(1+i)}{i} \div (1+i)^{38}$

$$= \dfrac{(1+i)^{39}-(1+i)}{i} \cdot \dfrac{1}{(1+i)^{38}}$$

$$= \dfrac{\dfrac{(1+i)^{39}}{(1+i)^{38}} - \dfrac{(1+i)}{(1+i)^{38}}}{i}$$

$$= \dfrac{(1+i)-(1+i)^{-37}}{i}$$

Extra Dividends – Percent Change

1. percent change $= \dfrac{\text{new value} - \text{old value}}{\text{old value}} \times 100$

$$= \dfrac{50-40}{40} \times 100$$

$$= 25\%$$

3. $NV = OV(1 + p)$

$$= 60(1 + 0.20)$$

$$= 72$$

The stock's new price is $72.

5. $OV = \dfrac{NV}{1+p}$

$$= \dfrac{600}{1+0.40}$$

$$= 428.57143$$

The old price of one ounce of gold at the beginning of the month is $428.57.

7. Fidelity Magellan

$NV = 10,000(1 + 0.480)$

$= 14,800$

new value: $14,800

9. Mutual Shares

$NV = 10,000(1 + 0.605)$

$= 16,050$

new value: $16,050

11. United States from 1975 to 1980:

percent change $= \dfrac{NV - OV}{100} \times 100$

$$= \dfrac{477,771 - 213,992}{213,992} \times 100$$

$$= 123.3\%$$

13. Japan from 1975 to 1980:

percent change $= \dfrac{271,737 - 113,569}{113,569} \times 100$

$$= 139.3\%$$

15. Soviet Union from 1980 to 1985:

percent change $= \dfrac{65,967 - 67,059}{67,059} \times 100$

$$= -1.6\%$$

17. General Electric:

$OV = \dfrac{NV}{1 + p}$

$OV = \dfrac{100,000}{1 + 4.51}$

$= 18,148.82$

The value at the beginning of the decade was $18,148.82.

19. AT&T:

$$OV = \frac{100,000}{1 + 3.48}$$

$$= 22,321.429$$

The value at the beginning of the decade was

$22,321.43.

Extra Dividends – Using Index Numbers to Measure Change

1. $I_{70,88}$ $= \dfrac{V_{88}}{V_{70}} \times 100$

$$= \frac{140}{80} \times 100 = 175\%$$

3. $V_{70} = \dfrac{V_{88}}{I_{80,88}} \times 100$

$$= \frac{33,000,000}{280} \times 100 \approx 8,214,285.7; \; \$8,214,286$$

5. a. percent increase in nominal income

$$= \frac{30,000 - 15,000}{15,000} \times 100 = 100\%$$

   b. percent increase in real income

$$= \frac{\dfrac{30,000}{170} \times 100 - 15,000}{15,000} \times 100 = 17.6\%$$

7. 1.18

9. 1.35

11. 1.1

13. 1.32

15. $V_{84} = \dfrac{\$1000}{311.1} \times 100 = \$321.44$

17. $V_{80} = \dfrac{\$1000}{246.8} \times 100 = \$405.19$

CHAPTER
1
FUNCTIONS AND LINEAR MODELS

1–1 Functions

1. $y = 3x - 2$

   (a) $x = 0$ is associated with $y = -2$

   (b) $x = 4$ is associated with $y = 10$

3. If $f(x) = -2x + 7$,

   (a) $f(0) = 7$

   (b) $f(1) = -2(1) + 7 = 5$

   (c) $f(5) = -3$

   (d) $f(-3) = 13$

5. If $z(t) = \dfrac{5}{t + 7}$

   (a) $z(0) = \dfrac{5}{7}$

(continued)

5. *(continued)*

   (b) $z(1) = \dfrac{5}{(1 + 7)} = \dfrac{5}{8}$

   (c) $z(-6) = \dfrac{5}{(-6 + 7)} = 5$

   (d) $z(8) = \dfrac{5}{(8 + 7)} = \dfrac{1}{3}$

7.

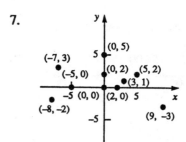

9. $f(x) = 3x + 2$

x	$f(x)$
0	2
1	5
2	8
3	1 1

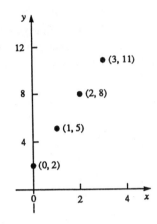

11. If $f(x) = \dfrac{5}{(x-2)(x+7)}$, the domain of f is the set of real numbers x such that $x \neq 2$ and $x \neq -7$.

13. If $f(x) = \sqrt{x - 2}$, the domain of f is the set of real numbers x such that $x \geq 2$.

15. If $h(x) = \dfrac{8}{(x - 3)^2}$, the domain of h is the set of real numbers x such that $x \neq 3$.

17. Graphs a, c, and d are those of functions. In graphs b, e, and f, at least one value of x corresponds to more than one value of y.

19. The equation $y^2 = x + 5$ does not define a function, since its graph fails the vertical line test. If $x = -1$, for example, $y = \pm 2$

21. $f(x) = x^2 - 4x + 5$

    (a) $\begin{aligned} f(x + h) &= (x + h)^2 - 4(x + h) + 5 \\ &= x^2 + 2xh - 4x + h^2 - 4h + 5 \\ &= x^2 + x(2h - 4) + h^2 - 4h + 5 \end{aligned}$

    (b) $\begin{aligned} f(x + h) - f(x) &= f(x + h) - (x^2 - 4x + 5) \\ &= 2hx + h^2 - 4h \\ &= h(2x - 4) + h^2 \end{aligned}$

    (c) $\dfrac{f(x + h) - f(x)}{h} = 2x - 4 + h$

23. $f(x) = 5x^2 - 2x + 4$

    $\dfrac{f(x + h) - f(x)}{h}$

    $= \dfrac{5(x + h)^2 - 2(x + h) + 4 - (5x^2 - 2x + 4)}{h}$

    $= \dfrac{10xh + 5h^2 - 2h}{h} = 10x - 2 + 5h$

25. $g(x) = x^3 - 4x^2 + 5x - 9$

    $\dfrac{g(x + h) - g(x)}{h}$

    $= \dfrac{(x + h)^3 - 4(x + h)^2}{h}$

    $+ \dfrac{5(x + h) - 9 - (x^3 - 4x^2 + 5x - 9)}{h}$

    $= \dfrac{3x^2h + 3xh^2 + h^3 - 8xh - 4h^2 + 5h}{h}$

    $= 3x^2 - 3xh + h^2 - 8x - 4h + 5$

    $= 3x^2 - 8x + 5 + h(3x - 4) + h^2$

27. $h(x) = \begin{cases} 2 \text{ if } x < 1 \\ 4 \text{ if } 1 \le x < 6 \\ x \text{ if } x \ge 6 \end{cases}$

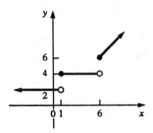

29. (a) $c(w) = \begin{cases} 1.25 \text{ if } w < 8 \\ 2.00 \text{ if } 8 \le w \le 16 \\ 5.00 \text{ if } w > 16 \end{cases}$

(b)

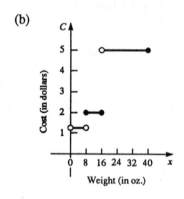

Weight (in oz.)

31. Let x = price per share

$$P(x) = \begin{cases} -7 \text{ if } x \ge 35 \\ 35 - (x + 7) \text{ if } x < 35 \end{cases}$$

or,

$$P(x) = \begin{cases} 28 - x \text{ if } x < 35 \\ -7 \text{ if } x \ge 35 \end{cases}$$

Price per share (in dollars)

1–2 Slope and Equations of Straight Lines

1. (a) $\Delta y = 8 - 5 = 3$

(b) $\Delta x = 6 - 2 = 4$

(c) slope $= \dfrac{\Delta y}{\Delta x} = \dfrac{3}{4} = 0.75$

(d) The slope is the increase in the distance of the ant from the x–axis, measured upward, per unit increase in distance from the y–axis, measured to the right.

3. $m = \dfrac{16 - 5}{7 - 4} = \dfrac{11}{3}$

5. $m = \dfrac{5 - 3}{7 - (-1)} = \dfrac{2}{8} = \dfrac{1}{4}$

7. $m = \dfrac{-7 - 3}{-5 - (-4)} = \dfrac{-10}{-1} = 10$

9. $m = \dfrac{-9 - (-2)}{-3 - (-8)} = \dfrac{-7}{5} = -\dfrac{7}{5}$

11. (a)

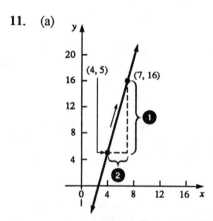

1 $16 - 5 = 11$ units vertical change

2 $7 - 4 = 3$ units horizontal change

Slope $= m = \dfrac{\Delta y}{\Delta x} = \dfrac{11}{3}$

(continued)

11. *(continued)*

(b)

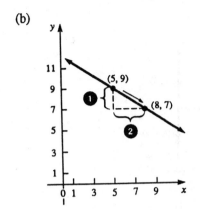

$7 - 9 = -2$ units
vertical change

$8 - 5 = 3$ units
horizontal change

Slope $= m = \dfrac{\Delta y}{\Delta x} = \dfrac{-2}{3}$

(c)

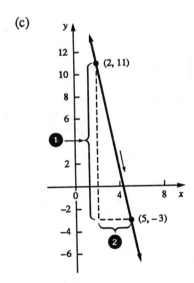

$-3 - 11 = -14$ units
vertical change

$5 - 2 = 3$ units
horizontal change

Slope $= m = \dfrac{\Delta y}{\Delta x} = \dfrac{-14}{3}$

13. $m = -\dfrac{1}{3}, \ (0,0)$

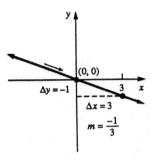

15. $m = -2, \ (2,1)$

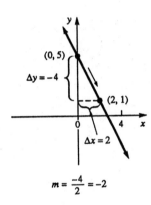

17. $m = -\dfrac{3}{4}, \ (4,-5)$

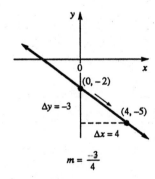

19. $m = -\dfrac{5}{3}$, $(0,2)$

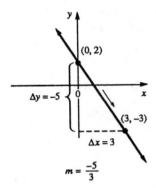

21. $m = \dfrac{-3 - (-3)}{-8 - (-9)} = 0$

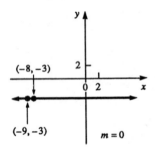

23. $m = \dfrac{-2 - (-2)}{-6 - 4} = 0$

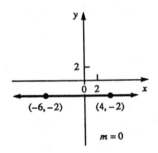

25. $m = \dfrac{-4 - 4}{6 - 6} = \dfrac{-8}{0}$, undefined

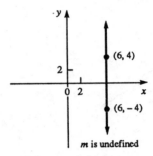

27. $m = \dfrac{9 - (-6)}{4 - 4} = \dfrac{15}{0}$, undefined

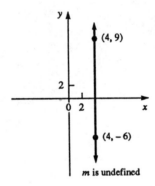

29. $(4,-1)$, $(9,-8)$

Point–slope form:

$$\frac{y - (-1)}{x - 4} = \frac{-8 - (-1)}{9 - 4} = -\frac{7}{5}; \quad y + 1 = -\frac{7}{5}(x - 4)$$

Slope–intercept form:

$$y = -1 - \frac{7}{5}(x - 4); \quad y = -\frac{7}{5}x + \frac{23}{5}$$

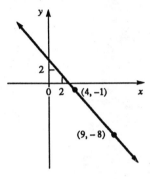

$$y + 1 = \frac{-7}{5}(x - 4)$$

$$y = \frac{-7}{5}x + \frac{23}{5}$$

31. $(3,2)$, $(7,10)$

Point–slope form:

$$\frac{y - 2}{x - 3} = \frac{10 - 2}{7 - 3} = 2; \quad y - 2 = 2(x - 3)$$

Slope–intercept form:

$$y = 2 + 2(x - 3); \quad y = 2x - 4$$

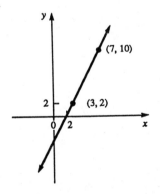

$$y - 2 = 2(x - 3)$$

$$y = 2x - 4$$

33. $(5,5)$, $(7,7)$

Point–slope form:

$$\frac{y - 5}{x - 5} = \frac{7 - 5}{7 - 5} = 1; \quad y - 5 = x - 5$$

$$y = x$$

Slope–intercept form:

$$y = -5 + x - 5; \quad y = x - 10$$

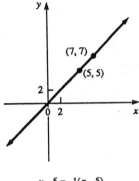

$$y - 5 = 1(x - 5)$$

$$y = x$$

35. $(5,-3)$, $(4,2)$

Point–slope form:

$$\frac{y - 2}{x - 4} = \frac{2 - (-3)}{4 - 5} = -5; \quad y - 2 = -5(x - 4)$$

Slope–intercept form:

$$y = 2 - 5(x - 4); \quad y = -5x + 22$$

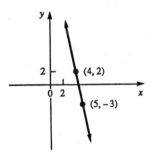

$$y - 2 = -5(x - 4)$$

$$y = -5x + 22$$

37. $(5,-1)$, $m = -3$

Point–slope form:
$$\frac{y - (-1)}{x - 5} = -3; \quad y + 1 = -3(x - 5)$$

Slope–intercept form:
$$y = -1 - 3(x - 5); \quad y = -3x + 14$$

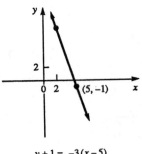

$$y + 1 = -3(x - 5)$$
$$y = -3x + 14$$

39. $(-4,-9)$, $m = 6$

Point–slope form:
$$\frac{y - (-9)}{x - (-4)} = 6; \quad y + 9 = 6(x + 4)$$

Slope–intercept form:
$$y = -9 + 6(x + 4); \quad y = 6x + 15$$

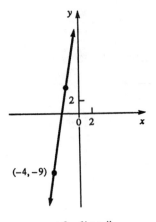

$$y + 9 = 6(x + 4)$$
$$y = 6x + 15$$

41. $(0,6)$, $m = 5$

Point–slope form:
$$\frac{y - 6}{x - 0} = 5; \quad y - 6 = 5x$$

Slope–intercept form:
$$y = 5x + 6$$

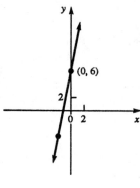

$$y - 6 = 5(x - 0)$$
$$y = 5x + 6$$

43. $(-4,7)$, $m = 1$

Point–slope form:
$$\frac{y - 7}{x - (-4)} = 1; \quad y - 7 = x + 4$$

Slope–intercept form:
$$y = 7 + x + 4; \quad y = x + 11$$

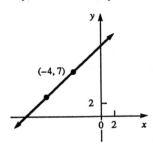

$$y - 7 = 1(x + 4)$$
$$y = x + 11$$

45. $2x + 7y = 21$; $(7,1)$, $(3,0)$, $(4,-3)$, $(14,-1)$, $(7,1)$ and $(14,-1)$ lie on the given line.

47. $y = 6x + 3$; (0,6), (1,9), $\left[-\frac{1}{2},0\right]$, (–1,–3)

(1,9), $(-\frac{1}{2},0)$, and (–1,–3) lie on the given line.

49. $g(x) = -3x$; (0,0), (1,–3), (2,–6), (5,14), (0,0), (1,–3), and (2,–6) lie on the given line.

51. (–3,1), (–3,9)
Slope undefined
$x = -3$

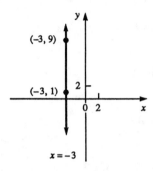

53. (4,1), (6,1)
Slope = 0
$y = 1$

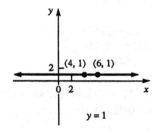

55. (–8,1), $m = 0$
$y - 1 = 0(x + 8)$; $y = 1$

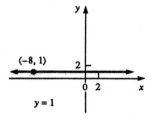

57. (7,2), m is undefined
$x = 7$

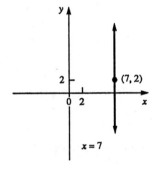

59. $y = 3x - 2;\ m_1 = 3$

$(5,6),\ m_2 = -\dfrac{1}{3};\ \ y - 6 = -\dfrac{1}{3}(x - 5)$

$y = -\dfrac{1}{3}x + \dfrac{23}{3}$

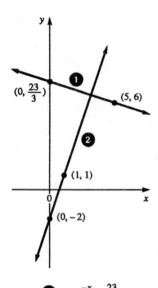

① $y = \dfrac{-x}{3} + \dfrac{23}{3}$

② $y = 3x - 2$

61. $y = 6x - 14,\ y = 6x + 13$

$m_1 = m_2 = 6;\ \ \therefore\ \dfrac{3}{2}$ the lines are parallel.

63. $y = -4x + 8,\ y = 4x + 16$

$m_1 = -4,\ m_2 = 4;\ \ \therefore\ $ the lines are not parallel.

65. $y = 3x + 5,\ y = -\dfrac{1}{3}x + 16$

$m_1 = 3,\ m_2 = -\dfrac{1}{3};\ \ \therefore\ $ the lines are perpendicular.

67. $y = 2x - 3,\ 2y + x = 10$

$m_1 = 2,\ m_2 = -\dfrac{1}{2};\ \ \therefore\ $ the lines are perpendicular

69. $3x - 5y = 11$

$y = \dfrac{3}{5}x - \dfrac{11}{5}$

71. $3x - 2y = 0$

$y = \dfrac{3}{2}x$

73. $y = 6x - 54$ is linear.

75. $y = 3x^2 - 4$ is not linear.

77. $3x - 4y = 5$ is not linear.

79. $y = x^3 - 5$ is not linear.

81.   (a) $m = \dfrac{642 - 110.7}{1989 - 1985} = \dfrac{531.3}{4} \approx 312.825$ billion dollars/year

(b) U.S. Overseas IOU's are increasing at the rate 132.825 billions dollars per year during the indicated time interval.

(c) $y - 642 = 132.825(x - 1989)$
$y = 642 + 132.825(x - 1989)$
$= 132.825x - 263,546.925$

(d) at $x = 1992,\ y = 642 + 132.825(1992 - 1989)$
$= 1040.475$ billion dollars

83.   (a) $m = \dfrac{670 - 149}{1979 - 1989} = -52.1$ defects/year

(b) Defects per 100 vehicles decreased at the rate of 52.1 per year during the indicated time interval.

(continued)

83. *(continued)*

(c) $y - 670 = -52.1(x - 1979)$ or
  $y - 149 = -52.1(x - 1989)$
  $y = 670 - 52.1(x - 1979) = -52.1x + 103,775.9$

(d) at $x = 1991$, $y = 670 - 52.1(1991 - 1979)$
  $= 44.8 \approx 45$ defects

85. (a) $m = \dfrac{22,000 - 5000}{13.3 - 6.5} = 2500$

(b) The value of the investment increased by 2500 per percentage point increase in inflation.

87. (a) $T(x) = \begin{cases} 0.15x & \text{if } 0 < x \le 15,475 \\ \\ 2321.25 + 0.28(x - 15,475) \\ \quad \text{if } 15,475 < x \le 37,425 \\ \\ 8467.25 + 0.33(x - 37,425) \\ \quad \text{if } 37,425 < x \le 117,895 \end{cases}$

(b)

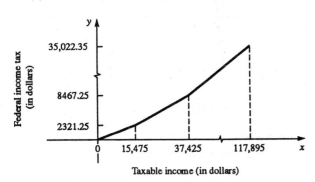

1–3 Graphing Linear Equations

1. $4x - 6y = 24$
  x–intercept: $(6,0)$
  y–intercept: $(0,-4)$

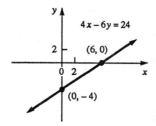

3. $-3x + 2y = 8$

  x–intercept: $\left[-\dfrac{8}{3}, 0\right]$

  y–intercept: $(0,4)$

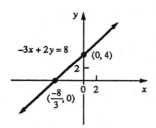

5. $2x + 7y = 11$

  x–intercept: $\left[\dfrac{11}{2}, 0\right]$

  y–intercept: $\left[0, \dfrac{11}{7}\right]$

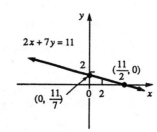

7. $y = -4x + 13$

x–intercept: $\left[\dfrac{13}{4}, 0\right]$

y–intercept: $(0, 13)$

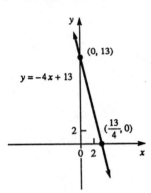

9. $f(x) = 6x + 4$

x–intercept: $\left[-\dfrac{2}{3}, 0\right]$

y–intercept: $(0, 4)$

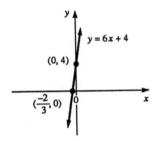

11. $y = 2x$

$(0,0)$ is one point; $m = 2$

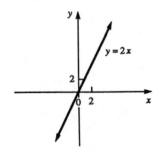

13. $3x + 2y = 0$

$y = -\dfrac{3}{2}x$

$(0,0)$ is one point; $m = -\dfrac{3}{2}$

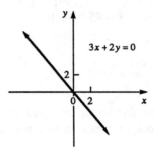

15. $g(x) = 7x$

$(0,0)$ is one point; $m = 7$

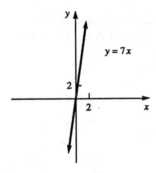

17. $f(x) = x$

$(0,0)$ is one point; $m = 1$

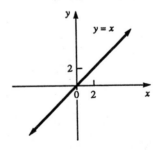

19. $f(x) = 3x + 5, x \geq 0$
y–intercept: $(0,5)$; $m = 3$

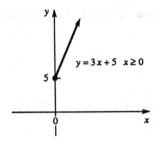

21. $f(x) = 2x + 3, x \geq 1$
$(1,5)$ is one point; $m = 2$

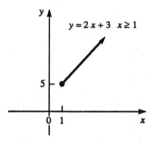

1–4 Linear Models

1. (a) $V = \dfrac{1400 - 1200}{60 - 40} = \dfrac{200}{20} = \$10/\text{unit}$

(b) $1200 = 10(40) + F;\ \ F = 1200 - 400 = 800$
$C(x) = 10x + 800$

(c) $F = \$800$

(d)

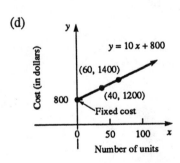

3. (a) $V = \dfrac{4200 - 2200}{70 - 30} = \dfrac{2000}{40} = \$50/\text{unit}$

(b) $4200 = 50(70) + F;\ \ F = 4200 - 3500 = 700$
$C(x) = 50x + 700$

(c) $F = \$700$

(d) $C(50) = 50(50) + 700 = \$3200$

5. $C(x) = 20x + 1000$
$C(100) = \$3000$

7. $C(x) = 15x + 8700$
$C(100) = \$10,200$

9. $C(x) = 40x + 900$
$R(x) = 70x$
$P(x) = 70x - (40x + 900) = 30x - 900$

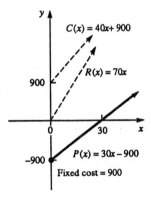

11. $C(x) = 100x + 8000$
      $R(x) = 140x$
      $P(x) = 140x - (100x + 8000) = 40x - 8000$

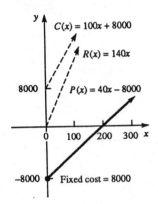

13. $C(x) = 25x + 7500$
      $R(x) = 40x$
      $P(x) = 40x - (25x + 7500) = 15x - 7500$

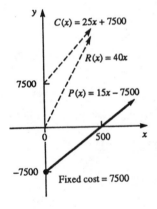

15. $m = \dfrac{\Delta p}{\Delta q} = \dfrac{10 - 5}{50 - 75} = \dfrac{5}{-25} = -\dfrac{1}{5}$

      $p = 5 - \dfrac{1}{5}(q - 75);\ \ 5p = -q + 100;\ \ 5p + q = 100$

      or $p = -\dfrac{1}{5}q + 20$

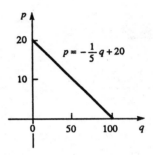

17. $m = \dfrac{\Delta p}{\Delta q} = \dfrac{120 - 60}{50 - 80} = \dfrac{60}{-30} = -2$

      $p = 60 - 2(q - 80) = -2q + 220;\ \ p + 2q = 220$
      $p = -2q + 220$

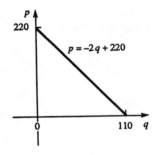

19. $m = \dfrac{\Delta p}{\Delta q} = \dfrac{600 - 200}{100 - 300} = \dfrac{400}{-200} = -2$

$p = 200 - 2(q - 300) = -2q + 800;\ p + 2q = 800$

or $p = -2q + 800$

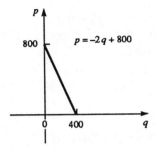

21. $6p + 5q = 30$

$q = 6 - \dfrac{6}{5}p$

23. $2p + q = 6$
$q = 6 - 2p$

25. $p + 3q = 6$
$p = 6 - 3q$

27. $8p + 5q = 40$

$p = 5 - \dfrac{5}{8}q$

29. $m = \dfrac{\Delta p}{\Delta q} = \dfrac{120 - 20}{30 - 10} = \dfrac{100}{20} = 5$

$p = 20 + 5(q - 10) = 5q - 30;\ p - 5q = -30$

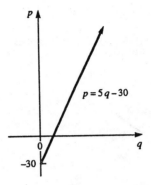

31. $m = \dfrac{\Delta p}{\Delta q} = \dfrac{120 - 100}{90 - 30} = \dfrac{20}{60} = \dfrac{1}{3}$

$p = 100 + \dfrac{1}{3}(q - 30) = \dfrac{1}{3}q + 90;\ 3p - q = 270$

or $p = \dfrac{1}{3}q + 90$

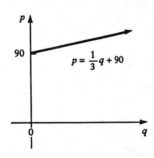

33. $4p - 3q = 12$

$q = \dfrac{4}{3}p - 4$

35. $3p - 5q = 30$

$p = \dfrac{5}{3}q + 10$

37. (a) $MPC = \dfrac{\Delta y}{\Delta x} = \dfrac{65-58}{90-80} = \dfrac{7}{10} = 0.70$

For each dollar increase in disposable income, consumption increases by $0.70 and $0.30 is saved.

(b) $y = 58 + \dfrac{7}{10}(x-80) = \dfrac{7}{10}x + 2 = 0.7x + 2$

(c)

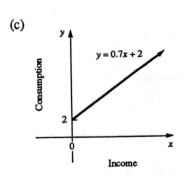

39. (a) $MPC = \dfrac{\Delta y}{\Delta x} = \dfrac{187-97}{200-100} = \dfrac{90}{100} = 0.90$

For each dollar increase in disposable income, consumption increases by $0.90 and $0.10 is saved.

(b) $y = 97 + \dfrac{9}{10}(x-100) = \dfrac{9}{10}x + 7$

(c)

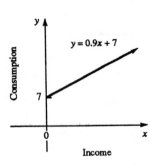

41. (a) $MPC = \dfrac{\Delta y}{\Delta x} = \dfrac{79-55}{90-60} = \dfrac{24}{30} = 0.80$

For each dollar increase in disposable income, consumption increases by $0.80 and $0.20 is saved.

(b) $MPS = 1 - MPC = 1 - 0.80 = 0.20$
An increase of one dollar in disposable income yields $0.20 in savings.

(c) $y = 55 + \dfrac{4}{5}(x-60) = \dfrac{4}{5}x + 7 = 0.8x + 7$

(d) $y(85) = \dfrac{4}{5}(85) + 7 = \$75$ billion

43. $y = 90,000 - \left[\dfrac{90,000 - 10,000}{4}\right]x$
$ = 90,000 - 20,000x$

45. $y = 45,000 - \left[\dfrac{45,000 - 0}{3}\right]x$
$ = 45,000 - 15,000x$

47. (a) $y = 30,000 - \left[\dfrac{30,000 - 1000}{10}\right]x$
$ = 30,000 - 2900x$

(b)

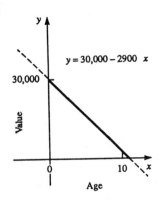

(c) $y(6) = 30,000 - 2900(6) = \$12,600$

1-5   Break-Even Analysis: Market Equilibrium

1. $C(x) = 5x + 1000$
   $R(x) = 9x$
   $P(x) = 9x - (5x + 1000) = 4x - 1000$
   Break-even point:
   $R(x) = C(x) \Rightarrow 9x = 5x + 1000$; $x = 250$
   $P(x) = 0 \Rightarrow x = 250$

3. $C(x) = 20x + 100,000$
   $R(x) = 120x$
   $P(x) = 120x - (20x + 100,000) = 100x - 100,000$
   Break-even point:
   $R(x) = C(x) \Rightarrow 120x = 20x + 100,000$; $x = 1000$
   $P(x) = 0 \Rightarrow x = 1000$

5. $C(x) = 25x + 80,000$
   $R(x) = 65x$
   $P(x) = 65x - (25x + 80,000) = 40x - 80,000$
   Break-even point:
   $R(x) = C(x) \Rightarrow 65x = 25x + 80,000$; $x = 2000$
   $P(x) = 0 \Rightarrow x = 2000$

7. (a) $C(x) = 5x + 2000$

   (b) $R(x) = 15x$

   (c) See below

   (d) $R(x) = C(x) \Rightarrow 15x = 5x + 2000$;
         $x = 200$ units, break-even point

   (e) $P(x) = R(x) - C(x) = 15x - (5x + 2000)$
         $\qquad\qquad = 10x - 2000$

   (continued)

7. *(continued)*

   (f)

   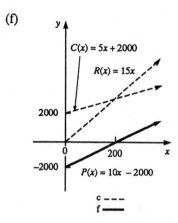

   (g) $P(300) = 10(300) - 2000 = \$1000$

   (h) $P(100) = 10(100) - 2000 = \$1000$

   (i) $P(x) = 40,000 = 10x = 2000 \Rightarrow x = 4200$ units

9. (a) $V = (50,000 + 70,000 + 10,000)/100$
         $\quad = \$1300/\text{unit}$

   (b) $C(x) = (200,000 + 50,000 + 10,000) + 1300x$
         $\qquad = 1300x + 260,000$

   (c) $R(x) = 1800x$

   (d) See below

   (e) $R(x) = C(x) \Rightarrow 1800x = 1300x + 260,000$;
         $x = 520$ units, break-even point

   (f) $P(x) = R(x) - C(x)$
         $\qquad = 1800x - (1300x + 260,000)$
         $\qquad = 500x - 260,000$

   (continued)

9. *(continued)*

(d) and (g)

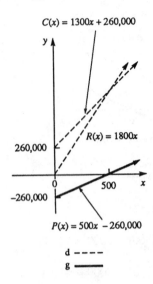

$C(x) = 1300x + 260,000$

$R(x) = 1800x$

$P(x) = 500x - 260,000$

d $----$

g $\underline{\quad\quad}$

(h) $P(1000) = 500(1000) - 260,000 = \$240,000$

(i) $P(x) = 104,000 = 500x - 260,000 \Rightarrow$
     $x = 728$ units

11. (a) $R(x) = 100x$

(b) Fixed charges = $500 + 1000 + 1000 = \$2500$
    Variable charges = $(1500 + 1400 + 1000 + 600 + 500)/100 = \$50/\text{units}$
    $C(x) = 50x + 2500$

(c) $R(x) = C(x) \Rightarrow 100x = 50x + 2500;$
    $x = 50$ units, break–even point

(d) $P(x) = R(x) - C(x) = 100x - (50x + 2500)$
    $= 50x - 2500$

(e) $P(x) = 4000 = 50x - 2500 \Rightarrow x = 130$ units

(f) $P(125) = 50(125) - 2500 = \$3750$

(g) If fixed charges increased by $1700,
    $P(x) = 50x - 2500 - 1700 = 50x - 4200$
    $P(x) = 0 \Rightarrow x = 84$ units, break–even point

13. (a) At the break–even point, $R(x) = C(x)$
    $R(x) =$
    $[0.6\, R(x)](0.6) + [0.4\, R(x)](0.85) + 180,000$
    VC for Stat I   VC for Stat II   FC
    $R(x) = 0.70\, R(x) + 180,000 \Rightarrow 0.3\, R(x)$
    $= 180,000$
    $R(x) = \$600,000$

(b) If the fixed charges become 1.4(180,000)
    $= 252,000$, let $P(x) = R(x) - C(x) = R(x) - (0.7\, R(x) + 252,000)$
    $105,000 = 0.3\, R(x) - 252,000$
    $0.3\, R(x) = 357,000$
    $R(x) = \$1,190,000$ to yield a profit of $105,000

15. $p = -\dfrac{7}{3}q + 31$ demand equation

$p = \dfrac{5}{3}q + 15$ supply equation

At equilibrium, $-\dfrac{7}{3}q + 31 = \dfrac{5}{3}q + 15$

$4q = 16 \Rightarrow q = 4$

$p = \dfrac{5}{3}q + 15 = \dfrac{20}{3} + 15 = \$21.67$

17. $p = -\dfrac{1}{3}q + 13$ demand equation

$p = \dfrac{1}{4}q + 6$ supply equation

At equilibrium, $-\dfrac{1}{3}q + 13 = \dfrac{1}{4}q + 6$

$\dfrac{7}{12}q = 7 \Rightarrow q = 12$

$p = \dfrac{1}{4}(12) + 6 = \$9$

19. (a)

q	p
150	35
300	20

$$m = \frac{\Delta p}{\Delta q} = \frac{35 - 20}{150 - 300} = \frac{15}{-150} = -\frac{1}{10}; \quad p = 35 - \frac{1}{10}(q - 150) = 50 - \frac{1}{10}q, \text{ demand equation}$$

(b)

q	p
125	25
350	40

$$m = \frac{\Delta p}{\Delta q} = \frac{40 - 25}{350 - 125} = \frac{15}{225} = \frac{1}{15}; \quad p = 25 + \frac{1}{15}(q - 125) = \frac{50}{3} + \frac{1}{15}q, \text{ supply equation}$$

(c)

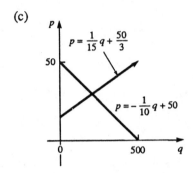

(d) At equilibrium, $50 - \frac{1}{10}q = \frac{50}{3} + \frac{1}{15}q$

$$\frac{100}{3} = \frac{5}{30}q \Rightarrow q = 200$$

(e) $p = 50 - \frac{1}{10}q = 50 - 20 = \$30$ each

EXTRA DIVIDENDS: Cost Accounting: Cost Segregation

1. (a)

Hours	Cost	
5000	$ 6500	low
5900	$ 7400	high

$$m = \frac{\Delta y}{\Delta x} = \frac{7400 - 6500}{5900 - 5000} = \frac{900}{900} = 1.0$$

$$y = 6500 + 1(x - 5000) = x + 1500$$

(b) Fixed cost = $1500

(c) Variable cost per hour = $1.00

3. (a)   High   Low
       Cost $ 39,000 $ 32,000
       Hours   24,000   15,000

$$m = \frac{\Delta y}{\Delta x} = \frac{39,200 - 32,000}{24,000 - 15,000} = \frac{7200}{9000} = 0.80$$

Variable cost per hour = $0.80

(b) $y = 32,000 + 0.8(x - 15,000) = 0.8x + 20,000$
     Fixed cost = $20,000/month

(c) Annual fixed cost = $240,000

EXTRA DIVIDENDS: Model Fitting: Goodness of Fit

1. (a) | x | 2 | 4 | 6 | 8 | 9 | |
    |---|---|---|---|---|---|---|
    | y | 7 | 9 | 13 | 16 | 25 | |
    | | | | | | | |
    | $y = 2x + 1$ | 5 | 9 | 13 | 17 | 19 | |
    | residual | 2 | 0 | 0 | −1 | 6 | $S = 4 + 1 + 36 = 41$ |
    | | | | | | | |
    | $y = 3x - 2$ | 4 | 10 | 16 | 22 | 25 | |
    | residual | 3 | −1 | −3 | −6 | 0 | $S = 9 + 1 + 9 + 36 = 55$ |

(b) $y = 2x + 1$ is the better–fitting linear model, since the sum–of–squares error is lower than that of the other model.

3. (a) | x | 2 | 3 | 7 | 8 | 9 | |
    |---|---|---|---|---|---|---|
    | y | 6 | 9 | 19 | 28 | 35 | |
    | | | | | | | |
    | $y = 4x - 3$ | 5 | 9 | 25 | 29 | 33 | |
    | residual | 1 | 0 | −6 | −1 | 2 | $S = 1 + 36 + 1 + 4 = 42$ |
    | | | | | | | |
    | $y = 3x + 2$ | 8 | 11 | 23 | 26 | 29 | |
    | residual | −2 | −2 | −4 | 2 | 6 | $S\ 4 + 4 + 16 + 4 + 36 = 64$ |

(b) $y = 4x - 3$ is the better–fitting linear model, since the sum–of–squares error is lower than that of the other model.

5. (a)

x	1	2	3	4	5	6	7
y	153.18	166.10	167.24	180.66	191.85	182.08	211.28

$y = 13x + 130$	143	156	169	182	195	208	221
residual	10.18	10.10	−1.76	−1.34	−3.15	−25.92	−9.72

$y = 20x + 70$	90	110	130	150	170	190	210
residual	63.18	56.10	37.24	30.66	21.85	−7.92	−1.28

x	8	9	10	11	12	13	14
y	238.90	250.84	231.32	242.17	291.70	304.00	337.00

$y = 13x + 130$	234	247	260	273	286	299	312
residual	4.90	3.84	−28.68	−30.83	5.70	5.00	25.00

$y = 20x + 70$	230	250	270	290	310	330	350
residual	8.90	0.84	38.68	−47.83	−18.30	−26.00	−13.00

For $y = 13x + 130$, $S = 3481.0598$
For $y = 20x + 70$, $S = 15{,}051.2198$

(b) $y = 13x + 130$ is the better–fitting linear model, since the sum–of–squares error is lower than that of the other model.

7. (a)

x	1	2	3	4	5	6	7	8	9
y	125	128	130	129	135	138	140	146	145

$y = 3x + 120$	123	126	129	132	135	138	141	144	147
residual	2	2	1	−3	0	0	−1	2	−2

$y = 4x + 115$	119	123	127	131	135	139	143	147	151
residual	6	5	3	−2	0	−1	−3	−1	−6

For $y = 3x + 120$, $S = 27$
For $y = 4x + 115$, $S = 121$

(b) $y = 3x + 120$ is the better–fitting linear model, since the sum–of–squares error is lower than that of the other model.

9. (a)

x	20	35	27	40	49	55	
y	210	279	230	190	252	287	

$y = 2x + 170$	210	240	224	250	268	280	
residual	0	39	6	−60	−16	7	$S = 5462$

$y = 3x + 100$	160	205	181	220	247	265	
residual	50	74	49	−30	5	22	$S = 11{,}786$

(b) $y = 2x + 170$ is the better–fitting linear model, since the sum–of–squares error is lower than that of the other model.

Review Exercises

1. $f(x) = 4x - 2$
$f(0) = -2, f(1) = 2, f(3) = 10$

3. $f(x) = x^2 + 2x - 1$
$f(0) = -1, f(1) = 2, f(3) = 14$

5. $f(x) = 4x + 8$

$$\frac{f(x + h) - f(x)}{h} = \frac{4(x + h) + 8 - (4x + 8)}{h} = \frac{4h}{h} = 4$$

7. $f(x) = x^2 - 4x + 3$

$$\frac{f(x + h) - f(x)}{h}$$

$$= \frac{(x + h)^2 - 4(x + h) + 3 - (x^2 - 4x + 3)}{h}$$

$$= \frac{2xh + h^2 - 4h}{h} = 2x - 4 + h$$

9. $f(x) = \sqrt{x - 4}$
domain of f: $x \geq 4$

11. $f(x) = \dfrac{x + 6}{(x - 7)(x + 5)}$

domain of f: all real values x such that $x \neq 7$ and $x \neq -5$

13. The graph is not that of a function, since it does not satisfy the vertical lines test.

15. $f(x) = \begin{cases} 2 \text{ if } x \leq 3 \\ 5 \text{ if } 3 < x \leq 8 \\ x \text{ if } x > 8 \end{cases}$

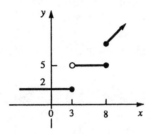

17. $P(x) = \begin{cases} -4 \text{ if } x \leq 20 \\ x - 24 \text{ if } x > 20 \end{cases}$
Let x = price per share of stock

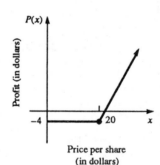

Price per share
(in dollars)

19. (6,5), (4,–3)
$$m = \frac{\Delta y}{\Delta x} = \frac{5 - (-3)}{6 - 4} = \frac{8}{2} = 4$$

21. (3,–2), (5,8)
$$m = \frac{\Delta y}{\Delta x} = \frac{8 - (-2)}{5 - 3} = \frac{10}{2} = 5$$
$$y = 8 + 5(x - 5) = 5x - 17$$

23. $4x - 2y = 10$
(1,–3) and (0,–5) lie on the line.
(2,5), (3,0) and (–1,–2) do not lie on the line.

25. Answers will vary
(0,0), $m < 0$

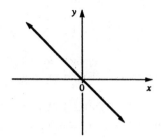

27. (3,4), m undefined

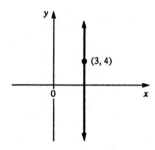

29. (3,4), m undefined
$x = 3$

31. $2x + 6y = 30$ $-4x - 12y = 24$
$m = -\dfrac{2}{6} = -\dfrac{1}{3}$ $m = -\dfrac{4}{12} = -\dfrac{1}{3}$
The lines are parallel.

33. $5x - 2y = 20$
x–intercept: (4,0), y–intercept: (0,−10)

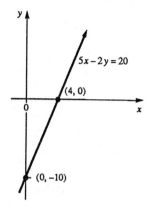

35. $f(x) = 6x$
x–intercept: (0,0), y–intercept: (0,0)

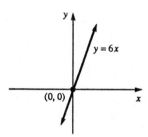

37. $C(x) = 30x + 1200$
$R(x) = 50x$
$P(x) = R(x) - C(x) = 20x - 1200$
Break–even point: $P(x) = 0 \Rightarrow x = 60$ units

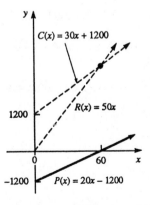

39. $P(x) = \dfrac{10,000}{10,000}x - \left[\dfrac{6000}{10,000}x + 2000\right]$

  $= 0.4x - 2000$

41. $P(x) = \dfrac{10,000}{7500}x - \left[\dfrac{3000}{7500}x + 5600\right]$

  $= \dfrac{14}{15}x - 5600$

43. variable cost per unit $= 4\left[\dfrac{6000}{10,000}\right] + 3\left[\dfrac{3000}{7500}\right]$

  $= 2.40 + 1.20 = \$3.60$

  selling price per unit $= 4\left[\dfrac{10,000}{10,000}\right] + 3\left[\dfrac{10,000}{7500}\right]$

  $= 4.00 + 4.00 = \$8.00$

  $P(x) = R(x) - C(x) = 8x - (3.60x + 2000 + 5600)$

  $= 4.40x - 7600$

  Break–even point: $4.40x = 7600 \Rightarrow x = 1727$ units

  (rounded to nearest unit)

45. Ford Motor Company:

  $y = 20.8 + \dfrac{22.0 - 20.8}{1989 - 1979}(x - 1979)$

  $= 20.8 + 0.12(x - 1979) = 0.12x - 216.68$

  Japanese companies:

  $y = 15.2 + \dfrac{25.6 - 15.2}{1989 - 1979}(x - 1979)$

  $= 15.2 + 1.04(x - 1979) = 1.04x - 2042.96$

  Break–even year:

  $20.8 + 0.12(x - 1979) = 15.2 + 1.04(x - 1979)$

  $\Rightarrow 5.6 = 0.92(x - 1979) \Rightarrow x - 1979 = \dfrac{5.6}{0.92}$

  $x = 1979 + \dfrac{5.6}{0.92} = 1985$ (rounded to nearest year)

47.

x	y
50	85
60	94

(a) $\dfrac{\Delta y}{\Delta x} = \dfrac{94 - 85}{60 - 50} = \dfrac{9}{10} = 0.9 = \text{MPC}$

  For each dollar increase in income, \$0.90 goes into consumption and \$0.10 into savings.

(b) MPS $= 1 - \text{MPC} = 1 - 0.90 = \$0.10$

  For each dollar increase in income, \$0.10 goes into savings.

(c) $y = 85 + 0.9(x - 50) = 0.9x + 40$

(d)

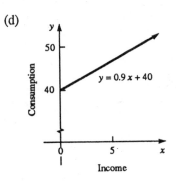

49. (a) $y = 80,000 - \left[\dfrac{80,000 - 5000}{5}\right]x$

  $= 80,000 - 15,000x$

(b)

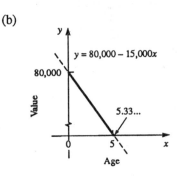

(c) $y(3) = 80,000 - 15,000(3) = \$35,000$

CHAPTER

2

GRAPHING POLYNOMIAL
AND RATIONAL FUNCTIONS

2–1 Graphing Concepts

1. $y = -|x - 2|$
Shift graph of $y = |x|$ 6 units to the left; reflect graph in the x–axis.

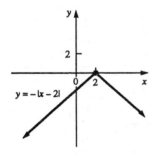

3. $y = |x - 8| + 2$
Shift graph of $y = |x|$ 8 units to the right and 2 units upwards.

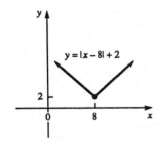

5. $y = -|x + 3| - 8$
Shift graph of $y = |x|$ 3 units to the left; reflect graph in the line $y = -8$

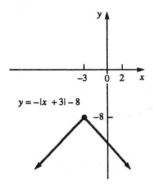

7. $f(x) = (x - 3)^{2/3}$
Shift graph of $y = x^{2/3}$ 3 units to the right.

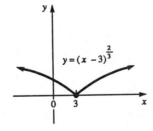

9. $f(x) = (x + 7)^{2/3}\ 5$
    Shift graph of $y = x^{2/3}$ 7 units to the left and 5 units upwards.

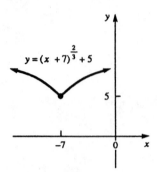

11. $f(x) = -(x - 5)^{2/3}$
     Shift graph of $y = x^{2/3}$ 5 units to the right; reflect graph in the line x–axis.

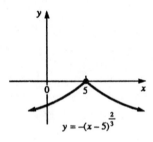

13. $f(x) = -(x - 3)^{2/3} - 1$
     Shift graph of $y = x^{2/3}$ 3 units to the right; reflect graph in the line $y = -1$.

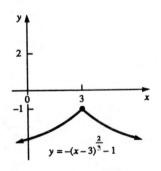

15. $f(x) = (x + 2)^{1/3}$
     Shift graph of $y = x^{1/3}$ 2 units to the left.

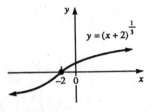

17. $f(x) = (x - 2)^{1/3} + 9$
     Shift graph of $y = x^{1/3}$ 2 units to the right and 9 units upwards.

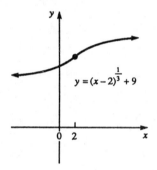

19. $f(x) = -(x - 1)^{1/3}$
     Shift graph of $y = x^{1/3}$ 1 unit to the right; reflect graph in the x–axis

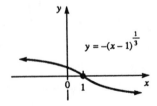

21. $f(x) = -(x - 5)^{1/3} + 2$

Shift graph of $y = x^{1/3}$ 5 units to the right; reflect graph in the line $y = 2$.

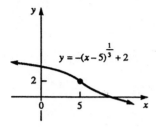

23. $f(x) = \sqrt{x - 5}$

Shift graph of $y = \sqrt{x}$ 5 units to the right.

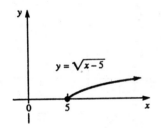

25. $f(x) = \sqrt{x + 9} + 2$

Shift graph of $y = \sqrt{x}$ 9 units to the left and 2 units downwards.

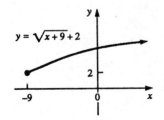

27. $f(x) = -\sqrt{x}$

Reflect graph of $y = \sqrt{x}$ in the x–axis.

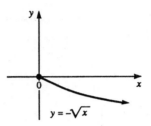

29. $f(x) = -\sqrt{x - 3}$

Shift graph of $y = \sqrt{x}$ 3 units to the right; reflect graph in the x–axis.

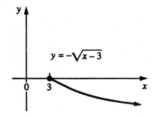

31. $y = f(x - 2)$

Shift graph of f 2 units to the right.

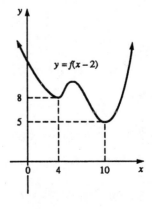

33. $y = -f(x)$

Reflect graph of f in the x–axis

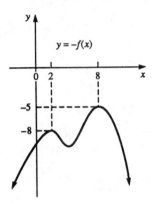

35. $y = f(x - 1) + 4$

Shift graph of f right 1 unit and upwards 4 units.

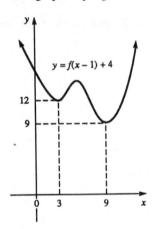

37. $y = -f(x - 2) + 1$

Shift graph of f 2 units to the right; reflect graph in the line $y = 1$.

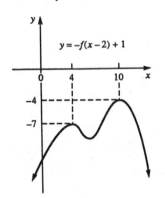

39. $y = -f(x) - 6$

Reflect fraph of f in the line $y = -6$.

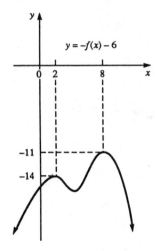

41. f is odd, since $f(-x) = -f(x)$.

43. f is odd, since $f(-x) = -f(x)$.

45. $f(x) = x^3$

(a)

x	-3	-2	-1	0	1	2	3
$f(x)$	-27	-8	-1	0	1	8	27

f is odd, since $f(-x) = -f(x)$.

(b)

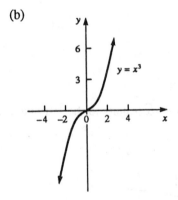

47. $f(x) = \dfrac{1}{x}$

(a)

x	-3	-2	-1	$-\frac{1}{2}$	0	$\frac{1}{2}$	1	2	3
$f(x)$	$-\frac{1}{3}$	$-\frac{1}{2}$	-1	-2	$*$	2	1	$\frac{1}{2}$	$\frac{1}{3}$

* Undefined.
f is odd since $f(-x) = -f(x)$.

(b)

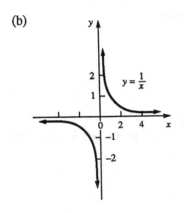

49. An increase of \$100 in fixed cost will raise the graph of the cost function by \$100. Shift upward 100 units.

51. The new equilibrium price is \$10.

2–2 Quadratic Functions and Their Graphs

1. (a) $y = 5x^2$

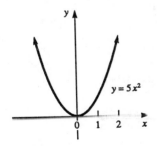

(*continued*)

1. (*continued*)

(b) $y = -5x^2$
Reflect graph of $5x^2$ in the x–axis.

(c) $y = \frac{1}{2}x^2$

(d) $y = -\frac{1}{2}x^2$
Reflect graph of $\frac{1}{2}x^2$ in the x–axis.

3. $y = -x^2$, $y = -4x^2$, $y = -\frac{1}{3}x^2$

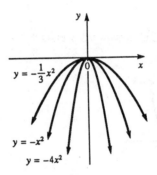

5. $y = x^2 + 7$
Shift graph of $y = x^2$ upward 7 units.

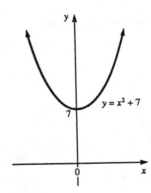

7. $y = x^2 - 9$
    Shift graph of $y = x^2$ downward 9 units.

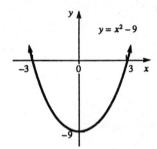

13. $y = 4(x - 1)^2$
     Shift graph of $y = 4x^2$ to the right 1 unit.

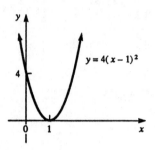

9. $f(x) = 2x^2 - 8$
    Shift graph of $y = 2x^2$ downward 8 units.

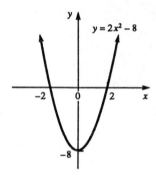

15. $f(x) = -2(x + 5)^2$
     Shift graph of $y = 2x^2$ to the left 5 units; reflect in the x–axis.

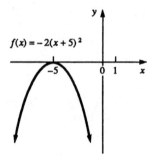

11. $y = (x - 3)^2$
     Shift graph of $y = x^2$ to the right 3 units.

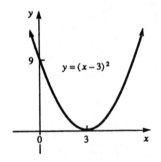

17. $y = 2(x - 1)^2 + 3$
     Shift graph of $y = 2x^2$ to the right 1 unit and upward 3 units.

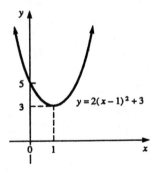

19. $f(x) = 3(x - 2)^2 - 1$
Shift graph of $3x^2$ to the right 2 units and downward 1 unit.

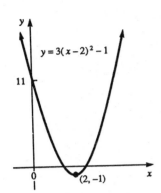

21. $y = -3(x + 2)^2 - 5$
Shift graph of $y = 3x^2$ to the left 2 units; reflect in the line $y = -5$.

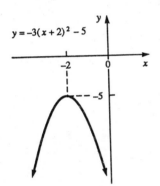

23. $f(x) = (x + 3)^2 - 1$
Shift graph of $y = x^2$ to the left 3 units and downward 1 unit.

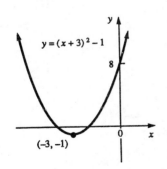

25. $f(x) = -(x + 2)^2 + 1$
Shift the graph of $y = x^2$ to the left 2 units; reflect in the line $y = 1$.

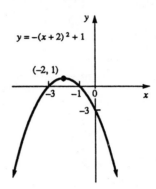

27. $y = 2x^2 + 8x = 2(x^2 + 4x + 4) - 8$
$\qquad = 2(x + 2)^2 - 8$
Shift graph of $y = 2x^2$ to the left 2 units and downward 8 units.

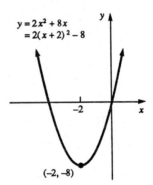

29. $f(x) = x^2 - 5x = \left[x - \dfrac{5}{2}\right]^2 - \dfrac{25}{4}$
Shift graph of $y = x^2$ to the right $\dfrac{5}{2}$ units and downward $\dfrac{25}{4}$.

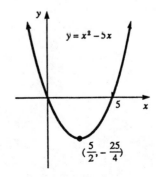

31. $y = -2x^2 + 6x$
      $\quad = -2(x^2 - 3x)$

$\quad = -2\left[x^2 - 3x + \dfrac{9}{4}\right] + \dfrac{9}{2}$

$\quad = -2\left[x - \dfrac{3}{2}\right]^2 + \dfrac{9}{2}$

Shift graph of $y = x^2$ to the right $\dfrac{3}{2}$ units; reflect in

the line $y = \dfrac{9}{2}$.

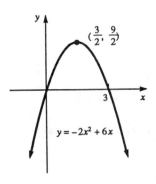

33. $y = x^2 - 6x + 5 = x^2 - 6x + 9 - 4 = (x - 3)^2 - 4$
      Shift graph of $y = x^2$ to the right 3 units and
      downward 4 units.

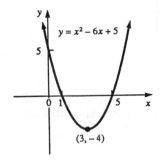

35. $f(x) = x^2 - 6x - 16 = x^2 - 6x + 9 - 25$
      $\quad\quad\quad = (x - 3)^2 - 25$
      Shift graph of $y = x^2$ to the right 3 units and
      downward 25 units.

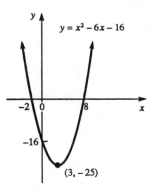

37. $y = x^2 - 2x - 15 = x^2 - 2x + 1 - 16 = (x - 1)^2 - 16$
      Shift graph of $y = x^2$ to the right 1 unit and
      downward 16 units.

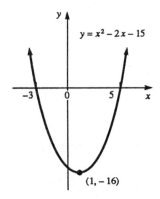

39. $y = x^2 - 4x - 5 = x^2 - 4x + 4 - 9 = (x - 2)^2 - 9$
      Shift graph of $y = x^2$ to the right 2 units and
      downward 9 units.

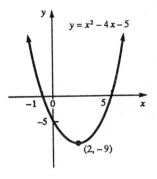

41. $f(x) = -5x^2 + 6x + 4$

$$= -5\left[x^2 - \frac{6}{5}x + \frac{9}{25}\right] + 4 + \frac{9}{5}$$

$$= -5\left[x - \frac{3}{5}\right]^2 + \frac{29}{5}$$

Shift graph of $y = 5x^2$ to the right $\frac{3}{5}$ unit; reflect in the line $y = \frac{29}{5}$.

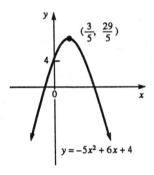

43. $y = -x^2 + 8x - 16 = -(x^2 - 8x + 16) = -(x - 4)^2$

Shift graph of $y = x^2$ to the right 4 units; reflect in the x–axis.

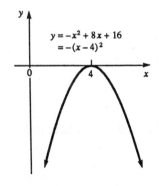

45. $f(x) = x^2 - 10x + 26 = (x - 5)^2 + 1$

Shift graph of $y = x^2$ to the right 5 units and upward 1 unit.

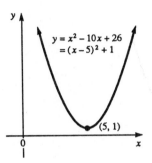

47. $f(x) = 2x^2 + 4x + 1$

$$= 2(x^2 + 2x + 1) - 1$$

$$= 2(x + 1)^2 - 1$$

Shift graph of $y = 2x^2$ to the left 1 unit and downward 1 unit.

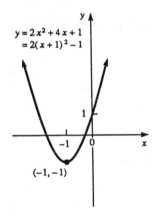

49. $f(x) = x^2 - 10x + 25 = (x - 5)^2$

Shift graph of $y = x^2$ to the right 5 units.

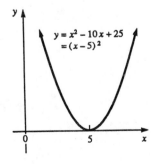

51. $y = x^2 - 6x + 9 = (x - 3)^2$
Shift graph of $y = x^2$ to the right 3 units.

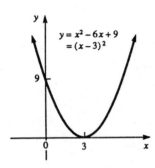

53. $y = (x - 2)(x + 3)$
x–intercepts: 2, –3

x–coordinate of vertex: $\dfrac{2 - 3}{2} = -\dfrac{1}{2}$

y–coordinate of vertex: $\left[-\dfrac{1}{2} - 2\right]\left[-\dfrac{1}{2} + 3\right] = -\dfrac{25}{4}$

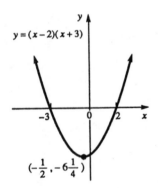

55. $y = 4(x - 2)(x + 3)$
x–intercepts: 2, –3

x–coordinate of vertex: $\dfrac{2 - 3}{2} = -\dfrac{1}{2}$

y–coordinate of vertex: $4\left[-\dfrac{1}{2} - 2\right]\left[-\dfrac{1}{2} + 3\right] = -25$

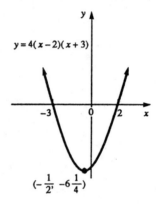

57. $f(x) = (x - 6)^2$
Shift graph of $y = x^2$ to the right 6 units.

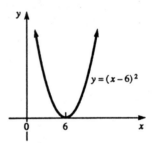

59. $y = -5(x + 1)^2$
Shift graph of $y = 5x^2$ to the left 1 unit; reflect in the x–axis.

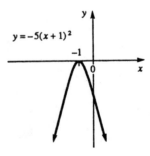

61. $y = (x - 2)(x + 2)$
x–intercepts: 2, –2
x–coordinate of vertex: 0
y–coordinate of vertex: $(-2)(2) = -4$

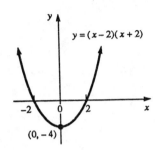

63. $y = -3(x - 2)(x + 2)$
x–intercepts: 2, –2
x–coordinate of vertex: 0
y–coordinate of vertex: $-3(-2)(2) = 12$; opens downward

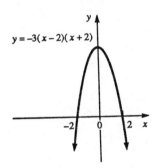

2–3 Applications of Quadratic Functions

1. $p = -2x + 100$

(a) $R(x) = xp = -2x^2 + 100x = -2x(x - 50)$
(b) x–intercepts: 0, 50
     x–coordinates of vertex: 25
     R–coordinate of vertex:
       $-2(25)(25 - 50) = 1250$

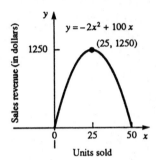

(c) $R_{max} = \$1250$
(d) $x = 25$ units
(e) $p(25) = -2(25) + 100 = \$50$

3. $p = -4x + 320$

(a) $R(x) = xp = -4x^2 + 320x = -4x(x - 80)$
(b) x–intercepts: 0, 80
     x–coordinate of vertex: 40
     R–coordinate of vertex:
       $-4(40)(40 - 80) = 6400$

(c) $R_{max} = \$6400$
(d) $x = 40$ units
(e) $p(40) = -4(40) + 320 = \$160$

5. $p = -6x + 1800$

(a) $R(x) = xp = -6x^2 + 1800x = -6x(x - 300)$

(b) x–intercepts: 0, 300

x–coordinate of vertex: 150

R–coordinate of vertex:

$-6(150)(150 - 300) = 135,000$

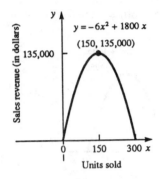

(c) $R_{max} = \$135,000$

(d) $x = 150$ units

(e) $p(150) = -6(150) + 1800 = \$900$

7. $p = -x + 1200$

$C(x) = 200x + 160,000$

(a) $R(x) = xp = -x^2 + 1200x = -x(x - 1200)$

(b) x–intercepts of R: 0, 1200

x–coordinate of vertex: 600

R–coordinate of vertex:

$-600(600 - 1200) = 360,000$

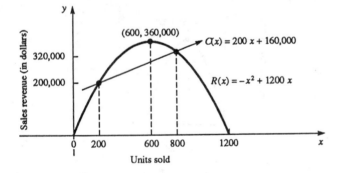

(c) $R(x) = C(x) \Rightarrow -x^2 + 1200x = 200x + 160,000$

$x^2 - 1000x + 160,000 = (x - 200)(x - 800) = 0$

Break–even points: $x = 200, 800$ units

(d) profit > 0 if $200 < x < 800$

(continued)

7. *(continued)*

(e) $P(x) = R(x) - C(x)$

$= -x^2 + 1200x - 200x - 160,000$

$= -x^2 + 1000x - 160,000$

(f) $P(x) = -(x - 200)(x - 800)$

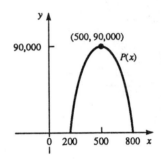

(g) $P(x) = 0 \Rightarrow x = 200, 800$

(h) $P(x) > 0$ if $200 < x < 800$, as determined in part (d).

(i) $P_{max} = P(500)$

$= -(500 - 200)(500 - 800) = \$90,000$

(j) $x = 500$ units

9. $p = -2x + 2700$

$C(x) = 300x + 540,000$

(a) $R(x) = xp = -2x^2 + 2700x = -2x(x - 1350)$

(b) x–intercepts of R: 0, 1350

x–coordinate of vertex: 675

R–coordinate of vertex:

$-2(675)(675 - 1350) = 911,250$

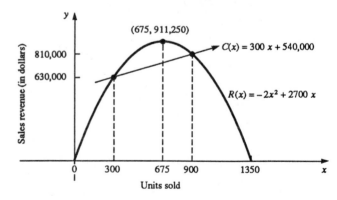

(continued)

9. *(continued)*

(c) $R(x) = C(x) \Rightarrow -2x^2 + 2700x = 300x + 540{,}000$
$x^2 - 1200x + 270{,}000 = (x - 300)(x - 900) = 0$
Break–even points: $x = 300, 900$ units

(d) profit > 0 if $300 < x < 900$

(e) $P(x) = R(x) - C(x)$
$= -2x^2 + 2700x - 300x - 540{,}000$
$= -2x^2 + 2400x - 540{,}000$

(f) $P(x) = -2(x - 300)(x - 900)$

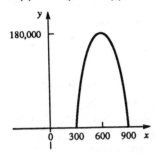

(g) $P(x) = 0 \Rightarrow x = 300, 900$

(h) $P(x) > 0$ if $300 < x < 900$, as determined in part (d).

(i) $P_{max} = P(600) = -2(600 - 300)(600 - 900)$
$= \$180{,}000$

(j) $x = 600$ units.

11. Supply function: $p = x^2 + 12$ $(x > 0)$
Demand function: $p = (x - 6)^2$ $(0 < x \le 6)$

(a) Equilibrium $\Rightarrow x^2 + 12 = (x - 6)^2$
$= x^2 - 12x + 36$
$\Rightarrow x = 2$ units

(b) $p(2) = \$16$

(c)

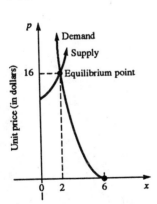

Quantity (in millions of units)

13. Supply function: $p = x^2 + 16$ $(x > 0)$
Demand function: $p = (x - 8)^2$ $(0 < x \le 8)$

(a) Equilibrium $\Rightarrow x^2 + 16 = (x - 8)^2$
$= x^2 - 16x + 64$
$\Rightarrow x = 3$ units

(b) $p(3) = \$25$

(c)

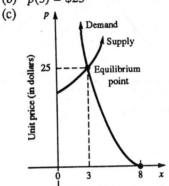

Quantity (in millions of units)

15. (a) $C(x) = 200x + 50\left[\dfrac{x}{500} \cdot \dfrac{x}{20}\right]$ Note: There are

$= 200x + \dfrac{x^2}{200}$ $\dfrac{x}{20}$ defectives

per lot and $\dfrac{x}{500}$ lots per month.

$R(x) = 270x$

$P(x) = R(x) - C(x) = 270x - 200x - \dfrac{x^2}{200}$

$= 70x - \dfrac{x^2}{200} = \dfrac{x}{200}(14{,}000 - x)$

(b)

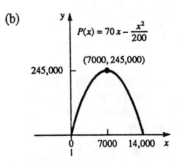

(c) $P(x) = 0 \to x = 0$ or $14{,}000$ units. P_{max} occurs at $x = 7{,}000$.

$P_{max} = \dfrac{7000}{200}(14{,}000 - 7000) = \$245{,}000$ per month

(d) $x = 7000$ units/month

17. $P(t) = 0.8(t - 2)^2 + 40$ $(t > 0)$

(a)
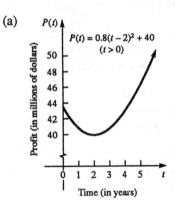

(b) $P(2) = 0.8(2 - 2)^2 + 40 = 40$ million dollars
(c) $P(4) = 0.8(4 - 2)^2 + 40 = 43.2$ million dollars
(d) $P(5) = 0.8(5 - 2)^2 + 40 = 47.2$ million dollars

19. $y = -0.10x^2 + 30$ $(0 \le x \le 17)$

(a)
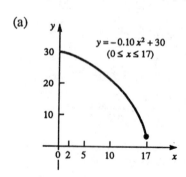

(b) $x = 5 \Rightarrow y = -0.10(25) + 30 = 27.5$
(c) $x = 10 \Rightarrow y = -0.10(100) + 30 = 20.0$
(d) $x = 2y \Rightarrow y = -0.10(4y^2) + 30$

$0.4y^2 + y - 30 = 0$

$$y = \frac{-1 \pm \sqrt{1 + 48}}{0.8} = \frac{-1 \pm 7}{0.8} = 7.5; \ x = 15$$

21. $S(t) = -16t^2 + 192t$ $(0 \le t \le 12)$

(a)
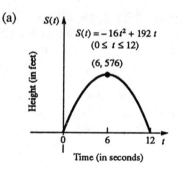

(b) $S(t) = -16(t^2 - 12t + 36) + 576$
$= -16(t - 6)^2 + 576$
S is a maximum when $t = 6$. The maximum height is 576 feet.

(c) $S = 0 \Rightarrow t = 12$ seconds.

23. $y = 0.5(x - 7)^2$
focus is at $(7, 0.5)$

25. $y = 2(x - 1)^2 + 3; \ y - 3 = 2(x - 1)^2$
focus is at $(1, 3.125)$

27. $y = 0.1(x - 4)^2 + 5; \ y - 5 = 0.1(x - 4)^2$
focus is at $(4, 7.5)$

29. $y = 0.25x^2$
focus is at $(0, 1)$

31. $y = f(t) = \begin{cases} (t-5)^2 + 40 & \text{if } 1 < t \le 5 \\ (t-10)^2 + 12 & \text{if } 5 < t \le 10 \end{cases}$

(a)

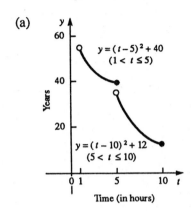

(b) $y(3) = (3-5)^2 + 40 = 44$
(c) $y(4) = (4-5)^2 + 40 = 41$
(d) $y(6) = (6-10)^2 + 12 = 28$
(e) $y(8) = (8-10)^2 + 12 = 16$

2–4 Some Special Polynomial and Rational Functions

1. $y = -6x + 7$
    degree = 1

3. $f(x) = 9x^8 + 4x^6 - 8x^2 + 6$
    degree = 8

5. $y = 4x^3 - 8x^2 + 7x + 3$
    degree = 3

7. There are 5 local extreme points and 2 x–intercepts.
    The degree must be at least $5 + 1 = 6$.

9. If $f(x) = ax^n$, where $n \ge 3$ is an odd positive integer and $a > 0$, then $\lim\limits_{x \to -\infty} f(x) = -\infty$ and $\lim\limits_{x \to +\infty} f(x) = +\infty$.

The number of x–intercepts is at most equal to n and may be n, $n - 2$, $n - 4$,..., 1. The number of extreme points is at most $n - 1$ and may be $n - 1$, $n - 3$, ..., 0.

For $y = f(x) = x^3$:

x	-4	-3	-2	-1	0	1	2	3	4
$f(x)$	-64	-27	-8	-1	0	1	8	27	64

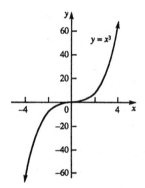

11. $f(x) = x^2$

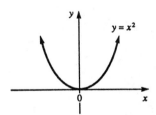

13. $f(x) = x^4$

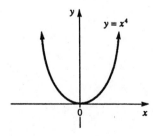

15. $y = -4x^2$

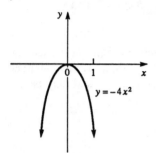

17. $y = x^3$

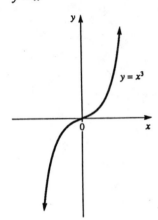

19. $f(x) = x^5$

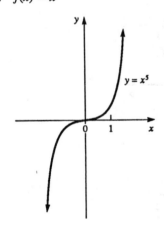

21. $f(x) = -2x^5$

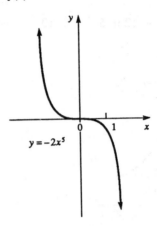

23. If $y = f(x) = \dfrac{k}{x^n}$, where n is an odd positive integer and $k > 0$, then $\lim\limits_{x \to 0+} f(x) = +\infty$ and $\lim\limits_{x \to 0-} f(x) = -\infty$. Also, $\lim\limits_{x \to +\infty} f(x) = 0$ and $\lim\limits_{x \to -\infty} f(x) = 0$. Hence, the coordinate axes are asymptotes of the graph.

For $f(x) = \dfrac{1}{x^3}$.

x	-4	-3	-2	-1	0	1	2	3	4
$f(x)$	$-\dfrac{1}{64}$	$-\dfrac{1}{27}$	$-\dfrac{1}{8}$	-1	$*$	1	$\dfrac{1}{8}$	$\dfrac{1}{27}$	$\dfrac{1}{64}$

*Not defined.

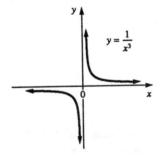

25. $y = \dfrac{3}{x^2}$

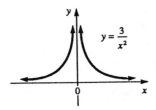

27. $f(x) = \dfrac{5}{x^7}$

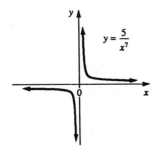

29. $f(x) = \dfrac{2}{x^5}$

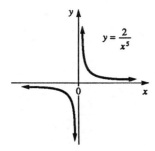

31. $y = \dfrac{5}{x^3}$

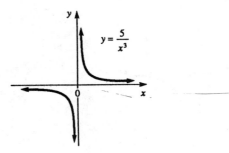

33. $y = \dfrac{6}{x^8}$

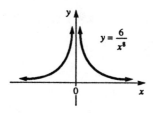

35. $f(x) = \dfrac{4}{x^5}$

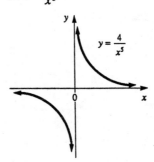

37. $y = -\dfrac{6}{x^8}$

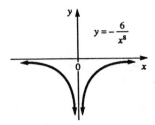

39. $y = -\dfrac{5}{x^3}$

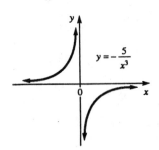

41. $y = -\dfrac{7}{x^6}$

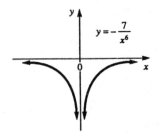

43. $y = 2(x - 5)^4$

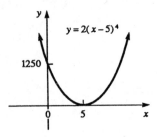

45. $y = 5(x + 2)^3$

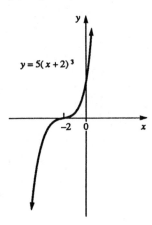

47. $y = (x - 2)^4 + 5$

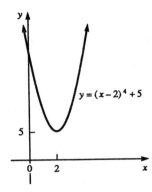

49. $y = (x + 5)^3 - 2$

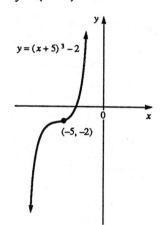

51. $y = \dfrac{3}{x - 5}$

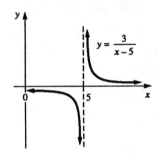

53. $y = \dfrac{2}{(x + 5)^6}$

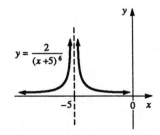

55. $y = \dfrac{7}{(x - 2)^4} + 3$

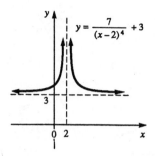

57. $y = \dfrac{2}{(x - 3)^5} - 1$

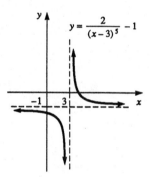

59. $xy = 30,000$

$y = \dfrac{30,000}{x}$

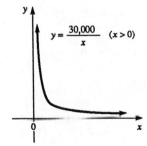

61. $y = \dfrac{200,000}{x}$ $(1000 \le x \le 5000)$

(a)

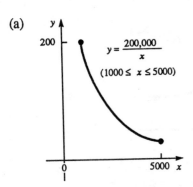

(b) $y(1000) = \dfrac{200,000}{1000} = 200$

(c) $y(2000) = 100$
(d) $y(4000) = 50$
(e) $y(5000) = 40$

63. $p = \dfrac{20,000}{x - 10}$ $(10 < x \le 5010)$

(a)

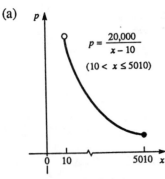

(b) $p(1010) = \dfrac{20,000}{(1010 - 10)} = 20$

(c) $p(2010) = \dfrac{20,000}{(2010 - 10)} = 10$

(d) $p(4010) = \dfrac{20,000}{(4010 - 10)} = 5$

(e) $p(5010) = \dfrac{20,000}{(5010 - 10)} = 4$

65. $C(x) = 50x + 40,000$

(a) $\bar{C}(x) =$ average cost $= \dfrac{C(x)}{x} = 50 + \dfrac{40,000}{x}$

(b)

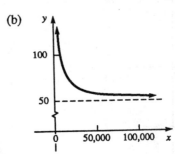

(c) $\bar{C}(1000) = 50 + \dfrac{40,000}{1000} = \$90$/unit

$\bar{C}(10,000) = 50 + \dfrac{40,000}{10,000} = \$54$/unit

$\bar{C}(100,000) = 50 + \dfrac{40,000}{100,000} = \$50.40$/unit

Each value, $\bar{C}(x)$, is the average cost, in dollars, of producing each of x units.

(d) $\lim_{x \to +\infty} \bar{C}(x) = \$50$/unit

67. $\bar{C}(x) = 20x + 60,000$

(a) $\bar{C}(x) = \text{average cost} = \dfrac{C(x)}{x} = 20 + \dfrac{60,000}{x}$

(b)
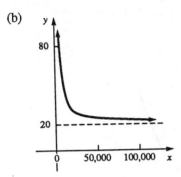

(c) $\bar{C}(1000) = 20 + \dfrac{60,000}{1000} = \$80/\text{unit}$

$\bar{C}(10,000) = 20 + \dfrac{60,000}{10,000} = \$26/\text{unit}$

$\bar{C}(100,000) = 20 + \dfrac{60,000}{100,000} = \$20.60/\text{unit}$

Each value, $\bar{C}(x)$, is the average cost, in dollars of producing each of x units.

(d) $\lim\limits_{x \to \infty} \bar{C}(x) = \$20/\text{unit}$

2–5 Graphing Polynomial Functions (Optional)

1. $f(x) = x^3 - 7x + 6 = (x - 1)(x + 3)(x - 2)$

	$-\infty$	-3		1		2		$+\infty$
sign of $(x - 1)$	$-$	$-$	$-$	0	$+$	$+$	$+$	$+$
sign of $(x + 3)$	$-$	$-$	0	$+$	$+$	$+$	$+$	$+$
sign of $(x - 2)$	$-$	$-$	$-$	$-$	$-$	$-$	0	$+$	$+$
sign of $f(x)$	$-$	$-$	0	$+$	0	$-$	0	$+$	$+$

x–intercepts: $(-3,0)$, $(1,0)$, $(2,0)$
y–intercept: $(0,6)$

(continued)

1. (continued)
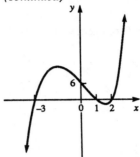

$f(x) = x^3 - 7x + 6$
$\quad = (x - 1)(x + 3)(x - 2)$

3. $f(x) = x^3 - 7x^2 + 11x - 5 = (x - 1)^2(x - 5)$

x–intercepts: $(1,0)$, $(5,0)$ (tangency at $(1,0)$)
y–intercept: $(0,-5)$
$\lim\limits_{x \to +\infty} f(x) = +\infty$; $\lim\limits_{x \to -\infty} f(x) = -\infty$

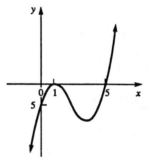

$f(x) = x^3 - 7x^2 + 11x - 5$
$\quad = (x - 1)^2(x - 5)$

5. $y = 3x^4 - 3x^3 - 9x^2 + 15x - 6 = 3(x - 1)^3(x + 2)$

x–intercepts: $(1,0)$, $(-2,0)$
y–intercept: $(0,-6)$
$\lim\limits_{x \to +\infty} f(x) = +\infty$; $\lim\limits_{x \to -\infty} f(x) = +\infty$

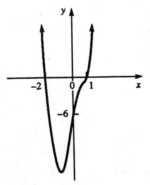

$y = 3x^4 - 3x^3 - 9x^2 + 15x - 6$
$\quad = 3(x - 1)^3(x + 2)$

7. $f(x) = x^3 - 3x^2 - 9x + 5 = (x - 1)^2(x + 5)$

x–intercepts: $(1,0)$, $(-5,0)$(tangency at $(1,0)$)
y–intercept: $(0,5)$
$\lim\limits_{x \to +\infty} f(x) = +\infty$; $\lim\limits_{x \to -\infty} f(x) = -\infty$

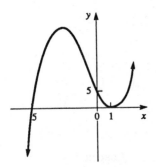

$f(x) = x^3 + 3x^2 - 9x + 5$
  $= (x - 1)^2(x + 5)$

9. $y = -x^3 + 36x = -x(x^2 - 36) = -x(x + 6)(x - 6)$

x–intercept: $(0,0)$, $(-6,0)$, $(6,0)$
y–intercept: $(0,0)$
$\lim\limits_{x \to +\infty} f(x) = -\infty$; $\lim\limits_{x \to -\infty} f(x) = +\infty$

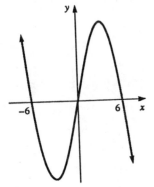

$f(x) = -x^3 + 36x$
  $= -x(x^2 - 36)$
  $= -x(x - 6)(x + 6)$

11. $f(x) = x^3 - 19x - 30 = (x + 2)(x - 5)(x + 3)$

x–intercepts: $(-2,0)$, $(5,0)$, $(-3,0)$
y–intercept: $(0,-30)$
$\lim\limits_{x \to +\infty} f(x) = +\infty$; $\lim\limits_{x \to -\infty} f(x) = -\infty$

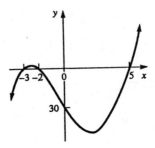

$f(x) = x^3 - 19x - 30$
  $= (x + 2)(x - 5)(x + 3)$

13. $f(x) = x^3 + x^2 - 5x + 3 = (x - 1)^2(x + 3)$

x–intercepts: $(1,0)$, $(-3,0)$ (tangency at $(1,0)$)
y–intercept: $(0,3)$
$\lim\limits_{x \to +\infty} f(x) = +\infty$; $\lim\limits_{x \to -\infty} f(x) = -\infty$

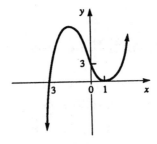

$f(x) = x^3 + x^2 - 5x + 3$
  $= (x - 1)^2(x + 3)$

15. $y = x^3 + 3x^2 - 16x + 12 = (x - 1)(x + 6)(x - 2)$

x–intercepts: (1,0), (2,0), (–6,0)
y–intercept: (0,12)
$\lim\limits_{x \to +\infty} f(x) = +\infty$; $\lim\limits_{x \to -\infty} f(x) = -\infty$

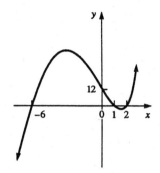

$$y = x^3 + 3x^2 - 16x + 12$$
$$= (x - 1)(x + 6)(x - 2)$$

17. $f(x) = \dfrac{1}{2}(x - 2)^3(x + 5)$

x–intercepts: (2,0), (–5,0)
y–intercept: (0,–20)
$\lim\limits_{x \to +\infty} f(x) = +\infty$; $\lim\limits_{x \to -\infty} f(x) = +\infty$

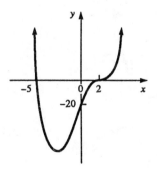

$$f(x) = \frac{1}{2}(x - 2)^3(x + 5)$$

19. $f(x) = -3x^5(x - 2)^3(x + 4)$

x–intercepts: (0,0), (2,0), (–4,0)
y–intercept: (0,0)
$\lim\limits_{x \to +\infty} f(x) = -\infty$; $\lim\limits_{x \to -\infty} f(x) = +\infty$

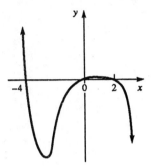

$$f(x) = -3x^5(x - 2)^3(x + 4)$$

21. $f(x) = 2x^3(x - 5)(x + 6)^2$

x–intercepts: (0,0), (5,0), (–6,0) (tangency at
(–6,0))

y–intercept: (0,0)
$\lim\limits_{x \to +\infty} f(x) = +\infty$; $\lim\limits_{x \to -\infty} f(x) = +\infty$

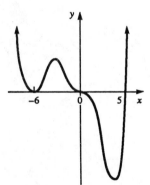

$$f(x) = 2x^3(x - 5)(x + 6)^2$$

23. $y = (x - 2)(x + 3)^4(x - 1)$

x–intercepts: (2,0), (–3,0), (1,0) (tangency at
(–3,0))

y–intercept: (0,162)

$\lim\limits_{x \to +\infty} f(x) = +\infty$; $\lim\limits_{x \to -\infty} f(x) = +\infty$

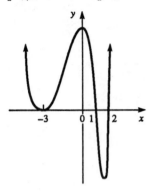

$y = (x - 2)(x + 3)^4(x - 1)$

25. $p(x) = -0.0001x^4 + 0.005x^3 - 0.07x^2 + 0.3x$
$(0 \le x \le 30)$
$= -0.0001x(x - 10)^2(x - 30)$

(a) x–intercepts: (0,0), (10,0), (30,0) (tangency
at (10,0))

y–intercept: (0,0)

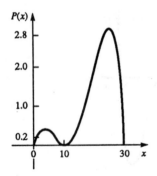

$P(x) = -0.0001\, x(x - 10)^2(x - 30)$

(b) $p(x) = 0$ at $x = 0, 10, 30$

(c) $p(x) > 0$ if $0 < x < 10$

(d) $p(x) > 0$ if $10 < x < 30$

(e) $p(20) = -0.0001(20)(20 - 10)^2(20 - 30) = 2$

27.   (a) $V = x(10 - 2x)(20 - 2x)$

(b)

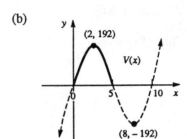

(c) $V \ge 0$ if $0 \le x \le 5$. This makes sense, since
one–half of 10, the length of the shorter side,
is 5.

(d) $V(x)$ is maximal at $x \approx 2$ inches.

29. $S(t) = t^3 - 16t^2 = t^2(t - 16)$

(a)

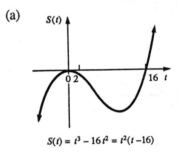

$S(t) = t^3 - 16t^2 = t^2(t - 16)$

(b) $S(5) = 25(5 - 16) = -275$

(c) $S(20) = 400(20 - 16) = 1600$

(d) $S(t) = 0 \Rightarrow t = 0, 16$

2–6 Graphing Rational Functions (Optional)

1. $y = \dfrac{(x-3)^2(x+8)}{(x-1)^2}$

y–intercept: (0,72)
x–intercepts: (3,0), (–8,0)(tangency at (3,0))
vertical asymptotes: $x = 1$; even exponent
$\lim\limits_{x \to +\infty} f(x) = +\infty$; $\lim\limits_{x \to -\infty} (x) = -\infty$

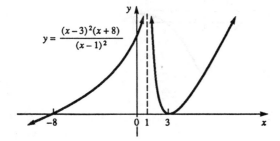

3. $f(x) = \dfrac{(x-1)(x+3)^2}{(x+1)^3(x+4)}$

y–intercept: $\left[0, -\dfrac{9}{4}\right]$

x–intercepts: (1,0), (–3,0)(tangency at (–3,0))
vertical asymptotes: $x = -1$, $x = -4$; odd exponents
$\lim\limits_{x \to +\infty} f(x) = 0$; $\lim\limits_{x \to -\infty} f(x) = 0$

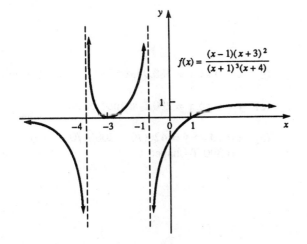

5. $y = \dfrac{1}{x-4}$

y–intercept: $\left[0, -\dfrac{1}{4}\right]$

vertical asymptote at $x = 4$
$\lim\limits_{x \to +\infty} f(x) = 0$; $\lim\limits_{x \to -\infty} f(x) = 0$

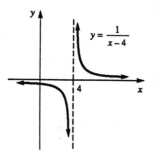

7. $y = \dfrac{3x+12}{x-2} = \dfrac{3(x+4)}{x-2}$
x–intercept: (–4,0)
y–intercept: (0,–6)
vertical asymptote at $x = 2$
$\lim\limits_{x \to +\infty} f(x) = 3$; $\lim\limits_{x \to -\infty} f(x) = 3$

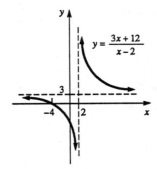

9. $f(x) = \dfrac{2x - 16}{x + 1} = \dfrac{2(x - 8)}{x + 1}$

x–intercept: (8,0)

y–intercept: (0,16)

vertical asymptote at $x = -1$

$\lim\limits_{x \to +\infty} f(x) = 2$; $\lim\limits_{x \to -\infty} f(x) = 2$

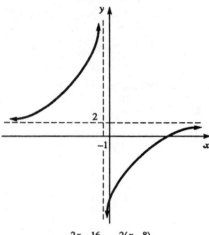

$$f(x) = \frac{2x - 16}{x + 1} = \frac{2(x - 8)}{x + 1}$$

11. $y = \dfrac{x^2 - 9x}{3x - 15} = \dfrac{x(x - 9)}{3(x - 5)}$

x–intercepts: (0,0), (9,0)

y–intercept: (0,0)

vertical asymptote: $x = 5$

$\lim\limits_{x \to +\infty} f(x) = +\infty$; $\lim\limits_{x \to -\infty} f(x) = -\infty$

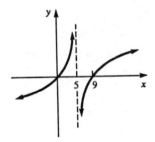

$$y = \frac{x^2 - 9x}{3x - 15} = \frac{x(x - 9)}{3(x - 5)}$$

13. $f(x) = \dfrac{(x - 3)(x + 3)}{x^2 + 1}$

There are no vertical asymptotes since the denominator of the function can never become zero.

x	-2	-1	$-\frac{1}{2}$	0	$\frac{1}{2}$	1	2
$f(x)$	-1	-4	-7	-9	-7	-6	-1

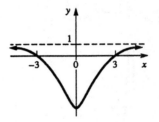

15. $y = \dfrac{420(x - 200)}{x - 210}$ $(0 \le x \le 200)$

(a)

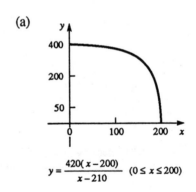

$$y = \frac{420(x - 200)}{x - 210} \quad (0 \le x \le 200)$$

(b) $x = 0 \Rightarrow y = 380.95$, or 381 T–shaped beams

(c) $y = 0 \Rightarrow x = 200$ I–shaped beams

(d) $x = 70 \Rightarrow y = 420(70 - 200)/(70 - 210)$
     $= 390$ T–shaped beams

17. $y = \dfrac{40x}{110 - x}$ $(0 \le x \le 300)$

(a)

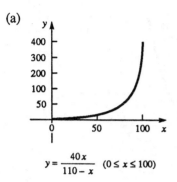

$y = \dfrac{40x}{110 - x}$ $(0 \le x \le 100)$

(b)

x, %	y, $ millions
10	4.000
20	8.889
50	33.333
80	106.667
100	400.000

Extra Dividends

Model Fitting: Goodness of Fit

1.
x	1	2	3	4	5
y	40	80	200	300	530

For the model $y = 20x^2 + 10$, the sum of squares
error is $[20(1)^2 + 10 - 40]^2 + [20(2)^2 + 10 - 80]^2 +$
$[20(3)^2 + 10 - 200]^2 + [20(4)^2 + 10 - 300]^2$
$+ [20(5)^2 + 10 - 530]^2 = 1600$
For the model $y = 100x - 90$, the sum of squares
error is $[100(1) - 90 - 40]^2 + [100(2) - 90 - 80]^2$
$+ [100(3) - 90) - 200]^2 + [100(4) - 90 - 300]^2$
$+ [100(5) - 90 - 530]^2 = 16,400$

Thus, the quadratic model, $y = 20x^2 + 10$, gives the
better fit to the given data.

3. $y = 48.4x + 768$ sum of squares error = 1,568,435
    $y = 1.2x^2 + 1111$ sum of squares error = 1,596,801

The linear model is slightly better than the
quadratic model.

Review Exercises

1. $y = |x - 7|$

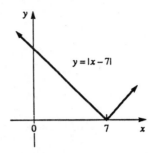

3. $y = -|x + 2| + 1$

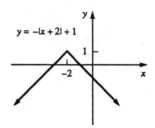

5. $f(x) = 1/x^3$

x	-3	-2	-1	0	1	2	3
f(x)	$-\frac{1}{27}$	$-\frac{1}{8}$	-1	*	1	$\frac{1}{8}$	$\frac{1}{27}$

*Not defined
f is odd, since $f(-x) = -f(x)$.

7. $f(x) = x^2 - 25$

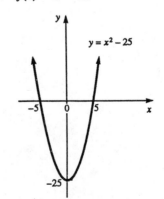

9. $f(x) = -3(x + 2)^2 + 4$

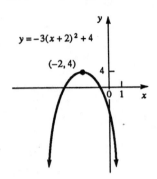

11. $y = -4x^2 + 24x = -4(x - 3)^2 + 36$

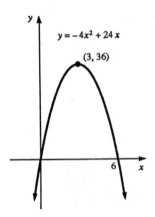

13. $f(x) = x^2 + 2x - 15 = (x + 1)^2 - 16$

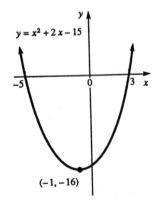

15. $f(x) = x^2 - 8x + 16 = (x - 4)^2$

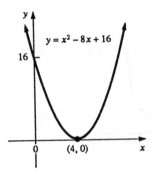

17. $y = (x + 4)(x - 6) = x^2 - 2x - 24 = (x - 1)^2 - 25$

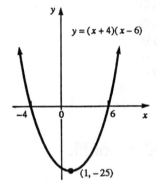

19. $p = -2x + 140$

(a) $R(x) = xp = -2x^2 + 140x$

(b) $R(x) = -2(x^2 - 70x) = -2(x - 35)^2 + 2450$

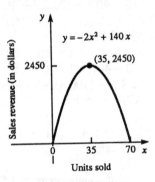

(c) $R_{max} = R(35) = \$2450$

(d) $R = 2450 \Rightarrow x = 35$ units

(e) $x = 35 \Rightarrow p = -2(35) + 140 = \$70$

21. $p = x^2 + 8$ supply function $(x > 0)$
$p = (x - 4)^2$ demand function $(0 < x \le 4)$

(a) $x^2 + 8 = (x - 4)^2 \Rightarrow 8 = -8x + 16 \Rightarrow x = 1$, equilibrium quantity

(b) $p = 1^2 + 8 = \$9$, equilibrium price

(c)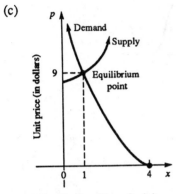

23. $y = x^6$

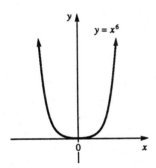

25. $y = 5x^8$

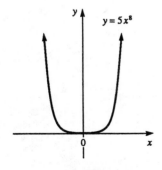

27. $y = 3/x^4$

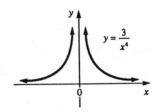

29. $y = -8/x^3$

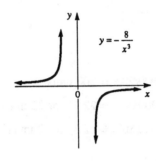

31. $y = -4(x - 2)^4 + 3$

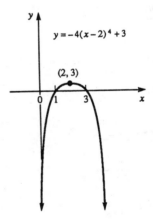

33. $y = 8,000,000/x^2$ $(200 \le x \le 800)$

(a)

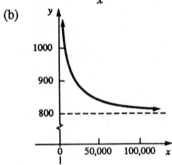

(b) $y(200) = 8,000,000/200^2 = 200$ units

(c) $y(600) = 8,000,000/600^2 = 22.22$, or 22 units

(d) $y(800) = 8,000,000/800^2 = 12.5$, or 12 units

35. $C(x) = 800x + 480,000$ $(x > 0)$

(a) average cost function $= \bar{C}(x) = \dfrac{C(x)}{x}$

$$= 800 + \dfrac{480,000}{x}$$

(b)

$\bar{C}(x) = 800 + \dfrac{480,000}{x}$ $(x > 0)$

(c) $\bar{C}(2000) = 800 + \dfrac{480,000}{2000} = \$1040/\text{unit}$

$\bar{C}(10,000) = 800 + \dfrac{480,000}{10,000} = \$848/\text{unit}$

$\bar{C}(100,000) = 800 + \dfrac{480,000}{100,000} = \$804.80/\text{unit}$

$\bar{C}(x)$ is the cost per unit when x units are produced.

(d) $\lim\limits_{x \to +\infty} \bar{C}(x) = \$800/\text{unit}$

37. $f(x) = (x + 2)^3(x - 6)^2$

x–intercepts: $(-2,0)$, $(-6,0)$ (tangency at $(-6,0)$)
y–intercept: $(0,288)$
$\lim\limits_{x \to +\infty} f(x) = +\infty$; $\lim\limits_{x \to -\infty} f(x) = -\infty$

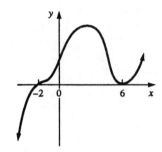

$f(x) = (x + 2)^3(x - 6)^2$

39. $f(x) = x^3 - 4x^2 + 4x = x(x^2 - 4x + 4) = x(x - 2)^2$

x–intercepts: $(0,0)$, $(2,0)$ (tangency at $(2,0)$)
y–intercept: $(0,0)$
$\lim\limits_{x \to +\infty} f(x) = +\infty$; $\lim\limits_{x \to -\infty} f(x) = -\infty$

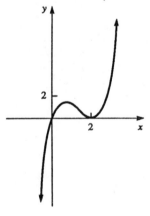

$f(x) = x^3 - 4x^2 + 4x$
$= x(x - 2)^2$

41. $f(x) = \dfrac{(x+2)(x-5)^2}{x-1}$

x–intercepts: $(-2,0)$, $(5,0)$ (tangency at $(5,0)$)
y–intercept: $(0,-50)$
vertical asymptote at $x = 1$
$\lim\limits_{x \to +\infty} f(x) = +\infty$; $\lim\limits_{x \to -\infty} f(x) = +\infty$

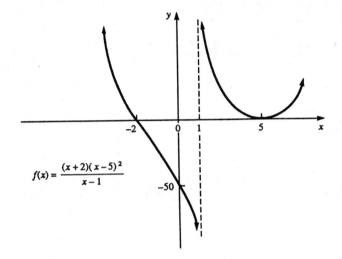

$f(x) = \dfrac{(x+2)(x-5)^2}{x-1}$

43. $f(x) = \dfrac{5x-40}{x-2}$

x–intercept: $(8,0)$
y–intercept: $(0,20)$
vertical asymptote at $x = 2$
$\lim\limits_{x \to +\infty} f(x) = 5$; $\lim\limits_{x \to -\infty} f(x) = 5$

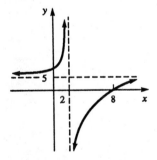

$f(x) = \dfrac{5x-40}{x-2} = \dfrac{5(x-8)}{x-2}$

45. $f(x) = \dfrac{x^2-36}{x-8} = \dfrac{(x-6)(x+6)}{x-8}$

x–intercepts: $(6,0)$, $(-6,0)$
y–intercept: $(0,9/2)$
vertical asymptote at $x = 8$
$\lim\limits_{x \to +\infty} f(x) = +\infty$; $\lim\limits_{x \to -\infty} f(x) = -\infty$

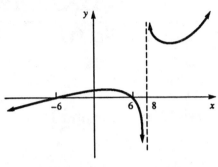

$f(x) = \dfrac{x^2-36}{x-8} = \dfrac{(x-6)(x+6)}{x-8}$

CHAPTER
3
EXPONENTIAL AND LOGARITHMIC FUNCTIONS

3–1 Exponential Functions and Their Graphs

1. $f(x) = 5^x$
   y–intercept: $(0,1)$
   horizontal asymptote: $y = 0$

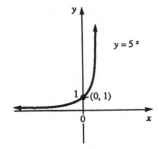

3. $y = 4^x$
   y–intercept: $(0,1)$
   horizontal asymptote: $y = 0$

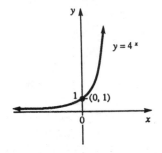

5. $y = 2 \cdot 3^x$
   y–intercept: $(0,2)$
   horizontal asymptote: $y = 0$

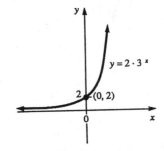

7. $f(x) = -4 \cdot 5^x$
   y–intercept: $(0,-4)$
   horizontal asymptote: $y = 0$

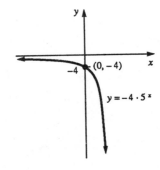

9. $y = -3 \cdot 4^x$
    y–intercept: $(0,-3)$
    horizontal asymptote: $y = 0$

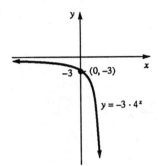

11. $f(x) = 7e^x$
     y–intercept: $(0,7)$
     horizontal asymptote: $y = 0$

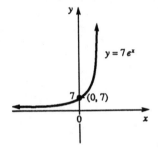

13. $y = -3e^x$
     y–intercept: $(0,-3)$
     horizontal asymptote: $y = 0$

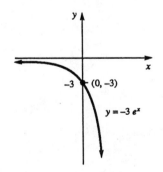

15. $y = -10e^x$
     y–intercept: $(0,-10)$
     horizontal asymptote: $y = 0$

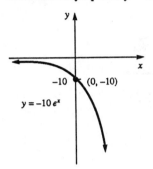

17. $f(x) = -3 \cdot 4^x + 8$
     y–intercept: $(0,5)$
     horizontal asymptote: $y = 8$

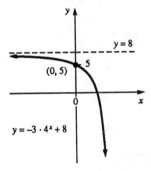

19. $y = 4e^x - 1$
     y–intercept: $(0,3)$
     horizontal asymptote: $y = -1$

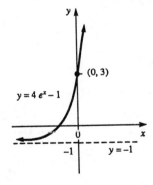

21. $f(x) = -2e^x - 5$
    y–intercept: $(0,-7)$
    horizontal asymptote: $y = -5$

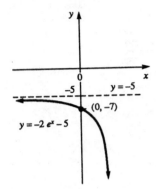

23. $f(x) = 3^{-x}$
    y–intercept: $(0,1)$
    horizontal asymptote: $y = 0$

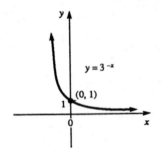

25. $y = 4 \cdot 5^{-x}$
    y–intercept: $(0,4)$
    horizontal asymptote: $y = 0$

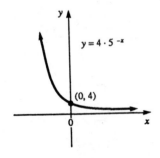

27. $y = 3 \cdot 4^{-x}$
    y–intercept: $(0,3)$
    horizontal asymptote: $y = 0$

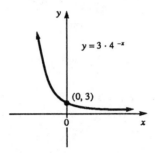

29. $f(x) = -2 \cdot 3^{-x}$
    y–intercept: $(0,-2)$
    horizontal asymptote: $y = 0$

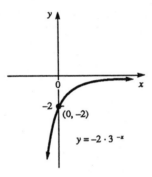

31. $y = 3e^{-x}$
    y–intercept: $(0,3)$
    horizontal asymptote: $y = 0$

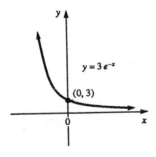

33. $y = 10e^{-x}$
 y–intercept: $(0,10)$
 horizontal asymptote: $y = 0$

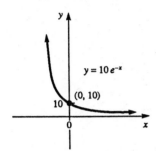

35. $f(x) = -7e^{-x}$
 y–intercept: $(0,-7)$
 horizontal asymptote: $y = 0$

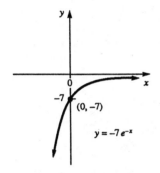

37. $y = 2 \cdot 3^{-x} + 5$
 y–intercept: $(0,7)$
 horizontal asymptote: $y = 5$

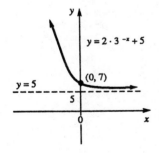

39. $f(x) = 4e^{-x} + 1$
 y–intercept: $(0,5)$
 horizontal asymptote: $y = 1$

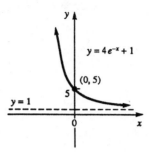

41. $y = -2e^{-x} + 30$
 y–intercept: $(0,28)$
 horizontal asymptote: $y = 30$

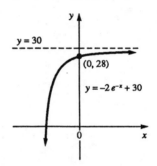

43. $y = 30e^{-x} + 30$
 y–intercept: $(0,60)$
 horizontal asymptote: $y = 30$

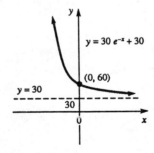

45. $y = 10(1 - e^{-x})$
y–intercept: $(0,0)$
horizontal asymptote: $y = 10$

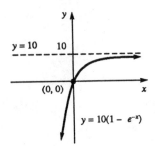

47. $y = \left[\dfrac{1}{2}\right]^x$
y–intercept: $(0,1)$
horizontal asymptote: $y = 0$

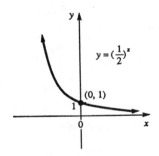

49. $y = \left[\dfrac{1}{2}\right]^{-x}$
y–intercept: $(0,1)$
horizontal asymptote: $y = 0$

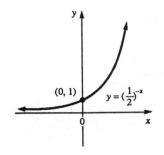

51. $y = 5 \cdot \left[\dfrac{1}{5}\right]^x + 1$
y–intercept: $(0,6)$ horizontal asymptote: $y = 1$

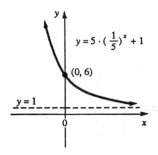

53. $y = 3e^{2x} + 1$
y–intercept: $(0,4)$
horizontal asymptote: $y = 1$

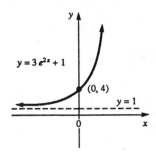

55. $f(x) = -2e^{0.05x} + 1$
y–intercept: $(0,-1)$
horizontal asymptote: $y = 1$

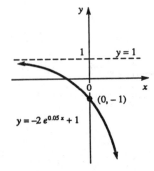

57. $y = 10e^{0.07x}$
y–intercept: $(0,10)$
horizontal asymptotes: $y = 0$

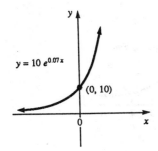

59. $f(x) = 7e^{-0.10x}$
y–intercept: $(0,7)$
horizontal asymptote: $y = 0$

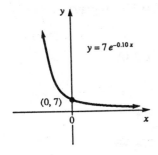

61. $y = 10(1 - e^{-0.20x})$
y–intercept: $(0,0)$
horizontal asymptote: $y = 10$

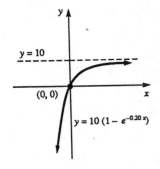

63. $y = -8e^{-0.40x}$
y–intercept: $(0,-8)$
horizontal asymptote: $y = 0$

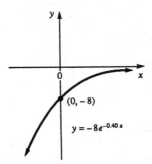

65. (a) $y = 500,000(3^t)$

(b)

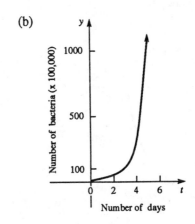

(c) $y(4) = 500,000(3^4) = 40,500,000$

(d) $y(6) = 500,000(3^6) = 364,500,000$

67. $y = 3 \cdot 2^t$

(a)

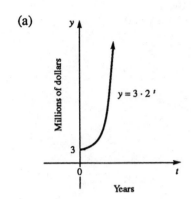

(b) 19×0: \$3,000,000; 19×1: \$6,000,000;
19×2: \$12,000,000

69. $y = 1000e^{0.05t}$ $(t \geq 0)$

(a)

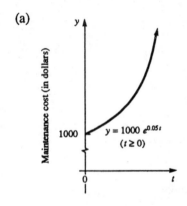

(b) $y(2) = 1000e^{0.10} \approx 1105.17$

71. $y(t) = 2000e^{-0.60t}$ $(t \geq 0)$

(a)

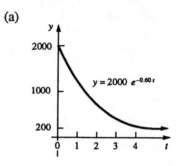

(b) $y(0) = 2000e^0 = 2000$ grams

(c) $y\left[\dfrac{1}{2}\right] = 2000e^{-0.30} \approx 1481.64$ grams

(d) $y(3) = 2000e^{-1.80} \approx 330.60$ grams

73. $N(x) = 50 - 50e^{-0.3x}$ $(x \geq 0)$

(a)

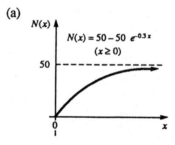

(b) $N(5) = 50(1 - e^{-1.5}) \approx 39$

(c) $\lim\limits_{x \to \infty} N(x) = 50$

75. $T = 38.6e^{-0.05t} + 60 \ (t \geq 0)$

(a)

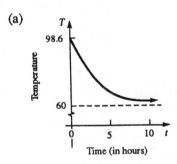

(b) The room temperature is 60°.

(c) $T(2) = 38.6e^{-0.10} + 60 \approx 94.9°$

77. $y = 80 - 60e^{-0.70x}$

(a)

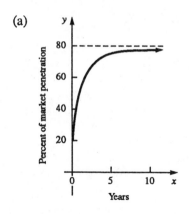

(b) $y(2) = 80 - 60e^{-1.4} \approx 65.20\%$

(c) $y(10) = 80 - 60e^{-7} \approx 79.95\%$

(d) The saturation level is 80%.

3–2 Logarithmic Functions

1. $5^2 = 25 \Rightarrow 2 = \log_5 25$

3. $2^6 = 64 \Rightarrow 6 = \log_2 64$

5. $10^{-2} = 0.01 \Rightarrow -2 = \log_{10} 0.01$

7. $t^w = s \Rightarrow w = \log_t s$

9. $b^{x+y} = N \Rightarrow x + y = \log_b N$

11. $y = \log_9 81 \Rightarrow y = 2$

13. $y = \log_3 1 \Rightarrow y = 0$

15. $y = \log_7 7 \Rightarrow y = 1$

17. $y = \log_{10} 10 \Rightarrow y = 1$

19. $y = \log_{10} 1000 \Rightarrow y = 3$

21. $y = \log_{10} 100,000 \Rightarrow y = 5$

23. $y = \ln e^2 = 2$

25. $\log_2 32 = 5$, since $2^5 = 32$

27. $\log_3 1 = 0$

29. $\log_8 1 = 0$

31. $\log 10 = \log_{10} 10 = 1$

33. $\log 1000 = 3$

35. $\log 100,000 = 5$

37. $y = \log_3 x$
x–intercept $= (1,0)$

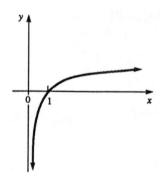

39. $\log 3.5 = \log 7 - \log 2 = 0.5441$

41. $\log 49 = \log 7^2 = 2 \log 7 = 1.6902$

43. $\log 56 = \log 2^3 + \log 7 = 3 \log 2 + \log 7 = 1.7481$

45. $\log \sqrt[3]{2} = \frac{1}{3} \log 2 = 0.1003$

47. $\log \sqrt{14} = \frac{1}{2} \log 14 = \frac{1}{2}(\log 2 + \log 7) = 0.5731$

49. $\log \sqrt{98} = \frac{1}{2} \log 98 = \frac{1}{2}(1.9912) = 0.9956$

51. $\log 371 = \log 100 + \log 3.71 = 2.5694$

53. $\log 37,100 = \log 3.71 + \log 10,000 = 4.5694$

55. $\log 0.371 = \log 3.71 - \log 10 = -0.4306$

57. $\log 0.00371 = \log 3.71 - \log 1000 = -2.4306$

59. $\ln 6 = \ln 3 + \ln 2 = 1.791759$

61. $\ln \frac{2}{3} = \ln 2 - \ln 3 = -0.405465$

63. $\ln 8 = 3 \ln 2 = 2.079441$

65. $\ln \sqrt{3} = \frac{1}{2} \ln 3 = 0.549306$

67. $\ln 0.75 = \ln \frac{3}{4} = \ln 3 - 2 \ln 2 = -0.287682$

69. $\log 8.73 = 0.9410142$

71. $\log 4760 = 3.677607$

73. $\log 0.80 = -0.09691$

75. $\ln 10 = 2.3025851$

77. $\ln 0.15 = -1.897120$

79. $\ln 80 = 4.3820266$

81. $\log x = 0.9047 \Rightarrow x = 8.0297$

83. $\log x = 3.9047 \Rightarrow x = 8029.7$

85. $\log x = 5.4099 \Rightarrow x = 256,980$

87. $\log x = 1.3610 \Rightarrow x = 22.9615$

89. $\ln x = 1.7047 \Rightarrow x = 5.4997$

91. $3.4 = e^x \Rightarrow x = \ln 3.4 = 1.2238$

93. $5.5 = e^x \Rightarrow x = \ln 5.5 = 1.7047$

95. $\log \left[\dfrac{xy}{z}\right] = \log x + \log y - \log z$

97. $\log (x^2 y) = 2 \log x + \log y$

99. $\log \sqrt{xy} = \dfrac{1}{2}(\log x + \log y)$

101. $S(x) = 100,000 + 8000 \ln (x + 2)$

    (a) $S(1) = 100,000 + 8000 \ln (1 + 2)$
           $= \$108,788.90$ thousand
        $S(3) = 100,000 + 8000 \ln (3 + 2)$
           $= \$112,875.50$ thousand
       $S(10) = 100,000 + 8000 \ln (10 + 2)$
           $= \$119,879.25$ thousand
       $S(20) = 100,000 + 8000 \ln (20 + 2)$
           $= \$124,728.34$ thousand

    (b) $S(21) - S(20) = \$125,083.95 - 124,728.34$
                  $= \$355.61$ thousand
    Spending an additional \$1000 on advertising yields \$355.61 thousand in increased revenue.

    (c) $S(31) - S(30) = \$127,972.06 - \$127,725.89$
                  $= \$246.17$
    Corresponding increase in sales is \$246.17 thousand.

103. (a) $R(1) = 13.22219 \Rightarrow$
      revenue $= \$13,222.19 \approx \$13.22$ thousand
      revenue from the sale of 1 unit
      $R(2) = 16.12784 \Rightarrow$
      revenue $= \$16,127.84 \approx \$16.13$ thousand
      revenues from the sale of 2 units
      $R(5) = 20.04321 \Rightarrow$
      revenue $= \$20,043.21 \approx \$20.04$ thousand
      revenue from the sale of 5 units
      $R(10) = 23.03196 \Rightarrow$
      revenue $= \$23,031.96 \approx \$23.03$ thousand
      revenue from the sale of 10 units

    (b) $R(11) - R(10) = 23.44392 - 23.03196$
                    $= 0.41196$
    $\Rightarrow$ increase in revenue
      $\$411.96 \approx \$0.41$ thousand
    If the company increases its sales from 10 to 11 units, revenues will increase by \$0.41 thousand.

    (c) $R(21) - R(20) = 26.24282 - 26.03144$
                    $= 0.21138$
    $\Rightarrow$ increase in revenue $= \$211.38 \approx \$0.21$ thousand.

105. $y = y_0 e^{kt}$
    If $y = 3y_0$, $3 = e^{kt}$; $\ln 3 = kt$; $t = \dfrac{\ln 3}{k}$

    $y = 2000 e^{0.20t}$

    (a) initial value of $y = 2000$

    (b) tripling time $= \dfrac{\ln 3}{k} = \dfrac{\ln 3}{0.20} \approx 5.493$ days

    (c) quadrupling time

      $= t = \dfrac{\ln 4}{k} = \dfrac{\ln 4}{0.20} \approx 6.931$ days

107. $P = 1000e^{0.05t}$

(a) $P = 2P_0 \Rightarrow 2 = e^{0.05t}$; $\ln 2 = 0.05t$;

$$t = \frac{\ln 2}{0.05} \approx 13.86 \text{ years}$$

(b) $P = 3P_0 \Rightarrow 3 = e^{0.05t}$;

$$t = \frac{\ln 3}{0.05} \approx 21.97 \text{ years}$$

109. $p = e^{1+0.3y}$

(a) $\ln p = 1 + 0.3y$; $y = \dfrac{\ln p - 1}{0.3}$

(b) $y(10) = \dfrac{\ln 10 - 1}{0.3} \approx 4.342 \text{ million units}$

3-3 Fitting Exponential Models

1. (2,6) and (8,40)

$$\text{slope} = \frac{\ln y_2 - \ln y_1}{x_2 - x_1} = \frac{\ln(40/6)}{8 - 2} \approx 0.3162$$

$\ln y - \ln y_1 = 0.3162(x - x_1)$
$\ln y = 1.7918 + 0.3162x - 0.6324$
$ = 1.1594 + 0.3162x$
$y = ab^x = e^{1.1594}(e^{0.3162x}) = 3.188(1.3719^x)$

3. (3,20) and (6,70)

$$\text{slope} = \frac{\ln(70/20)}{6 - 3} = 0.4176$$

$\ln y - \ln 20 = 0.4176(x - 3)$
$\ln y = 1.7430 + 0.4176x$
$y = 5.7143(1.5183^x)$

5. (1,15) and (5,85)

$$\text{slope} = \frac{\ln(85/15)}{5 - 1} = 0.4337$$

$\ln y - \ln 15 = 0.4337(x - 1)$
$\ln y = 2.2744 + 0.4337x$
$y = 9.7220(1.5430^x)$

7. (a) (2,1.35) and (9,3.90)

$$\text{slope} = \frac{\ln(3.90/1.35)}{9 - 2} = 0.1516$$

$\ln y - \ln 1.35 = 0.1516(x - 2)$
$\ln y = -0.0030 + 0.1516x$
$y = 0.9970(1.1637^x)$

(b) $y(11) = 5.28$ EPS

9. (a) (1,3.43) and (7,8.22)

$$\text{slope} = \frac{\ln(8.22/3.43)}{7 - 1} = 0.1457$$

$\ln y - \ln 3.43 = 0.1457(x - 1)$
$\ln y = 1.0869 + 0.1457x$
$y = 2.965(1.1568^x)$

(b) $y(8) = 9.51$ EPS

Review Exercises

1. $y = 7^x$

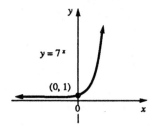

3. $y = e^x$

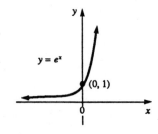

5. $y = 8e^x$

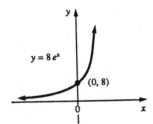

7. $y = -5e^{-x}$

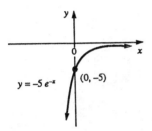

9. $f(x) = 4 + 2e^x$

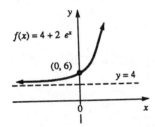

11. $f(x) = 9 - 2e^{-x}$

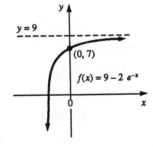

13. $f(x) = -1 + 5e^{-x}$

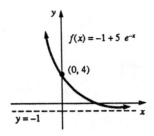

15. $f(x) = 5 + e^{-2x}$

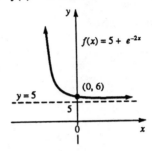

17. $y = 5e^t$

(a)

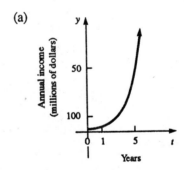

(b) $y(2) = 5e^2 \approx 36.95$ million dollars
$y(3) = 5e^3 \approx 100.43$ million dollars
$y(5) = 5e^5 \approx 742.07$ million dollars

19. $y = 80 - 74e^{-0.27x}$

(a)

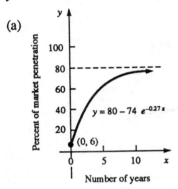

(b) $y(3) = 80 - 74e^{-0.81} \approx 47.08\%$

(c) $y(10) = 80 - 74e^{-2.7} \approx 75.03\%$

(d) The saturation level is 80%.

21. $10^4 = 10,000$
$4 \log 10 = \log 10,000; \; 4 = \log 10,000$

23. $y = e^x$
$\ln y = x$

25. $\log_3 9 = 2$

27. $\log 1 = 0$

29. $\log 10 = 1$

31. $\ln 0.987 \approx -0.0130852$

33. $\log x = 0.8875 \Rightarrow x = 10^{0.8875} \approx 7.7179$

35. $\log x = 4.56 \Rightarrow x = 10^{4.56} \approx 36.307$

37. $4.6 \approx e^{1.5260563}$

39. $\log (st/r) = \log s + \log t - \log r$

41. $\log (uv)^5 = 5(\log u + \log v)$

43. $\log uv = \log u + \log v$

45. $y = \ln x$

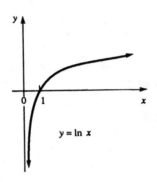

47. $(2,8)$ and $(5,60)$

$\text{slope} = \dfrac{\ln \, (60/8)}{5 - 2} = 0.6716$

$\ln \, y = \ln 8 + 0.6716(x - 2) = 0.7362 + 0.6716x$
$y = 2.0879(1.9574^x)$

CHAPTER
4
Mathematics of Finance

4–1 Simple Interest and Discount

1. $I = Prt = 1000(0.07)(3) = \$210$
   $S = P + I = 1000 + 210 = \$1210$

3. $I = Prt = 5000(0.08)(6/12) = \$200$
   $S = P + I = 5000 + 200 = \$5200$

5. $I = Prt = 2000(0.09)(4/12) = \$60$
   $S = P + I = 2000 + 60 = \$2060$

7. (a) $S = P + Prt = P(1 + rt) = 10,000(1 + 0.09t)$
   $= 10,000 + 900t$

   (b)

   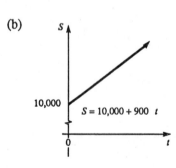

   $S = 10,000 + 900\ t$

   (c) Annual increase $= Pr =$ slope of graph

9. $P = \dfrac{S}{1 + rt} = \dfrac{8000}{1 + 0.09(2)} = \$6779.66$

11. $P = \dfrac{S}{1 + rt} = \dfrac{15,000}{1 + 0.08(2.5)} = \$12,500$

13. $P = \dfrac{S}{1 + rt} = \dfrac{20,000}{1 + 0.08(3/12)} = \$19,607.84$

15. $B = S(1 - dt) = 10,000[1 - 0.09(4/12)] = \$9700$

17. $B = S(1 - dt) = 5000[1 - 0.06(6/12)] = \$4850$

19. $B = S(1 - dt) = 30,000[1 - 0.09(2/3)] = \$28,200$

21. $r = \dfrac{I}{Pr} = \dfrac{300}{9700(4/12)} = 0.0927835$, or 9.27835%

23. $r = \dfrac{I}{Pt} = \dfrac{150}{4850(6/12)} = 0.0618557$, or 6.18557%

25. $r = \dfrac{I}{Pt} = \dfrac{1800}{28,200(2/3)} = 0.0957447$, or 9.57447%

27. $S = \dfrac{B}{1 - dt} = \dfrac{8000}{1 - 0.08(9/12)} = \$8510.64$

29. $S = \dfrac{B}{1 - dt} = \dfrac{12,000}{1 - 0.12(10/12)} = \$13,333.33$

31. $S = \dfrac{B}{1 - dt} = \dfrac{9000}{1 - 0.10(1/2)} = \$9473.68$

33. $S = P(1 + rt) = 10,000[1 + 0.08(9/12)]$
    $\qquad = \$10,600$, maturity value
    $B = S(1 - dt) = 10,600[1 - 0.06(2/12)]$
    $\qquad = \$10,494$, proceeds

35. $S = P(1 + rt) = 14,000[1 + 0.08(3/12)]$
    $\qquad = \$14,280$, maturity value
    $B = S(1 - dt) = 14,280[1 - 0.06(1/12)]$
    $\qquad = \$14,208.60$, proceeds

37. $S = P(1 + rt) = 30,000[1 + 0.09(18/12)]$
    $\qquad = \$34,050$, maturity value
    $B = S(1 - dt) = 34,050[1 - 0.08(9/12)]$
    $\qquad = \$32,007$, proceeds

39. $P = \dfrac{S}{1 + rt} = \dfrac{10,000}{1 + 0.08(9/12)} = \$9433.96$

41. $P = \dfrac{S}{1 + rt} = \dfrac{10,000}{1 + 0.10(5)} = \$6666.67$

43. (a) $D = Sdt = 6000(0.09)(5) = \$2700$, discount

    (b) $B = S(1 - dt) = 6000 - 2700 = \$3300$, proceeds

    (c) Equivalent simple interest rate $= \dfrac{2700}{3300(5)}$
       $= 0.163634$, or 16.3634%

45. (a) $S = \dfrac{B}{1 - dt} = \dfrac{5000}{1 - 0.07(4/12)}$
       $= \$5119.45$, maturity value

    (b) Equivalent simple interest rate
       $= \dfrac{5119.45 - 5000}{5000(4/12)}$
       $= 0.07167236$, or 7.16736%

47. $S = P(1 + rt) = 600(1 + 0.06(4/12)$
    $\qquad = \$612$, maturity value

49. (a) $S = 20,000(1 + 0.1(1.5))$
       $\qquad = \$23,000$, maturity value
       $B = S(1 - dt) = 23,000(1 - 0.09(6/12))$
       $\qquad = 21,965$, paid by investment firm

    (b) Amount earned $= 23,000 - 21,965 = \$1035$

51. $S = 30,000(1 + 0.12(6/12))$
    $\qquad = \$31,800$, maturity value
    $B = S(1 - dt) = 31,800(1 - 0.15(6/12))$
    $\qquad = \$29,415$
    (Answer c)

4–2 Compound Interest

In Problems 1–7, the formulas $S = P(1 + i)^n$ and $i = r/m$ are used to calculate a compound amount.
Interest $I = S - P$.

1. $S = 1000\left[1 + \dfrac{0.08}{4}\right]^{40} = \$2208.04$
   $I = \$1208.04$

3. $S = 8000\left[1 + \dfrac{0.06}{1}\right]^{10} = \$14{,}326.78$

   $I = \$6326.78$

5. $S = 5000\left[1 + \dfrac{0.04}{4}\right]^{(6)(4)} = \$6348.67$

   $I = \$1348.67$

7. $S = 3000\left[1 + \dfrac{0.12}{12}\right]^{(4)(12)} = \$4836.68$

   $I = \$1836.68$

In problems 9–11, the formula $P\ \dfrac{S}{\left[1 + \dfrac{r}{m}\right]^{mt}}$ is used to calculate a present value.

9. $P = \dfrac{8000}{(1 + 0.08/2)^{(2)(5)}} = \$5404.51$

11. $P = \dfrac{6000}{(1 + 0.08/4)^{(4)(9)}} = \$2941.34$

13. $P = \dfrac{9000}{(1 + 0.08)^{20}} = \$1930.93$

15. $P = \dfrac{8000}{(1 + 0.08/4)^{(4)(4)}} = \$5827.57$

17. $P = \dfrac{20{,}000}{(1 + 0.08/2)^{(2)(20)}} = \$4165.78$

19. (a) $S = 10{,}000\left[1 + \dfrac{0.08}{4}\right]^{(10)(4)} = \$22{,}080.40$

    (b) $I = \$12{,}080.40$

21. $S = 500(1 + 0.05)^{20} = \$1326.65$

23. $S = 5000\left[1 + \dfrac{0.08}{2}\right]^{(3)(2)}\left[1 + \dfrac{0.10}{2}\right]^{(7)(2)}$

    $\qquad + 6000\left[1 + \dfrac{0.10}{2}\right]^{(7)(2)}$

    $\qquad = 12{,}526.23 + 11{,}879.59 = \$24{,}405.82$

25. $S = 600\left[1 + \dfrac{0.08}{4}\right]^{(11)(4)} + 300\left[1 + \dfrac{0.08}{4}\right]^{(5)(4)}$

    $\qquad = 1434.03 + 445.79 = \$1879.82$

27. $P = \dfrac{1000}{(1 + 0.06/2)^{(2)(5)}} = \$744.09$

29. (a) $P = \dfrac{10{,}000}{(1 + 0.12/4)^{(4)(3)}} = \$7013.80$

    (b) $P = \dfrac{10{,}000}{(1 + 0.12/4)^{4}} = \$8884.87$

In Problems 31–33, the formula $r = \left[1 + \dfrac{i}{m}\right]^{m} - 1$ is used to calculate effective annual interest rate.

31. $r = \left[1 + \dfrac{0.06}{2}\right]^{2} - 1 = 0.0609$, or 6.09%

33. $r = \left[1 + \dfrac{0.12}{12}\right]^{12} - 1 = 0.126825$, or 12.6825%

In Problems 35–37, the formula $t = \dfrac{\ln 2}{m\,\ln(1 + i/m)}$ is used to calculate doubling time. In Problems 39–41 ln 2 is replaced by $\ln 3$ or $\ln 4$.

35. $t = \dfrac{\ln 2}{2\,\ln(1 + 0.06/2)} \approx 11.72 \approx 12$ years

37. $t = \dfrac{\ln 2}{\ln(1 + 0.08)} \approx 9.0065 \approx 9$ years

39. $t = \dfrac{ln\ 3}{4\ ln(1 + 0.08/4)} \approx 13.87 \approx 14$ years

41. $t = \dfrac{ln\ 4}{4\ ln(1 + 0.08/4)} \approx 17.50$ years

43. $\left[1 + \dfrac{0.07}{12}\right]^{(12)(3)} = 1.232926$

45. $\left[1 + \dfrac{0.10}{12}\right]^{(12)(4)} = 1.489354$

47. $\left[1 + \dfrac{0.09}{52}\right]^{(52)(2)} = 1.197031$

49. $\left[1 + \dfrac{0.09}{365}\right]^{(365)(3)} 1.309921$

51. $\left[1 + \dfrac{0.08}{365}\right]^{(365)(4)} = 1.377079$

53. $S = 10,000\left[1 + \dfrac{0.08}{365}\right]^{(365)(5)} = \$14,917.59$

55. $P = \dfrac{50,000}{(1 + 0.07/365)^{(365)(3)}} = \$40,530.03$

57. $\left[1 + \dfrac{r}{2}\right]^2 = \left[1 + \dfrac{0.08}{4}\right]^4 = 1 + \dfrac{r}{2} = 1.0404$

$r = 0.0808$, or 8.08%

59. $\left[1 + \dfrac{r}{12}\right]^{12} = \left[1 + \dfrac{0.06}{2}\right]^2$

$1 + \dfrac{r}{12} = (1 + 0.03)^{1/6} \approx 1.00493862$

$r = 0.0592635$, or 5.92635%

61. $1 + 0.088 = 1.088$

$\left[1 + \dfrac{0.085}{365}\right]^{365} = 1.088706$

Since $1.088706 > 1.088$, choose 8.5% compounded daily.

63. $\left[1 + \dfrac{0.0875}{12}\right]^{12} = 1.0911$

Since $1.0911 > 1.09$, choose 8.75% compounded monthly.

65. $S = Pe^{rt} = 5000\ e^{(0.05)(8)} = \$7459.12$

67. $S = Pe^{rt} = 6000\ e^{(0.06)(5)} = \$8099.15$

69. $P = Se^{-rt} = 10,000\ e^{-(0.10)(7)} = \$4965.85$

71. $P = Se^{-rt} = 4000\ e^{-(0.07)(2)} = \$3477.43$

73. $r = e^{0.06} - 1 \approx 0.0725082$, or 7.25082%

75. $r = e^{0.09} - 1 \approx 0.0941743$, or 9.41743%

4–3 Geometric Series and Annuities

1. Sum $= 2 + 2 \cdot 5 + 2 \cdot 5^2 + 2 \cdot 5^3 = 312$

$S_n = \dfrac{a(1 - r^n)}{1 - r} = \dfrac{2(1 - 5^4)}{1 - 5} = 312$

3. Sum
$= 7 + 7 \cdot 4 + 7 \cdot 4^2 + 7 \cdot 4^3 + 7 \cdot 4^4 + 7 \cdot 4^5 + 7 \cdot 4^6$
$= 38,227$

$S_n = \dfrac{a(1 - r^n)}{1\ r} = \dfrac{7(1 - 4^7)}{1 - 4} = 38,227$

5. $S = R \cdot s_{\overline{n}|i} = 1000 \cdot s_{\overline{40}|0.03}$

$\quad = 1000\left[\dfrac{(1 + 0.03)^{40} - 1}{0.03}\right]$

$\quad = 1000(75.401260) = \$75,401.26$

7. $S = R \cdot s_{\overline{n}|i} = 5000 \cdot s_{\overline{20}|0.05}$

$\quad = 5000\left[\dfrac{(1 + 0.05)^{20} - 1}{0.05}\right]$

$\quad = 5000(33.065954) = \$165,329.77$

9. $S = R \cdot s_{\overline{n}|i} = 1500 \cdot s_{\overline{24}|0.02}$

$\quad = 1500\left[\dfrac{(1 + 0.02)^{24} - 1}{0.02}\right]$

$\quad = 1500(30.421863) = \$45,632.79$

11. $S = R\left[s_{\overline{n+1}|i} - 1\right] = 1000\left[s_{\overline{41}|0.03} - 1\right]$

$\quad = 1000(77.663298) = \$77,663.30$

13. $S = R\left[s_{\overline{n+1}|i} - 1\right] = 5000\left[s_{\overline{21}|0.05} - 1\right]$

$\quad = 5000(34.719252) = \$173,596.26$

15. $S = R\left[s_{\overline{n+1}|i} - 1\right] = 1500\left[s_{\overline{25}|0.02} - 1\right]$

$\quad = 1500(31.0302998) = \$46,545.45$

17.   (a) $S = R \cdot s_{\overline{n}|i} = 1500 \cdot s_{\overline{20}|0.05}$

$\quad = 1500\left[\dfrac{(1 + 0.05)^{20} - 1}{0.05}\right]$

$\quad = 1500(33.065954) = \$49,598.93$

(b) $I = 49,598.93 - 20(1500) = \$19,598.93$

19. $S = R\left[s_{\overline{n+1}|i} - 1\right]$

$\quad = 100\left[\dfrac{(1 + 0.01)^{49} - 1}{0.01} - 1\right]$

$\quad = 100(61.834834) = \$6183.48$

21. $S = R \cdot s_{\overline{n}|i} = 100 \cdot s_{\overline{48}|0.01}$

$\quad = 100(61.222608) = \$6122.26$

23. $S = R \cdot s_{\overline{n}|i} = 500 \cdot s_{\overline{18}|0.06} + 100 \cdot s_{\overline{10}|0.06}$

$\quad = 500(30.905653) + 100(13.180795)$

$\quad = 15,452.83 + 1318.08 = \$16,770.91$

25. $s_{\overline{100}|0.015} = \dfrac{(1 + 0.015)^{100} - 1}{0.015} = 228.80304$

27. $n = 10 \times 12 = 120, \; i = \dfrac{0.20}{12} \approx 0.01666667$

$\quad s_{\overline{120}|0.01666667} = \dfrac{(1 + 0.01666667)^{120} - 1}{0.01666667}$

$\quad = 376.0952$

29. $S = R \cdot s_{\overline{n}|i} = 400 \cdot s_{\overline{84}|\frac{0.10}{12}}$

$\quad = 400\left[\dfrac{(1 + 0.00833333)^{84} - 1}{0.00833333}\right]$

$\quad = 400(120.950423) = \$48,380.17$

31. $S = R\left[s_{\overline{n+1}|i} - 1\right] = 400\left[s_{\overline{85}|\frac{0.10}{12}} - 1\right]$

$= 400\left[\dfrac{1.00833333^{85} - 1}{0.00833333} - 1\right] = 400(121.95832)$

$= \$48,783.33$

4–4 Present Value of an Annuity

In Problems 1–5 use the formula

$A = R \cdot a_{\overline{n}|i} = R\left[\dfrac{1 - (1+i)^{-n}}{i}\right]$ to calculate the present

value A of an annuity.

1. $n = 6 \times 4 = 24, i - 0.08/4 = 0.02 = R = 3000$

$A = 3000\left[\dfrac{1 - (1+0.02)^{-24}}{0.02}\right]$

$= 3000(18.913926) = \$56,741.78$

3. $n = 10 \times 2 = 20, i = 0.06/2 = 0.03, R = 2000$

$A = 2000\left[\dfrac{1 - (1+0.03)^{-20}}{0.03}\right]$

$= 2000(14.877475) = \$29,754.95$

5. $n = 9 \times 4 = 36, i = 0.12/4 = 0.03, R = 5000$

$A = 5000\left[\dfrac{1 - (1+0.03)^{-36}}{0.03}\right]$

$= 5000(21.832253) = \$109,161.26$

In Problems 7–11 use the formula

$A = R\left[1 + a_{\overline{n+1}|i}\right]$ to calculate the present value A

of an annuity due.

7. $n = 24, i = 0.02, R = 3000$

$A = 3000\left[1 + a_{\overline{23}|0.02}\right]$

$= 3000(19.292204) = \$57,876.61$

9. $n = 20, i = 0.03, R = 2000$

$A = 2000\left[1 + a_{\overline{19}|0.03}\right]$

$= 2000(15.323799) = \$30,647.60$

11. $n = 36, i = 0.03, R = 5000$

$A = 5000\left[1 + a_{\overline{35}|0.03}\right]$

$= 5000(22.487220) = \$112,436.10$

13. $n = 10 \times 2 = 20, i = 0.10/2 = 0.05, R = 2000$

$A = 2000 \cdot a_{\overline{20}|0.05}$

$= 2000(12.462210) = \$24,924.42$

15. $n = 5 \times 4 = 20, i = 0.08/4 = 0.02, R = 400$

$A = 400 \cdot a_{\overline{20}|0.02} = 400(16.351433) = \$6540.57$

17. $n = 3 \times 12 = 36, i = 0.12/12 = 0.01, R = 500,$ payment due

$A = 500\left[1 + \cdot a_{\overline{35}|0.01}\right]$

$= 500(30.408580) = \$15,204.29$

19. $n = 28, i = 0.03, R = 1000,$ payment due

$A = 1000\left[1 + a_{\overline{27}|0.03}\right]$

$= 1000(19.327031) = \$19,327.03$

21. $n = 20, i = 0.02, R = 400,$ payment due

$A = 400\left[1 + a_{\overline{19}|0.02}\right]$

$= 400(16.678462) = \$6671.38$

23. $a_{\overline{365}|\,0.024} = \dfrac{1-(1+0.024)^{-365}}{0.024} = 41.659417$

25. $n = 20 \times 12 = 240,\ i = 0.10/12 = 0.00833333$

$a_{\overline{240}|\,0.010/12} = \dfrac{1-(1+0.00833333)^{-240}}{0.00833333}$

$\qquad\qquad = 103.62462$

27. $n = 4 \times 12 = 48,\ i = 0.07/12 = 0.005833333$
$\quad R = 600$

$A = 600 \cdot a_{\overline{48}|\,0.07/12} = 600(41.760201)$

$\quad = \$25,056.12$

29. $n = 48,\ i = 0.07/12,\ R = 600,$ payment due

$A = 600 \left[1 + a_{\overline{47}|\,0.07/12} \right] = 600(42.003803)$

$\quad = \$25,202.28$

31. $n = 6 \times 12 = 72,\ i = 0.05/12,\ R = 600$
$\quad A = 600 \cdot a_{\overline{72}|\,0.07/12} = 600(58.654444)$

$\quad = \$35,192.67$

4–5 Sinking Funds and Amortization

1. $S = \$35,000,\ n = 20,\ i = 0.02$
$\quad S = R \cdot s_{\overline{n}|\,i} = R \cdot s_{\overline{20}|\,0.02}$

$s_{\overline{20}|\,0.02} = \dfrac{(1+0.02)^{20}-1}{0.02} = 24.29737$

$R = \dfrac{35,000}{24.29737} = \$1440.49$

3. $S = \$10,000,\ n = 12,\ i = 0.03$
$\quad S = R \cdot s_{\overline{12}|\,0.03}$

$s_{\overline{48}|\,0.01} = \dfrac{(1+0.03)^{12}-1}{0.03} = 14.19203$

$R = \dfrac{10,000}{14.19203} = \$704.62$

5. $S = \$28,000,\ n = 6,\ i = 0.09$
$\quad S = R \cdot s_{\overline{6}|\,0.09}$

$s_{\overline{6}|\,0.09} = \dfrac{(1+0.09)^{6}-1}{0.09} = 7.5233345$

$R = \dfrac{28,000}{7.5233345} = \$3721.75$

7. $S = \$145,000,\ n = 120,\ i = \dfrac{0.102}{12} = 0.0085$
$\quad S = R \cdot s_{\overline{120}|\,0.0085}$

$s_{\overline{120}|\,0.0085} = \dfrac{(1+0.0085)^{120}-1}{0.0085}$

$R = \dfrac{145,000}{207.20781} = \$699.78$

9. $A = \$30,000,\ n = 5,\ i = 0.09$
$\quad A = R \cdot a_{\overline{5}|\,0.09}$

$a_{\overline{5}|\,0.09} = \dfrac{1-(1+0.09)^{-5}}{0.09} = 3.8896512$

$R = \dfrac{30,000}{3.8896512} = \$7712.77$

11. $A = \$95,000,\ n = 60,\ i = 0.005$
$\quad A = R \cdot a_{\overline{60}|\,0.005}$

$a_{\overline{60}|\,0.005} = \dfrac{1-(1+0.005)^{-60}}{0.005} = 51.72556$

$R = \dfrac{95,000}{51.72556} = \$1836.62$

13. $A = \$80,000$, $n = 20$, $i = 0.04$

   $A = R \cdot a_{\overline{20}|\,0.04}$

   $a_{\overline{20}|\,0.04} = \dfrac{1 - (1 + 0.04)^{-20}}{0.04} = 13.590326$

   $R = \dfrac{80,000}{13.590326} = \$5886.54$

15. $A = \$230,000$, $n = 300$, $i = \dfrac{0.1068}{12} = 0.0089$

   $A = R \cdot a_{\overline{300}|\,0.0089}$

   $a_{\overline{300}|\,0.0089} = \dfrac{1 - (1 + 0.0089)^{-300}}{0.0089} = 104.48595$

   $R = \dfrac{230,000}{104.48595} = \$2201.25$

17. $S = \$10,000$, $n = 20$, $i = 0.02$

   $S = R \cdot s_{\overline{20}|\,0.02}$

   $s_{\overline{20}|\,0.02} = \dfrac{(1 + 0.02)^{20} - 1}{0.02} = 24.29737$

   $R = \dfrac{10,000}{24.29737} = \$411.57$

19. $S = \$5,000$, $n = 12$, $i = 0.03$

   $S = R \cdot s_{\overline{12}|\,0.03}$

   $s_{\overline{12}|\,0.03} = \dfrac{(1 + 0.03)^{12} - 1}{0.03} = 14.19203$

   $R = \dfrac{5,000}{14.19203} = \$352.31$

21. $A = \$80,000 - \$20,000 = \$60,000$

   $n = 48$, $i = 0.02$

   $A = R \cdot a_{\overline{48}|\,0.02}$

   $a_{\overline{48}|\,0.02} = \dfrac{1 - (1 + 0.02)^{-48}}{0.02} = 30.67312$

   $R = \dfrac{60,000}{30.67312} = \$1956.11$

23. $S = \$60,000,\ n = 4,\ i = 0.10$

$S = R \cdot s_{\overline{4}|\,0.10}$

$s_{\overline{4}|\,0.10} = \dfrac{(1 + 0.10)^4 - 1}{0.10} = 4.641$

(a) $R = \dfrac{60{,}000}{4.641} = \$12{,}928.25$ annual payment

(b) Total interest $= 60{,}000 - 4(12{,}928.25) = \$8287$

(c) <u>Sinking</u> <u>Fund</u> <u>Schedule</u>

Payment Number	Payment	Interest	Total
1	12,928.25	0	12,928.25
2	12,928.25	1292.83	27,149.33
3	12,928.25	2714.93	42,792.51
4	12,928.25	4279.25	60,000.01

25. $A = \$90,000,\ n = 5,\ i = 0.09$

$A = R \cdot a_{\overline{5}|\,0.09}$

$a_{\overline{5}|\,0.09} = \dfrac{1 - (1 + 0.09)^{-5}}{0.05} = 3.8896512$

(a) $R = \dfrac{90{,}000}{3.8896512} = \$23{,}128.32$ annual payment

(b) Total interest $= 5(23{,}138.32) - 90{,}000 = \$25{,}691.60$

(c) <u>Amortization</u> <u>Schedule</u>

Payment Number	Payment R	Interest	Principal Reduction	Balance
0				90,000.00
1	23,138.32	8100.00	15,038.55	74,961.68
2	23,138.32	6746.55	16,391.77	58,569.91
3	23,138.32	5271.29	17,867.03	40,702.88
4	23,138.32	3663.26	19,475.06	21,227.82
5	23,138.32	1910.50	21,227.82	0.00
		25,691.60		

(d) Balance after 2 years $= \$58,569.91$

27. $A = \$100,000$, $n = 360$, $i = 0.0075$

$$A = R \cdot a_{\overline{360}|0.0075} \qquad a_{\overline{360}|0.0075} = \frac{1 - (1 + 0.0075)^{-360}}{0.0075} = 124.28187$$

(a) $R = \dfrac{100,000}{124.28187} = \$804.62$ monthly payment

(b) Balance after 48 payments (4 years) = present value of remaining 312 payments $= 804.62 \cdot a_{\overline{312}|0.0075}$

$$= (804.62)(120.37701) = \$96,857.75$$

29. $A = \$90,000$, $n = 15 \times 12 = 180$, $i = 0.01$

$$A = R \cdot a_{\overline{180}|0.01} \qquad a_{\overline{180}|0.01}$$

$$= \frac{1 - (1 + 0.01)^{-180}}{0.01} = 83.321664$$

(a) $R = \dfrac{90,000}{83.321664} = \$1080.15$ monthly payment

(b) Total payments $= 180(1080.15) = \$194,427.00$
Total interest $= \$104,427.90$

(c) Balance after 10 years = present value of
remaining 60 payments $= 1080.15 \cdot a_{\overline{60}|0.01}$

$$= (1080.15)(44.955039) = \$48,558.19$$

31. Let $A = R \cdot a_{\overline{n}|i}$, with fixed values of n and i.

Then, $R = \dfrac{A}{a_{\overline{n}|i}} = cA$, where $c = \dfrac{1}{a_{\overline{n}|i}}$ is a constant.

If $A = kL$, $R = c(kL) = k(cL)$
Hence, the periodic payment is proportional to the amount of the loan.

33. $A = \$100,000$, $n = 360$, $i = \dfrac{0.0975}{12} = 0.008125$

$$A = R \cdot a_{\overline{360}|0.008125} = R(116.39351)$$

$R = \$859.15$ monthly payment
Total interest $= 360(859.15) - 100,000 = \$209,294$
For $i = \dfrac{0.1075}{12} = 0.00895833$, $a_{\overline{360}|0.00895833}$

$$= 107.12626$$

$R = \dfrac{100,000}{107.12626} = \$933.48$ monthly payment

Total interest $= 360(933.48) - 100,000$
$$= \$236,052.80$$

35. $S = \$2,000,000,\ n = 4,\ i = 0.10$

$$S = R \cdot s_{\overline{4}|\,0.10}$$

$$s_{\overline{4}|\,0.10} = \frac{(1 + 0.10)^4 - 1}{0.10} = 4.641$$

$$R = \frac{2,000,000}{4.641} = \$430,941.61 \text{ annual deposit.}$$

37. Exercise 10.

Interest $= 7(0.07)(25,000) = \$12,250$

Total loan $= 25,000 + 12,250 = 37,250$

Quarterly payment $= 37,250 \div 28 = \$1330.36$

39. Exercise 12

Interest $= 6(0.12)(130,000) = \$93,600$

Total loan $= 130,000 + 93,600 = 223,600$

Monthly payment $= 223,600 \div 72 = \$3105.56$

4–6 Equations of Value; Deferred Annuities; Complex Annuities

1.

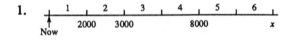

Choose Now as the comparison point and write an e

$8000(1 + 0.1)^{-4} = 2000(1 + 0.1)^{-1}$
$$+ 1000(1 + 0.1)^{-3} + x(1 + 0.1)^{-6}$$

Multiply by $(1 + 0.1)^6$:

$8000(1.1)^2 = 2000(1.1)^5 + 1000(1.1)^3 + x$

$9680 = 3221.02 + 1331 + x$

$x = \$5127.98$, last payment

3.

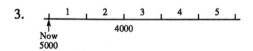

Find the present value of the loans:

$5000 + 4000(1 + 0.10)^{-2} = 5000 + 3305.79$
$$= 8305.79$$

(continued)

3. *(continued)*

Amortize over 5 years:

$A = \$8305.79,\ n = 5,\ i = 0.10$

$$A = R \cdot a_{\overline{5}|\,0.10}$$

$$a_{\overline{5}|\,0.10} = \frac{1 - (1 + 0.10)^{-5}}{0.10} = 3.7907868$$

$$R = \frac{8305.79}{3.7907868} = \$2191.05, \text{ annual payment}$$

5.

Present value $= 6000 + 5000(1 + 0.08)^{-3}$
$$= \$9969.16$$

Amortize over 7 years: $A = R \cdot a_{\overline{7}|\,0.08}$

$$a_{\overline{7}|\,0.08} = \frac{1 - (1 + 0.08)^{-7}}{0.08} = 5.20637$$

$$R = \frac{9969.16}{5.20637} = \$1914.80, \text{ annual payment}$$

7.

Present value of \$500 payments $= 500 \cdot a_{\overline{4}|\,0.10}$

$$= 500(3.1698654) = \$1584.93$$

Bring \$800 payments to end of year 4:

Value $= 800 \cdot a_{\overline{4}|\,0.10} = 800(3.7907868)$

$$= \$3032.63$$

Present value $= (3032.63)(1 + 0.1)^{-4} = 2071.33$

Total present value $= 1584.93 + 2071.33$
$$= \$3656.26$$

9.

1	2	3	4	5	6

8000   5000

$$\text{Present value} = 8{,}000 + 5000\left[1 + \frac{0.07}{12}\right]^{-36}$$

$$= 8{,}000 + 4055.39 = \$12{,}055.39$$

Monthly payment to amortize \$12,055.39 in 72 mon

11.

5	10	15	20	25

150,000   x   x   1.5x   1.5x   1.5x

↑
Comparison point

Bring payments forward to comparison point:
$i = 0.1008/12 = 0.0084$
$$X \cdot s_{\overline{24}|\,0.0084} = x(26.46773)$$

Bring payments back:
$$1.5x \; a_{\overline{36}|\,0.0084} = 1.5x(30.95519)$$

$$150{,}000(1.0084)^{24} = (26.46773 + 46.432786)x$$
$$183{,}349.34 = 72.900516x$$
$$x = \$2515.06 \text{ monthly payment, ye}$$
$$1.5x = \$3772.59 \text{ monthly payment, ye}$$

13.

1	2	3	4	5	6	7	8

↑ Now   ↑ Payment begins

$d = 7$ quarters of deferment

15.

1 - 12	13	14	15	16	17	18

↑ Now   ↑ Payment begins

$d = 17$ monthly periods of deferment

17.

1 - 20	21	22

↑ Now   ↑ Payment begins

$d = 21$ quarterly periods of deferment

19.

1	2	3	4	5 - 43

$d = 3$ deferred periods
$i = \dfrac{0.16}{4} = 0.04, \; n = 40$

Value of money at end of 3rd quarter
$$= 500 \cdot a_{\overline{40}|\,0.04}$$
$$= 500(19.792774)$$
$$= 9896.39$$

Present value $= 9896.39(1.04)^{-3} = \$8797.85$

21.

1 - 2	3	4 - 20

↑ Payment begins

$$i = \frac{0.16}{2} = 0.08$$

Periods deferred = 2
Value of money at end of 1 year $= 1000 \cdot a_{\overline{20}|\,0.08}$

$$= 1000(9.8181474) = 9818.15$$
Present value $= 9818.15(1.08)^{-2} = \$8417.48$

23.

1 - 4	5	6	7 - 25

↑ Payment begins

$$i = \frac{0.20}{4} = 0.05$$

Periods deferred = 5
Value of \$50,000 at end of 5 quarters
$$= 50{,}000(1 + 0.05)^5$$
$$= \$63{,}814.08$$

Amortization: $A = R \cdot a_{\overline{20}|\,0.05} = R(12.46221)$

$$R = \frac{63{,}814.08}{12.46221} = \$5120.61, \text{ quarterly payment.}$$

25. $i = \dfrac{0.12}{12} = 0.01$ (monthly)

Find equivalent quarterly rate:

$(1 + r) = (1 + 0.01)^3 = 1.030301$

$r = 0.030301$

$n = 20,\ R = \$1000$

Future value of annuity $= 1000\ s_{\overline{20}|\,0.030301}$

$$= 1000\left[\frac{1.030301^{20} - 1}{0.030301}\right]$$

$$= 1000(26.952797)$$
$$= \$26,952.80$$

Present value $= 1000\ a_{\overline{20}|\,0.030301}$

$$= 1000\left[\frac{1 - 1.030301^{-20} - 1}{0.030301}\right]$$

$$= 1000(14.836157)$$
$$= \$14,836.16$$

27. $i = \dfrac{0.12}{4} = 0.03$ (quarterly)

Find equivalent monthly rate: $(1 + r)^3 = 1 + 0.03$

$1 + r = (1 + .03)^{1/3} \approx 1.0099016$

$r = 0.0099016$

$n = 72,\ R = \$5600$

Future value of annuity $= 5600\ s_{\overline{72}|\,0.0099016}$

$$= 5600\left[\frac{1.0099016^{72} - 1}{0.0099016}\right]$$

$$= 5600(104.30528)$$
$$= \$584,109.57$$

Present value $= 5600\ a_{\overline{72}|\,0.0099016}$

$$= 5600\left[\frac{1 - 1.0099016^{-20}}{0.0099016}\right]$$

$$= 5600(51.311411)$$
$$= \$287,343.90$$

29. $i = 0.01$ (monthly)

Find equivalent quarterly rate:

$1 + r = (1 + 0.01)^3 = 1.030301$

$r = 0.030301$

$n = 24,\ R = \$650$

Present value $= 650\ a_{\overline{24}|\,0.030301}$

$$= 650\left[\frac{1 - (1 + 0.030301)^{-24}}{0.030301}\right]$$

$$= 650(16.88076) = \$10,972.49$$

31. $i = 0.03$ (quarterly)

Find equivalent monthly rate: $(1 + r)^3 = 1 + 0.03$

$1 + r = (1 + 0.03)^{1/3} = 1.0099016$

$r = 0.0099016$

Periods deferred $= 9,\ n = 21,\ R = 1000$

Present value

$$= (1.0099016)^{-9}\left[1000 \cdot a_{\overline{21}|\,0.0099016}\right]$$

$$a_{\overline{21}|\,0.0099016} = \frac{1 - (1 + 0.0099016)^{-21}}{0.0099016}$$

$$= 18.876536$$

Present value $= (0.9151419)(1000)(18.876536)$

$$= \$17,274.71$$

33. $i = 0.005$ (monthly)

Find equivalent semiannual rate:

$1 + r = (1 + 0.005)^6 \approx 1.0303775$

$r = 0.0303775$

$n = 14,\ A = \$12,500$

$$A = R \cdot a_{\overline{14}|\,0.0303775} = R\left[\frac{1 - (1 + 0.0303755)^{-14}}{0.0303755}\right]$$

$12,500 = R(11.267214)$

$R = \$1109.41$, semiannual payment

35. $i = 0.03$ (semiannually)

Find equivalent monthly rate:

$$(1 + r)^6 = 1 + 0.03$$
$$1 + r = (1 + 0.03)^{1/6} \approx 1.0049386$$
$$r = 0.0049386$$

$n = 48, R = \$150$

Payment begins   Payment ends

$$\text{Future value} = 150 \left[a_{\overline{49}|\,0.0049386} - 1 \right]$$

$$= 150 \left[\frac{1.0049386^{49} - 1}{0.0049386} - 1 \right]$$

$$= 150(55.283851 - 1)$$

$$= \$8142.58$$

$$\text{Present value} = 150 \left[(1 + a_{\overline{47}|\,0.0049386} \right]$$

$$= 150 \left[1 + \frac{1 - 1.0049386^{-47}}{0.0049386} - 1 \right]$$

$$= 150(1 + 41.852218)$$

$$= \$6427.83$$

37. $i = 0.09$ (annually)

Find equivalent semiannual rate:

$$(1 + r)^2 = 1 + 0.09$$
$$1 + r = (1 + 0.09)^{1/2} = 1.0440307$$
$$r = 0.0440307$$

$n = 20, R = \$500$

Payment begins   Payment ends

$$\text{Future value} = 500 \left[s_{\overline{21}|\,0.0440307} - 1 \right]$$

$$= 500 \left[\frac{1.0440307^{21} - 1}{0.0440307} - 1 \right]$$

$$= 500(33.422195 - 1)$$

$$= \$16,211.10$$

(continued)

37. *(continued)*

$$\text{Present value} = 500 \left[1 + a_{\overline{19}|\,0.044307} \right]$$

$$= 500 \left[1 + \frac{1 - 1.044307^{-19}}{0.044307} \right]$$

$$= 500(1 + 12.695473)$$

$$= \$6,847.74$$

39. $i = 0.02$ (quarterly)

Find equivalent monthly rate:

$$(1 + r)^3 = 1 + 0.02$$
$$1 + r = (1 + 0.02)^{1/3} \approx 1.0066227, \quad r = 0.0066227$$

$n = 60, A = \$4600$

Payment begins   Payment ends

$$4600 = R \left[1 + a_{\overline{59}|\,0.0066227} \right]$$

$$= R \left[1 + \frac{1 - 1.0066227^{-59}}{0.0066227} \right]$$

$$= R(1 + 48.706983) = 49.706983R$$

$R = \$92.54$, monthly payment

41. $i = 0.015$ (quarterly)

Find equivalent annual rate:

$$1 + r = (1 + 0.015)^4 \approx 1.0613636$$
$$r = 0.0613636$$

$n = 4, s = \$8,000$, payments due

$$8,000 = R \left[s_{\overline{5}|\,0.0613636} - 1 \right]$$

$$= R \left[\frac{1.0613636^5 - 1}{0.0613636} - 1 \right]$$

$$= R(5.6524599 - 1) = 4.6524599R$$

$R = \$1719.52$, annual payment

Extra Dividends: Net Present Value (Capital Investment Decision)

1. *NPV* of cash inflows
   $= 70{,}000(1.1)^{-1} + 80{,}000(1.1)^{-2} + 100{,}000(1.1)^{-3}$
   $+ 110{,}000(1.1)^{-4} + 130{,}000(1.1)^{-5}$
   $= 63{,}636.36 + 66{,}115.70 + 75{,}131.48 + 75{,}131.48$
   $+ 80{,}719.77 = \$360{,}734.8$
   Net present value $= \$360{,}734.80 - \$350{,}000$
   $\qquad\qquad\qquad = \$10{,}734.80$
   The investment is earning at least 10% compounded

3. $NPV = 120{,}000\, a_{\overline{8}|\,0.014} - 520{,}000$
   $= 556{,}663.67 - 520{,}000 = \$36{,}663.67$

5. Present value of inflows $= 60{,}000(1.12)^{-1}$
   $+ 70{,}000(1.12)^{-2} = 53{,}571.43 + 55{,}803.57$
   $\qquad\qquad\qquad = \$109{,}375 = $ cost of project

7. Present value of inflows $= 10{,}000(3.791) = \$37{,}910$
   $NPV = 37{,}910 - 40{,}000 = -2{,}090$ (at 10%)
   $NPV = 50{,}000 - 40{,}000 = 10{,}000$ (at 0%)
   Hence, the rate of return is greater than 0% but less

9. Present value $= 100{,}000$
   $= 40{,}000(1.12)^{-1} + x(1.12)^{-2}$
   $100{,}000(1.12)^2 = 40{,}000(1.12) + x$
   $x = \$80{,}640$

Review Exercises

1. Simple interest $= 1{,}000(0.09)(6) = \$540$
   Future value $= \$1{,}540$

3. Simple interest $= 10{,}000(0.08)(10) = \$8{,}000$
   Future value $= \$18{,}000$

5. (a) $B = S - D = S - Sdt = S\left[1 - 0.08 \times \dfrac{6}{12}\right]$
   $8000 = S(1 - 0.04)$
   $S = \$8333.33$

   (b) In sale at 6% discount, $D = 0.06\left[\dfrac{2}{12}\right](8333.33)$
   $\qquad\qquad\qquad\qquad\qquad = \$83.33$
   $B = 8333.33 - 83.33 = \$8250$ paid by third
   third party
   Original holder earns $8250 - 8000 = \$250$

7. $S = P(1 + i)^n = 10{,}000\left[1 + \dfrac{0.08}{2}\right]^{(11)(2)}$
   $= 10{,}000(1.04)^{22}$
   $= \$23{,}699.19$

9. $S = P(1 + i)^n;\ 10{,}000 = P\left[1 + \dfrac{0.12}{4}\right]^{(5)(4)}$
   $\qquad\qquad\qquad\qquad = P(1.8061112)$
   $P = \dfrac{10{,}000}{1.8061112} = \$5{,}536.76$

11. $S = P(1 + i)^n;\ 80{,}000 = P\left[1 + \dfrac{0.12}{12}\right]^{(8)(12)}$
   $\qquad\qquad\qquad\qquad = P(2.5992729)$
   $P = \dfrac{80{,}000}{2.5992729} = \$30{,}777.84$

13. $S = Pe^{rt} = 8000e^{(0.06)(3)} = 8000(1.1972174)$
   $\qquad\qquad\qquad\qquad = \$9577.74$

15. $\left[1 + \dfrac{0.09}{12}\right]^{12} = 1.093807 = 1\,(1 + r);$
   effective rate $= 9.3807\%$

17. $\left[1 + \dfrac{0.079}{365}\right]^{365} = 1.082195 = (1 + r);$
   effective rate $= 8.2195\%$

19. $S = P(1 + i)^n = 10,000\left[1 + \dfrac{0.06}{2}\right]^{(3)(2)}$

$\qquad = 10,000(1.1940523)$
$\qquad = \$11,940.52$ at end of 3 years
Add \$5000 deposit: $S = \$16,940.52$
Use this amount as new P for next 2 years:
$\qquad S = 16,940.52\ e^{0.07 \times 2} = \$19,486.24$

21. $S = P(1 + i)^n$; $53,429.05 = P\left[1 + \dfrac{0.12}{4}\right]^{(11)(4)}$

$\qquad\qquad\qquad\qquad = P(3.6714523)$

$P = \dfrac{53,429.05}{3.6714523} = \$14,552.57$

23. $A = R\ a_{\overline{n}|i} = 400 \cdot a_{\overline{20}|0.03} = 400\left[\dfrac{1 - 1.03^{-20}}{0.03}\right]$

$\qquad\qquad = 400(14.877475) = \$5950.99$

25. $S = R\ s_{\overline{n}|i} = R\ s_{\overline{60}|0.005} = R\left[\dfrac{1.005^{60} - 1}{0.005}\right]$

$\qquad = R(69.770029)$

$R = \dfrac{10,000}{69.770029} = \$143.33$, monthly payment

27. $PV = 500\ a_{\overline{8}|0.03} + 6000(1 + 0.03)^{-20}$

$\qquad = 500\left[\dfrac{1 - 1.03^{-8}}{0.03}\right] + 6000(1.03)^{-20}$

$\qquad = 500(7.0196922) + 6000(0.5536758)$
$\qquad = 3509.85 + 3322.05 = \$6831.90$

29. $S = R\ s_{\overline{n}|i} = R\ s_{\overline{40}|0.02} = R\left[\dfrac{1.02^{40} - 1}{0.02}\right]$

$\qquad = R(60.401983)$

$R = \dfrac{20,000}{60.401983} = \$331.11$ quarterly payment

31. Present value of inflows $= 10,000\ a_{\overline{5}|0.08}$

$\qquad = 10,000(3.99271) = \$39,927.10 =$ investment

33. $i = 0.06$ (semiannually)
Find equivalent monthly rate:
$\qquad (1 + r)^6 = 1 + 0.06$
$\qquad 1 + r = (1.06)^{1/6} \approx 1.0097588$; $r = 0.0097588$
$A = R\ a_{\overline{n}|i} = 100\ a_{\overline{60}|0.0097588}$

$\qquad = 100\left[\dfrac{1 - 1.0097588^{-60}}{0.0097588}\right] = 4525.20$

Purchase price $= 1000 + 4525.20 = \$5525.20$

35. $A = R \cdot a_{\overline{n}|i} = R\ a_{\overline{40}|0.04} = R\left[\dfrac{1 - 1.04^{-40}}{0.04}\right]$

$\qquad = R(19.792774)$

$R = \dfrac{70,000 - 20,000}{19.792774} = \$2526.17$, semiannual payment

37. $S = R \cdot s_{\overline{n}|i} = R \cdot s_{\overline{60}|0.01} = R\left[\dfrac{1.01^{60} - 1}{0.01}\right]$

$\qquad = R(81.66967)$

$R = \dfrac{10,000}{81.66967} = \$122.44$, monthly payment

39. $A = R \cdot a_{\overline{n}|i} = 600 \cdot a_{\overline{20}|0.03}$

$\qquad = 600\left[\dfrac{1 - 1.03^{-20}}{0.03}\right]$

$\qquad = 600(14.877475)$
$A = \$8926.48$, initial deposit

41. $i = 0.02$, $n = 48$, $R = \$100$ due
Future value $= 100\left[s_{\overline{49}|0.02} - 1\right]$

$\qquad = 100\left[\dfrac{1.02^{49} - 1}{0.02} - 1\right]$

$\qquad = 100(80.940589) = \$8094.06$

43. Calculate present values:

$4000 + 5000(1.08)^{-1} = 3000(1.08)^{-2} + x(1.08)^{-3}$

$4000(1.08)^3 + 5000(1.08)^2 - 3000(1.08) = x$

$5038.85 + 5832.00 - 3240.00 = x$

$x = \$7630.85$, final payment

45. $i = 0.02$ (quarterly)

Find equivalent semiannual rate:

$1 + r = (1 + 0.02)^2 = 1.0404$

$r = 0.0404$

Periods deferred = 5

At end of 5th period $A = R \cdot a_{\overline{12}\,|\,0.0404}$

$$= R\left[\frac{1 - 1.0404^{-12}}{0.0404}\right]$$

$= R(9.363329)$

Present value $= R(9.363329)(1.0404)^{-5}$

$= R(7.681191)$

Example: If $R = 1000$, $PV = \$7681.19$

47. $i = \dfrac{0.08}{4} = 0.02$ (quarterly)

Find equivalent annual rate:

$1 + r = (1 + 0.02)^4 \approx 1.082432$

$r = 0.082432$

$n = 5$, $R = \$600$

Future value $= R \cdot s_{\overline{n}\,|\,i} = 600 \cdot s_{\overline{5}\,|\,0.082432}$

$$= 600\left[\frac{1.082432^5 - 1}{0.082432}\right]$$

$= 600(5.89513) = \$3537.08$

Present value $= R \cdot a_{\overline{n}\,|\,i} = 600 \cdot a_{\overline{5}\,|\,0.082432}$

$$= 600\left[\frac{1 - 1.082432^{-5}}{0.082432}\right]$$

$= 600(3.967246) = \$2380.35$

49. $i = \dfrac{0.12}{2} = 0.06$ (semiannually)

Find equivalent monthly rate:

$(1 + r)^6 = 1 + 0.06$

$1 + r = (1 + 0.06)^{1/6} \approx 1.00975879;$

$r = 0.00975879$

$n = 48$, $R = \$1000$

Future value $= R \cdot s_{\overline{n}\,|\,i}$

$$= 1000\left[\frac{1.00975879^{48} - 1}{0.009758795}\right]$$

$= 1000(60.852632) = \$60,852.60$

Present value $= R \cdot a_{\overline{n}\,|\,i}$

$$= 1000\left[\frac{1 - 1.00975879^{-48}}{0.00975879}\right]$$

$= 1000(38.179679) = \$38,179.68$

51. $i = \dfrac{0.06}{12} = 0.005$ (monthly)

Find equivalent quarterly rate:

$1 + r = (1 + 0.005)^3 \approx 1.0150751$

$r = 0.0150751$

$n = 20$, $A = \$35,000$, periods deferred = 3

$(35,000)(1 + 0.0150751)^3 = R \cdot a_{\overline{20}\,|\,0.0150751}$

$$= R\left[\frac{1 - 1.0150751^{-20}}{0.0150751}\right] = R(17.155931)$$

$R = \dfrac{36,606.87}{17.155931} = \$2,133.77$, quarterly payment

CHAPTER
5
LINEAR SYSTEMS AND MATRICES

5–1 Linear Systems

1. Multiply (2) by –2 and add to (1):

(1) $2x - 3y = 6$ $\rightarrow$ $2x - 3y = 6$

(2) $x - 4y = 25$ $\quad -2x + 14y = -50$

$\overline{\qquad\qquad\qquad}$

$11y = 44$

$\Rightarrow y = -4$

From (2), $x = 7y + 25 = -28 + 25 = -3$

$(x,y) = (-3,-4)$

3. Multiply (1) by 2 and add to (2):

(1) $4x + y = 8$ $\rightarrow$ $8x + 2y = 16$

(2) $6x - 2y = -9$ $\quad 6x - 2y = -9$

$\overline{\qquad\qquad\qquad}$

$14x = 7$

$\Rightarrow x = \dfrac{1}{2}$

From (1), $y = 8 - 4x = 8 - 2 = 6$

$(x,y) = \left[\dfrac{1}{2}, 6\right]$

5. Multiply (1) by 2, (2) by 3, and add:

(1) $-3x + 10y = 5$ $\rightarrow$ $-6x + 20y = 10$

(2) $2x + 7y = 24$ $\quad 6x + 21y = 72$

$\overline{\qquad\qquad\qquad}$

$41y = 82$

$\Rightarrow y = 2$

From (2), $x = \dfrac{1}{2}(24 - 7y) = 5$

$(x,y) = (5, 2)$

7. Multiply (2) by –1 and add to (1):

(1) $5x + 2y = 20$ $\rightarrow$ $5x + 2y = 20$

(2) $5x - 5y = -25$ $\quad -5x + 5y = 25$

$\overline{\qquad\qquad\qquad}$

$7y = 45$

$\Rightarrow y = \dfrac{45}{7}$

From (1), $x = \dfrac{1}{5}(20 - 2y) = \dfrac{10}{7}$

$(x,y) = \left[\dfrac{10}{7}, \dfrac{45}{7}\right]$

9. (1) $\frac{1}{3}x - \frac{3}{2}y = -4$ (1) $\times 6$ $2x - 9y = -24$

$\qquad$ (2) $5x - 4y = 14$ $\qquad\qquad$ $5x - 4y = 14$

$\qquad\qquad$ (1) $\times\ 5$ $10x - 45y = -120$
$\qquad\qquad$ (2) $\times -2$ $-10x + 8y = -28$

$\qquad\qquad\qquad\qquad\qquad -37y = -148 \Rightarrow y = 4$

$\qquad$ From (1), $x = 3\left[-4 + \frac{3}{2}y\right] = 6$

11. (1) $\frac{x}{2} + \frac{y}{5} = \frac{8}{5}$ (1) $\times 10$ $5x + 2y = 16$

$\qquad$ (2) $\frac{x}{3} + \frac{y}{4} = \frac{17}{12}$ (2) $\times 12$ $4x + 3y = 17$

$\qquad\qquad$ (1) $\times\ 3$ $15x + 6y = 48$
$\qquad\qquad$ (2) $\times -2$ $-8x - 6y = -34$

$\qquad\qquad\qquad\qquad 7x \qquad = 14 \Rightarrow x = 2$

$\qquad\qquad\qquad$ From (1), $y = 5\left[\frac{8}{5} - \frac{x}{2}\right] = 3$

$\qquad$ $(x,y) = (2,3)$

13. $2x + y = 21$

$\qquad\qquad$ $y = 5x \quad\rightarrow\quad 2x + 5x = 21 \quad\Rightarrow\quad \begin{array}{l} x = 21/7 = 3 \\ y = 5\cdot 3 = 15 \end{array}$

$\qquad$ $(x,y) = (3,15)$

15. $-5x + y = 13 \qquad\rightarrow\qquad -5x + 7x - 1 = 13 \quad\Rightarrow\quad 2x = 14 \quad\Rightarrow\quad x = 7$

$\qquad\qquad$ $y = 7x - 1 \qquad\qquad\qquad\qquad\qquad y = 7\cdot 7 - 1 = 48$

$\qquad$ $(x,y) = (7,48)$

17. $\qquad\quad x = 2y - 5 \quad\rightarrow\quad 3(2y - 5) + 4y = 5$

$\qquad$ $3x = 4y = 5$

$\qquad\qquad\qquad\qquad \Rightarrow -10y = 20 \Rightarrow y = 2$

$\qquad\qquad\qquad\qquad\qquad\qquad x = -1$

$\qquad$ $(x,y) = (-1,2)$

19.
$$
\begin{aligned}
x + 2y &= 1 \\
3x - 5y &= -8
\end{aligned}
\quad \longrightarrow \quad
\begin{aligned}
-3x - 6y &= -3 \\
3x - 5y &= -8 \\
\hline
-11y = 11 &\Rightarrow y = 1 \\
x = 1 - 2y &= 1 - 2 = -1
\end{aligned}
$$

$(x,y) = (-1,1)$

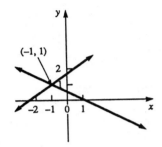

21.
$$
\begin{aligned}
3x + 5y &= -7 \\
2x - 6y &= 11
\end{aligned}
\quad \longrightarrow \quad
\begin{aligned}
6x + 10y &= 14 \\
-6x + 18y &= -33 \\
\hline
28y = -19 &\Rightarrow y = -\frac{19}{28} \\
x = \frac{1}{2}(11 + 6y) &= \frac{97}{28}
\end{aligned}
$$

$(x,y) = \left[\dfrac{97}{28},\ -\dfrac{19}{28}\right]$

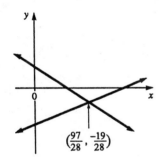

23. $2x + y = 11$ $-4x - 2y = -22$
       $3x + 2y = 18$ $\rightarrow$ $3x + 2y = 18$
                        $\overline{}$
                        $-x = -4 \Rightarrow x = 4$
                        $y = 11 - 2x = 3$

$(x,y) = (4,3)$

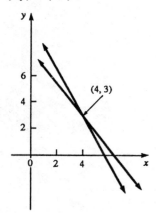

25. $-3x + 4y = 23$ $-6x + 8y = 46$
       $2x - 5y = -20$ $\rightarrow$ $6x - 15y = -60$
                        $\overline{}$
                        $-7y = -14 \Rightarrow y = 2$
                        $x = \frac{1}{2}(5y - 20) = \frac{1}{2}(-10) = -5$

$(x,y) = (-5,2)$

27. $2x + 3y = -13$ $-4x - 6y = 26$
       $4x - 5y = 29$ $\rightarrow$ $4x - 5y = 29$
                        $\overline{}$
                        $-11y = 55 \Rightarrow y = -5$
                        $x = \frac{1}{2}(-13 - 3y) = \frac{1}{2}(2) = 1$

$(x,y) = (1,-5)$

29. $x + y = -2$ $\rightarrow$ $-2x + 3(-2 - x) = -11 \Rightarrow -5x = -5$

       $-2x + 3y = -11$ $ \Rightarrow x = 1$
                        $y = -2 - x = -3$

$(x,y) = (1,-3)$

31. $4x - 3y = 2$ $\rightarrow$ $4(-5 - 2y) - 3y = 2 \Rightarrow -11y = 22 \Rightarrow y = -2$

       $x + 2y = -5$ $x = -5 - 2y = -1$
$(x,y) = (-1,-2)$

33. (1) $5x - 7y = 70$ $10x - 14y = 140$
   (2) $-10x + 14y = -120$ $-10x + 14y = 120$
$$\longrightarrow \quad \overline{}$$
$$0 = 260$$

Eq. (1) becomes $y = \dfrac{5}{7}x - 10$

Eq. (2) becomes $y = \dfrac{5}{7}x + \dfrac{60}{7}$

Lines of equal slope with different intercepts

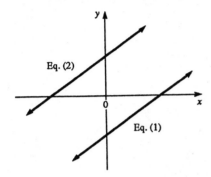

35. $3x - 8y = 10$ $-12x + 32y = -40$
    $12x - 32y = 75$ $12x - 32y = 75$
$$\longrightarrow \quad \overline{}$$
$$0 = 35 \implies \text{no solution}$$

37. $2x - y = 1$ $6x - 3y = 3$
    $-6x + 3y = 8$ $-6x + 3y = 8$
$$\longrightarrow \quad \overline{}$$
$$0 = 8 \implies \text{no solution}$$

39. $3x - y = 9$
$$\longrightarrow \; 3x - 3x = 9 \implies 0 = 9; \text{ no solution}$$
    $y = 3x$

41. Let x = lbs meat, y = lbs spinach
    $500x + 200y = 1200$ (protein)
    $100x + 800y = 1000$ (iron)
$$\downarrow$$
     $5x + 2y = 12$
   $-5x - 40y = -50$
$$\overline{}$$
$$-38y = -38 \implies y = 1$$
$$x = \frac{1}{5}(12 - 2y) = \frac{1}{5} \cdot 10 = 2$$

   $(x,y) = (2,1)$

43. Let x = no. of wagons, y = no. of cars

$$6x + 5y = 29{,}000 \quad \text{(profit)}$$
$$x = 4y$$
$$\downarrow$$
$$6(4y) + 5y = 29{,}000 \Rightarrow 29y = 29{,}000 \Rightarrow y = 1000$$
$$x = 4000$$

$(x,y) = (4000,\ 1000)$

45. $y = 25x$

$$\rightarrow \quad 25x = 10x + 6000 \Rightarrow 15x = 6\,000$$
$$y = 10x + 6000 \qquad\qquad \Rightarrow x = 400$$
$$y = 10{,}000$$

47. Let x = investment in corporate bonds, y = investment in U.S. Treasury bonds

$$0.12x + 0.08y = 0.09(100{,}000) \quad \text{(total yield)}$$
$$x + y = 100{,}000 \quad \text{(available funds)}$$
$$\downarrow$$
$$0.12x + 0.08(100{,}000 - x) = 9000$$
$$0.04x = 9000 - 8000 \Rightarrow x = 25{,}000$$
$$y = 75{,}000$$

$(x,y) = (\$25{,}000,\ \$75{,}000)$

49. x = lbs cashews, y = lbs peanuts

$$6x + 2y = 60 \quad \text{(cost)}$$
$$x + y = 20 \quad \text{(weight)}$$
$$\downarrow$$
$$6x + 2(20 - x) = 60 \Rightarrow 4x = 20 \Rightarrow x = 5$$
$$y = 15$$

$(x,y) = (5 \text{ lbs}, 15 \text{ lbs})$

51. (a) $300{,}000 + 5x = 400{,}000 + 4x$ (total cost for x units)
$$x = 100{,}000 \text{ units}$$
    (b) Machine I: Total cost $= 300{,}000 + 5(60{,}000) = \$600{,}000$
            Machine II: Total cost $= 400{,}000 + 4(60{,}000) = \$640{,}000$
            Hence, machine I results in a lower cost.
    (c) Machine I: Total cost $= 300{,}000 + 5(120{,}000) = \$900{,}000$
            Machine II: Total cost $= 400{,}000 + 4(120{,}000) = \$880{,}000$
            Machine II results in a lower total cost

5–2 Linear Systems; Tableaus; Problem Formulation

1. $\begin{bmatrix} 2 & 7 & | & 9 \\ -1 & 4 & | & 15 \end{bmatrix}$

3. $\begin{bmatrix} 3 & 0 & | & -1 \\ 4 & 3 & | & 7 \end{bmatrix}$

5. $\begin{bmatrix} 1 & 2 & 5 & | & 6 \\ 2 & -3 & -8 & | & 4 \\ -1 & 4 & 5 & | & 9 \end{bmatrix}$

7. $\begin{bmatrix} 5 & -7 & 0 & | & 4 \\ 0 & 3 & 0 & | & 9 \\ 1 & 0 & 1 & | & 15 \end{bmatrix}$

9. $4x_1 + 8x_2 = 5$
$-2x_1 + 6x_2 = 0$

11. $-x_1 + x_2 + 2x_3 = 3$
$4x_1 \quad - x_3 = 5$
$2x_1 + x_2 - x_3 = 6$

13. $x_1 \quad + x_3 = 2$
$x_2 \quad = 3$
$x_3 = 5$

15. $\begin{bmatrix} 1 & 0 & | & 5 \\ 0 & -1 & | & 6 \end{bmatrix}$ not a final tableau

17. $\begin{bmatrix} 1 & 0 & | & 8 \\ 0 & 1 & | & 3 \end{bmatrix}$ final tableau; $x_1 = 8, x_2 = 3$

19. $\begin{bmatrix} 1 & 0 & 0 & | & 2 \\ 0 & 1 & 0 & | & -4 \\ 0 & 0 & 1 & | & 5 \end{bmatrix}$ final tableau; $x_1 = 2, x_2 = -4, x_3 = 5$

21. $\begin{bmatrix} 1 & 0 & 0 & | & 3 \\ 0 & -1 & 0 & | & 2 \\ 0 & 0 & 1 & | & 4 \end{bmatrix}$ not a final tableau

23. Let x_1 = no. of conservative sweaters, x_2 = no. of sporty sweaters, x_3 = no. of practical sweaters

$$5x_1 + 2x_2 + 3x_3 = 660 \quad \text{(cutting)}$$
$$3x_1 + 2x_2 + 4x_3 = 480 \quad \text{(sewing)}$$
$$x_1 + 2x_2 + x_3 = 220 \quad \text{(inspection)}$$

25. $\left.\begin{array}{l} x_1 + x_2 \qquad\qquad = 300 \\ \qquad\quad x_3 + x_4 = 200 \end{array}\right\}$ production capacities

  $\left.\begin{array}{l} x_1 \qquad + x_3 \qquad = 150 \\ \quad\ x_2 \qquad + x_4 = 350 \end{array}\right\}$ demands

27. Let x_1 = investment in real estate, x_2 = investment in stocks,
  x_3 = investment in bonds

$$x_1 + x_2 + x_3 = 2{,}000{,}000 \qquad \text{(funds)}$$
$$0.12x_1 + 0.09x_2 + 0.08x_3 = 0.11(2{,}000{,}000) \quad \text{(return)}$$
$$4x_1 + 6x_2 + 3x_3 = 5(2{,}000{,}000) \qquad \text{(risk)}$$

5–3 Gauss–Jordan Method of Solving Linear Systems

1. $\begin{bmatrix} 1 & 2 & \vline & 4 \\ -2 & 3 & \vline & -5 \end{bmatrix}$ $(2)R_1 + R_2 \rightarrow R_2$ $\begin{bmatrix} 1 & 2 & \vline & 4 \\ 0 & 7 & \vline & 3 \end{bmatrix}$

3. $\begin{bmatrix} 2 & 4 & 16 & \vline & 10 \\ 0 & 1 & 8 & \vline & 9 \\ 3 & 1 & 2 & \vline & 4 \end{bmatrix}$ $(1/2)R_1 \rightarrow R_1$ $\begin{bmatrix} 1 & 2 & 8 & \vline & 5 \\ 0 & 1 & 8 & \vline & 9 \\ 3 & 1 & 2 & \vline & 4 \end{bmatrix}$

5. $\begin{bmatrix} 1 & 0 & 3 & \vline & 2 \\ 2 & -1 & 2 & \vline & 4 \\ -3 & 1 & -1 & \vline & 5 \end{bmatrix}$ $\begin{array}{c} (-2)R_1 + R_2 \rightarrow R_2 \\ \hline (3)R_1 + R_3 \rightarrow R_3 \end{array}$ $\begin{bmatrix} 1 & 0 & 3 & \vline & 2 \\ 0 & -1 & -4 & \vline & 0 \\ 0 & 0 & 8 & \vline & 11 \end{bmatrix}$

7. $\begin{bmatrix} 1 & 0 & -3 & \vline & -1 \\ 0 & 1 & 4 & \vline & -2 \\ 0 & 0 & 1 & \vline & 3 \end{bmatrix}$ $\begin{array}{c} (-4)R_3 + R_2 \rightarrow R_2 \\ \hline (3)R_3 + R_1 \rightarrow R_1 \end{array}$ $\begin{bmatrix} 1 & 0 & 0 & \vline & 8 \\ 0 & 1 & 0 & \vline & -14 \\ 0 & 0 & 1 & \vline & 3 \end{bmatrix}$

In Problems 9–25, a tableau representing the given linear system is used in the Gauss–Jordan method to solve the system.

9. $\begin{bmatrix} 2 & -3 & | & 6 \\ 1 & -7 & | & 25 \end{bmatrix}$ $\xrightarrow{R_1 \longleftrightarrow R_2}$ $\begin{bmatrix} 1 & -7 & | & 25 \\ 2 & -3 & | & 6 \end{bmatrix}$ $\xrightarrow{(-2)R_1 + R_2 \rightarrow R_2}$ $\begin{bmatrix} 1 & -7 & | & 25 \\ 0 & 11 & | & -44 \end{bmatrix}$

$\xrightarrow{(1/11)R_2 \rightarrow R_2}$ $\begin{bmatrix} 1 & -7 & | & 25 \\ 0 & 1 & | & -4 \end{bmatrix}$ $\xrightarrow{(7)R_3 + R_1 \rightarrow R_1}$ $\begin{bmatrix} 1 & 0 & | & -3 \\ 0 & 1 & | & -4 \end{bmatrix}$

$(x,y) = (-3, -4)$

11. $\begin{bmatrix} 2 & 3 & -5 & | & -13 \\ -1 & 2 & 3 & | & -7 \\ 3 & -4 & -7 & | & 15 \end{bmatrix}$ $\xrightarrow[\substack{\text{followed by} \\ (-1)R_1 \rightarrow R_1}]{R_1 \longleftrightarrow R_2}$ $\begin{bmatrix} 1 & -2 & -3 & | & 7 \\ 2 & 3 & -5 & | & -13 \\ 3 & -4 & -7 & | & 15 \end{bmatrix}$ $\xrightarrow[(-3)R_1 + R_3 \rightarrow R_3]{(-2)R_1 + R_2 \rightarrow R_2}$

$\begin{bmatrix} 1 & -2 & -3 & | & 7 \\ 0 & 7 & 1 & | & -27 \\ 0 & 2 & 2 & | & -6 \end{bmatrix}$ $\xrightarrow[\substack{\text{followed by} \\ (1/2)R_2 \rightarrow R_2}]{R_2 \longleftrightarrow R_3}$ $\begin{bmatrix} 1 & -2 & -3 & | & 7 \\ 0 & 1 & 1 & | & -3 \\ 0 & 7 & 1 & | & -27 \end{bmatrix}$ $\xrightarrow[(-7)R_2 + R_3 \rightarrow R_3]{(2)R_2 + R_1 \rightarrow R_1}$

$\begin{bmatrix} 1 & 0 & -1 & | & 1 \\ 0 & 1 & 1 & | & -3 \\ 0 & 0 & -6 & | & -6 \end{bmatrix}$ $\xrightarrow{(-1/6)R_3 \rightarrow R_3}$ $\begin{bmatrix} 1 & 0 & -1 & | & 1 \\ 0 & 1 & 1 & | & -3 \\ 0 & 0 & 1 & | & 1 \end{bmatrix}$ $\xrightarrow[(-1)R_3 + R_2 \rightarrow R_2]{R_3 + R_1 \rightarrow R_1}$ $\begin{bmatrix} 1 & 0 & 0 & | & 2 \\ 0 & 1 & 0 & | & -4 \\ 0 & 0 & 1 & | & 1 \end{bmatrix}$ $(x,y,z) = (2,-4,1)$

13. $\begin{bmatrix} 2 & 1 & 3 & | & 11 \\ 4 & 3 & -2 & | & -1 \\ 6 & 5 & -4 & | & -4 \end{bmatrix}$ $\xrightarrow[(-3)R_1 + R_3 \rightarrow R_3]{(-2)R_1 + R_2 \rightarrow R_2}$ $\begin{bmatrix} 2 & 1 & 3 & | & 11 \\ 0 & 1 & -8 & | & -23 \\ 0 & 2 & -13 & | & -37 \end{bmatrix}$ $\xrightarrow[(-2)R_2 + R_3 \rightarrow R_3]{(-1)R_2 + R_1 \rightarrow R_1}$

$\begin{bmatrix} 2 & 0 & 11 & | & 34 \\ 0 & 1 & -8 & | & -23 \\ 0 & 0 & 3 & | & 9 \end{bmatrix}$ $\xrightarrow{(1/3)R_3 \rightarrow R_3}$ $\begin{bmatrix} 2 & 0 & 11 & | & 34 \\ 0 & 1 & -8 & | & -23 \\ 0 & 0 & 1 & | & 3 \end{bmatrix}$ $\xrightarrow[(8)R_3 + R_2 \rightarrow R_2]{(-11)R_3 + R_1 \rightarrow R_3}$

$\begin{bmatrix} 2 & 0 & 0 & | & 1 \\ 0 & 1 & 0 & | & 1 \\ 0 & 0 & 1 & | & 3 \end{bmatrix}$ $\xrightarrow{(1/2)R_1 \rightarrow R_1}$ $\begin{bmatrix} 1 & 0 & 0 & | & 1/2 \\ 0 & 1 & 0 & | & 1 \\ 0 & 0 & 1 & | & 3 \end{bmatrix}$ $(x,y,z) = (1/2,1,3)$

15. $\begin{bmatrix} 1 & 10 & | & 34 \\ 3 & 2 & | & 18 \end{bmatrix}$ $\xrightarrow{(-3)R_1 + R_2 \rightarrow R_2}$ $\begin{bmatrix} 1 & 10 & | & 34 \\ 0 & -28 & | & -84 \end{bmatrix}$ $\xrightarrow{(-1/28)R_3 \rightarrow R_3}$ $\begin{bmatrix} 1 & 10 & | & 34 \\ 0 & 1 & | & 3 \end{bmatrix}$

$\xrightarrow{(-10)R_2 + R_1 \rightarrow R_1}$ $\begin{bmatrix} 1 & 0 & | & 4 \\ 0 & 1 & | & 3 \end{bmatrix}$ $(x_1, x_2) = (4,3)$

17. $\begin{bmatrix} 5 & 7 & 1 & | & 1 \\ 3 & 2 & 3 & | & 8 \\ 2 & 3 & 5 & | & 19 \end{bmatrix}$ $\xrightarrow[\;(-1)R_3 + R_2 \to R_2\;]{\;(-2)R_3 + R_1 \to R_1\;}$ $\begin{bmatrix} 1 & 1 & -9 & | & -37 \\ 1 & -1 & -2 & | & -11 \\ 2 & 3 & 5 & | & 19 \end{bmatrix}$ $\xrightarrow[\;(2)R_1 + R_3 \to R_3\;]{\;(-1)R_1 + R_2 \to R_2\;}$

$\begin{bmatrix} 1 & 1 & -9 & | & -37 \\ 0 & -2 & 7 & | & 26 \\ 0 & 1 & 23 & | & 93 \end{bmatrix}$ $\xrightarrow[\;(2)R_3 + R_2 \to R_2\;]{\;(-1)R_3 + R_1 \to R_1\;}$ $\begin{bmatrix} 1 & 0 & -32 & | & -130 \\ 0 & 0 & 53 & | & 212 \\ 0 & 1 & 23 & | & 93 \end{bmatrix}$ $\xrightarrow{\;(-1/53)R_2 \to R_2\;}$

followed by
$R_3 \longleftrightarrow R_2$

$\begin{bmatrix} 1 & 0 & -32 & | & -130 \\ 0 & 1 & 23 & | & 93 \\ 0 & 0 & 1 & | & 4 \end{bmatrix}$ $\xrightarrow[\;(-23)R_3 + R_2 \to R_2\;]{\;(32)R_3 + R_1 \to R_1\;}$ $\begin{bmatrix} 1 & 0 & 0 & | & -2 \\ 0 & 1 & 0 & | & 1 \\ 0 & 0 & 1 & | & 4 \end{bmatrix}$

$(x_1, x_2, x_3) = (-2, 1, 4)$

19. $\begin{bmatrix} 1 & 2 & 1 & | & 4 \\ 2 & 1 & -1 & | & -4 \\ 3 & 1 & 1 & | & 1 \end{bmatrix}$ $\xrightarrow[\;(-3)R_1 + R_3 \to R_3\;]{\;(-2)R_1 + R_2 \to R_2\;}$ $\begin{bmatrix} 1 & 2 & 1 & | & 4 \\ 0 & -3 & -3 & | & -12 \\ 0 & -5 & -2 & | & -11 \end{bmatrix}$ $\xrightarrow{\;(-1/3)R_2 \to R_2\;}$

$\begin{bmatrix} 1 & 2 & 1 & | & 4 \\ 0 & 1 & 1 & | & 4 \\ 0 & -5 & -2 & | & -11 \end{bmatrix}$ $\xrightarrow[\;(5)R_2 + R_3 \to R_3\;]{\;(-2)R_2 + R_1 \to R_1\;}$ $\begin{bmatrix} 1 & 0 & -1 & | & -4 \\ 0 & 1 & 1 & | & 4 \\ 0 & 0 & 3 & | & 9 \end{bmatrix}$ $\xrightarrow{\;(1/3)R_3 \to R_3\;}$

$\begin{bmatrix} 1 & 0 & -1 & | & -4 \\ 0 & 1 & 1 & | & 4 \\ 0 & 0 & 1 & | & 3 \end{bmatrix}$ $\xrightarrow[\;(-1)R_3 + R_2 \to R_2\;]{\;R_3 + R_1 \to R_1\;}$ $\begin{bmatrix} 1 & 0 & 0 & | & -1 \\ 0 & 1 & 0 & | & 1 \\ 0 & 0 & 1 & | & 3 \end{bmatrix}$ $(x_1, x_2, x_3) = (-1, 1, 3)$

21. $\begin{bmatrix} -1 & 1 & -1 & | & -6 \\ 1 & 3 & 1 & | & 10 \\ -2 & 1 & 0 & | & 6 \end{bmatrix}$ $\xrightarrow[\;(2)R_2 + R_3 \to R_3\;]{\;R_2 + R_1 \to R_1\;}$ $\begin{bmatrix} 0 & 4 & 0 & | & 4 \\ 1 & 3 & 1 & | & 10 \\ 0 & 7 & 2 & | & 26 \end{bmatrix}$ $\xrightarrow{\;(1/4)R_1 \to R_1\;}$

followed by
$R_1 \longleftrightarrow R_2$

$\begin{bmatrix} 1 & 3 & 1 & | & 10 \\ 0 & 1 & 0 & | & 1 \\ 0 & 7 & 2 & | & 26 \end{bmatrix}$ $\xrightarrow[\;(-7)R_2 + R_3 \to R_3\;]{\;(-3)R_2 + R_1 \to R_1\;}$ $\begin{bmatrix} 1 & 0 & 1 & | & 7 \\ 0 & 1 & 0 & | & 1 \\ 0 & 0 & 2 & | & 19 \end{bmatrix}$ $\xrightarrow{\;(1/2)R_3 \to R_3\;}$

$\begin{bmatrix} 1 & 0 & 1 & | & 7 \\ 0 & 1 & 0 & | & 1 \\ 0 & 0 & 1 & | & 19/2 \end{bmatrix}$ $\xrightarrow{\;(-1)R_3 + R_1 \to R_1\;}$ $\begin{bmatrix} 1 & 0 & 0 & | & -5/2 \\ 0 & 1 & 0 & | & 1 \\ 0 & 0 & 1 & | & 19/2 \end{bmatrix}$

$(x_1, x_2, x_3) = (-5/2, 1, 19/2)$

23. $\begin{bmatrix} 1 & -1 & -1 & | & -4 \\ 1 & -4 & 0 & | & -14 \\ 0 & 1 & 2 & | & 1 \end{bmatrix}$ $\xrightarrow{(-1)R_1 + R_2 \rightarrow R_2}$ $\begin{bmatrix} 1 & -1 & -1 & | & -4 \\ 0 & -3 & 1 & | & -10 \\ 0 & 1 & 2 & | & 1 \end{bmatrix}$ $\begin{array}{c} R_3 + R_1 \rightarrow R_1 \\ \hline (3)R_3 + R_2 \rightarrow R_2 \end{array}$

$\begin{bmatrix} 1 & 0 & 1 & | & -3 \\ 0 & 0 & 7 & | & -7 \\ 0 & 1 & 2 & | & 1 \end{bmatrix}$ $\begin{array}{c} (1/7)R_2 \rightarrow R_2 \\ \hline \text{followed by} \\ R_2 \longleftrightarrow R_3 \end{array}$ $\begin{bmatrix} 1 & 0 & 1 & | & -3 \\ 0 & 1 & 2 & | & 1 \\ 0 & 0 & 1 & | & -1 \end{bmatrix}$ $\begin{array}{c} (-1)R_3 + R_1 \rightarrow R_1 \\ \hline (-2)R_3 + R_2 \rightarrow R_2 \end{array}$

$\begin{bmatrix} 1 & 0 & 0 & | & -2 \\ 0 & 1 & 0 & | & 3 \\ 0 & 0 & 1 & | & -1 \end{bmatrix}$ $(x_1, x_2, x_3) = (-2, 3, -1)$

25. $\begin{bmatrix} 1 & 2 & 1 & -1 & | & 9 \\ 0 & 1 & -1 & 2 & | & -3 \\ 4 & 0 & 5 & 3 & | & 16 \\ 0 & 0 & 2 & 52 & | & -46 \end{bmatrix}$ $\xrightarrow{(-4)R_1 + R_3 \rightarrow R_3}$ $\begin{bmatrix} 1 & 2 & 1 & -1 & | & 9 \\ 0 & 1 & -1 & 2 & | & -3 \\ 0 & -8 & 1 & 7 & | & -20 \\ 0 & 0 & 2 & 52 & | & -46 \end{bmatrix}$

$\begin{array}{c} (-2)R_2 + R_1 \rightarrow R_1 \\ \hline (8)R_2 + R_3 \rightarrow R_3 \end{array}$ $\begin{bmatrix} 1 & 0 & 3 & -5 & | & 15 \\ 0 & 1 & -1 & 2 & | & -3 \\ 0 & 0 & -7 & 23 & | & -44 \\ 0 & 0 & 2 & 52 & | & -46 \end{bmatrix}$ $\xrightarrow{(1/2)R_4 \rightarrow R_4}$ $\begin{bmatrix} 1 & 0 & 3 & -5 & | & 15 \\ 0 & 1 & -1 & 2 & | & -3 \\ 0 & 0 & -7 & 23 & | & -44 \\ 0 & 0 & 1 & 26 & | & -23 \end{bmatrix}$

$\begin{array}{c} (-3)R_4 + R_1 \rightarrow R_1 \\ (7)R_4 + R_3 \rightarrow R_3 \\ \hline R_4 + R_2 \rightarrow R_2 \end{array}$ $\begin{bmatrix} 1 & 0 & 0 & 83 & | & 84 \\ 0 & 1 & 0 & 28 & | & -26 \\ 0 & 0 & 0 & 205 & | & -205 \\ 0 & 0 & 1 & 26 & | & -23 \end{bmatrix}$ $\begin{array}{c} (1/205)R_3 \rightarrow R_3 \\ \hline \text{followed by} \\ R_4 \rightarrow R_3 \end{array}$ $\begin{bmatrix} 1 & 0 & 0 & -83 & | & 84 \\ 0 & 1 & 0 & 28 & | & -26 \\ 0 & 0 & 1 & 26 & | & -23 \\ 0 & 0 & 0 & 1 & | & -1 \end{bmatrix}$

$\begin{array}{c} (83)R_4 + R_1 \rightarrow R_1 \\ \hline (-28)R_4 + R_2 \rightarrow R_2 \\ (-26)R_4 + R_3 \rightarrow R_3 \end{array}$ $\begin{bmatrix} 1 & 0 & 0 & 0 & | & 1 \\ 0 & 1 & 0 & 0 & | & 2 \\ 0 & 0 & 1 & 0 & | & 3 \\ 0 & 0 & 0 & 1 & | & -1 \end{bmatrix}$; $(x_1, x_2, x_3, x_4) = (1, 2, 3, -1)$

27. $\begin{bmatrix} 3 & -5 & | & 8 \\ -6 & 10 & | & 30 \end{bmatrix}$ $\xrightarrow{(2)R_1 + R_2 \rightarrow R_2}$ $\begin{bmatrix} 3 & -5 & | & 8 \\ 0 & 0 & | & 46 \end{bmatrix}$ $\Rightarrow$ no solution

29. $\begin{bmatrix} 8 & -2 & | & 10 \\ -4 & 1 & | & -5 \end{bmatrix}$ $\xrightarrow{(1/2)R_1 + R_2 \rightarrow R_2}$ $\begin{bmatrix} 8 & -2 & | & 10 \\ 0 & 0 & | & 0 \end{bmatrix}$ $\xrightarrow{(1/8)R_1 \rightarrow R_1}$

$\begin{bmatrix} 1 & -1/4 & | & 5/4 \\ 0 & 0 & | & 0 \end{bmatrix}$ $\Rightarrow x_1 = \frac{5}{4} + \frac{1}{4}x_2 = \frac{5}{4} + \frac{1}{4}t$, where t is any real number.

$x_2 = t$

31. $\begin{bmatrix} 1 & -2 & 1 & | & 3 \\ 3 & -7 & 2 & | & 4 \\ -2 & 4 & -2 & | & 8 \end{bmatrix}$ $\begin{array}{c} (-3)R_1 + R_2 \rightarrow R_2 \\ \hline (2)R_1 + R_3 \rightarrow R_3 \end{array}$ $\begin{bmatrix} 1 & -2 & 1 & | & 3 \\ 0 & -1 & -1 & | & -5 \\ 0 & 0 & 0 & | & 14 \end{bmatrix}$ $\Rightarrow$ no solution

33.
$$\begin{bmatrix} 1 & 1 & 1 & | & 4 \\ 1 & 2 & 3 & | & 2 \\ 2 & 4 & 6 & | & 5 \end{bmatrix} \xrightarrow{(-2)R_2 + R_3 \rightarrow R_3} \begin{bmatrix} 1 & 1 & 1 & | & 4 \\ 1 & 2 & 3 & | & 2 \\ 0 & 0 & 0 & | & 1 \end{bmatrix} \Rightarrow \text{ no solution}$$

35.
$$\begin{bmatrix} 1 & 2 & 1 & | & 4 \\ 2 & 1 & 5 & | & 6 \end{bmatrix} \xrightarrow{(-2)R_1 + R_2 \rightarrow R_2} \begin{bmatrix} 1 & 2 & 1 & | & 4 \\ 0 & -3 & 3 & | & -2 \end{bmatrix} \xrightarrow{(-1/3)R_2 \rightarrow R_2}$$

$$\begin{bmatrix} 1 & -2 & 1 & | & 4 \\ 0 & 1 & -1 & | & 2/3 \end{bmatrix} \xrightarrow{(-2)R_2 + R_1 \rightarrow R_1} \begin{bmatrix} 1 & 0 & 3 & | & 8/3 \\ 0 & 1 & -1 & | & 2/3 \end{bmatrix}$$

Infinitely many solutions: $x_1 = \dfrac{8}{3} - 3t,\ x_2 = \dfrac{2}{3} + t,\ x_3 = t$

37.
$$\begin{bmatrix} 1 & 2 & 1 & 0 & | & 4 \\ -1 & -1 & 0 & 1 & | & 5 \\ 3 & 6 & 1 & 1 & | & 6 \end{bmatrix} \begin{array}{c} R_1 + R_2 \rightarrow R_2 \\ \xrightarrow{\hspace{2cm}} \\ (-3)R_1 + R_3 \rightarrow R_3 \end{array} \begin{bmatrix} 1 & 2 & 1 & 0 & | & 4 \\ 0 & 1 & 1 & 1 & | & 9 \\ 0 & 0 & -2 & 1 & | & -6 \end{bmatrix}$$

$$\xrightarrow{(-2)R_2 + R_1 \rightarrow R_1} \begin{bmatrix} 1 & 0 & -1 & -2 & | & -14 \\ 0 & 1 & 1 & 1 & | & 9 \\ 0 & 0 & -2 & 1 & | & -6 \end{bmatrix} \xrightarrow{(-1/2)R_3 \rightarrow R_3} \begin{bmatrix} 1 & 0 & -1 & -2 & | & -14 \\ 0 & 1 & 1 & 1 & | & 9 \\ 0 & 0 & 1 & -1/2 & | & 3 \end{bmatrix}$$

$$\begin{array}{c} R_3 + R_1 \rightarrow R_1 \\ \xrightarrow{\hspace{2cm}} \\ (-1)R_3 + R_2 \rightarrow R_2 \end{array} \begin{bmatrix} 1 & 0 & 0 & -5/2 & | & -11 \\ 0 & 1 & 0 & 3/2 & | & 6 \\ 0 & 0 & 1 & -1/2 & | & 3 \end{bmatrix}$$

Infinitely many solutions: $x_1 = -11 + \dfrac{5}{2}t,\ x_2 = 6 - \dfrac{3}{2}t,$

$$x_3 = 3 + \dfrac{1}{2}t,\ x_4 = t$$

39.
$$\begin{bmatrix} 2 & 1 & 1 & -1 & | & 6 \\ 1 & 1 & 1 & -1 & | & 8 \\ -2 & -1 & -1 & 1 & | & 9 \end{bmatrix} \begin{array}{c} (-2)R_2 + R_1 \rightarrow R_1 \\ \xrightarrow{\hspace{2cm}} \\ (2)R_2 + R_3 \rightarrow R_3 \end{array} \begin{bmatrix} 0 & -1 & -1 & 1 & | & -10 \\ 1 & 1 & 1 & -1 & | & 8 \\ 0 & 1 & 1 & -1 & | & 25 \end{bmatrix}$$

$$\xrightarrow{R_1 + R_3 \rightarrow R_3} \begin{bmatrix} 0 & -1 & -1 & 1 & | & -10 \\ 1 & 1 & 1 & -1 & | & 8 \\ 0 & 0 & 0 & 0 & | & 15 \end{bmatrix} \Rightarrow \text{ no solution}$$

41.
$$\begin{bmatrix} 1 & 2 & -1 & -1 & | & 5 \\ 2 & 5 & 1 & -2 & | & 8 \end{bmatrix} \xrightarrow{(-2)R_1 + R_2 \rightarrow R_2} \begin{bmatrix} 1 & 2 & -1 & -1 & | & 5 \\ 0 & 1 & 3 & 0 & | & -2 \end{bmatrix}$$

$$\xrightarrow{(-2)R_2 + R_1 \rightarrow R_1} \begin{bmatrix} 1 & 0 & -7 & -1 & | & 9 \\ 0 & 1 & 3 & 0 & | & -2 \end{bmatrix} \quad \text{Infinitely many solutions:}\ \begin{array}{l} x_1 = 9 + 7s + t,\ x_2 = -2 - 3s \\ x_3 = s,\ x_4 = t \end{array}$$

43. $\begin{bmatrix} 1 & 1 & | & 1 \\ 2 & 1 & | & 4 \\ 4 & 2 & | & 8 \end{bmatrix} \xrightarrow[\;(-2)R_1 + R_2 \rightarrow R_2\;]{(-2)R_2 + R_3 \rightarrow R_3} \begin{bmatrix} 1 & 1 & | & 1 \\ 0 & -1 & | & 2 \\ 0 & 0 & | & 0 \end{bmatrix} \xrightarrow[\substack{\text{followed by} \\ (-1)R_2 \;\rightarrow\; R_2}]{R_2 + R_1 \;\rightarrow\; R_1} \begin{bmatrix} 1 & 0 & | & 3 \\ 0 & -1 & | & -2 \\ 0 & 0 & | & 0 \end{bmatrix}$

$x_1 = 3,\; x_2 = -2$

45. $\begin{bmatrix} 1 & 3 & | & 4 \\ 2 & 6 & | & 8 \\ -3 & -9 & | & -12 \end{bmatrix} \xrightarrow[\;(3)R_1 + R_3 \rightarrow R_3\;]{(-2)R_1 + R_2 \rightarrow R_2} \begin{bmatrix} 1 & 3 & | & 4 \\ 0 & 0 & | & 0 \\ 0 & 0 & | & 0 \end{bmatrix}$ Infinitely many solutions: $x_1 = 4 - 3t,\; x_2 = t$

47. $\begin{bmatrix} 2 & 1 & 3 & | & 11 \\ 6 & 5 & -4 & | & -4 \\ 4 & 3 & -2 & | & -1 \\ -6 & -5 & 4 & | & 4 \end{bmatrix} \xrightarrow[\substack{(-2)R_1 + R_3 \rightarrow R_3 \\ (3)R_1 + R_4 \rightarrow R_4}]{(-3)R_1 + R_2 \rightarrow R_2} \begin{bmatrix} 2 & 1 & 3 & | & 11 \\ 0 & 2 & -13 & | & -37 \\ 0 & 1 & -8 & | & -23 \\ 0 & -2 & 13 & | & 37 \end{bmatrix} \xrightarrow[\substack{(-2)R_3 + R_2 \rightarrow R_2 \\ (2)R_3 + R_4 \rightarrow R_4}]{(-1)R_3 + R_1 \rightarrow R_1} \begin{bmatrix} 2 & 0 & 11 & | & 34 \\ 0 & 0 & 3 & | & 9 \\ 0 & 1 & -8 & | & -23 \\ 0 & 0 & -3 & | & -9 \end{bmatrix}$

$\xrightarrow[\substack{\text{followed by} \\ R_2 \longleftrightarrow R_3}]{(1/3)R_2 \;\rightarrow\; R_2} \begin{bmatrix} 2 & 0 & 11 & | & 34 \\ 0 & 1 & -8 & | & -23 \\ 0 & 0 & 1 & | & 3 \\ 0 & 0 & -3 & | & -9 \end{bmatrix} \xrightarrow[\substack{(8)R_3 + R_2 \rightarrow R_2 \\ (3)R_3 + R_4 \rightarrow R_4}]{(-11)R_3 + R_1 \rightarrow R_1} \begin{bmatrix} 2 & 0 & 0 & | & 1 \\ 0 & 1 & 0 & | & 1 \\ 0 & 0 & 1 & | & 3 \\ 0 & 0 & 0 & | & 0 \end{bmatrix} \xrightarrow[\text{delete } R_4]{(1/2)R_1 \;\rightarrow\; R_1} \begin{bmatrix} 1 & 0 & 0 & | & 1/2 \\ 0 & 1 & 0 & | & 1 \\ 0 & 0 & 1 & | & 3 \end{bmatrix}$

$x_1 = \dfrac{1}{2},\; x_2 = 1,\; x_3 = 3$

49. $\begin{bmatrix} 2 & 3 & 5 & | & 780 \\ 1 & 2 & 1 & | & 320 \\ 4 & 1 & 3 & | & 500 \end{bmatrix} \xrightarrow{R_1 \longleftrightarrow R_2} \begin{bmatrix} 1 & 2 & 1 & | & 320 \\ 2 & 3 & 5 & | & 780 \\ 4 & 1 & 3 & | & 500 \end{bmatrix} \xrightarrow[\;(-4)R_1 + R_3 \rightarrow RR_3\;]{(-2)R_1 + R_2 \rightarrow R_2} \begin{bmatrix} 1 & 2 & 1 & | & 320 \\ 0 & -1 & 3 & | & 140 \\ 0 & -7 & -1 & | & -780 \end{bmatrix}$

$\xrightarrow[\;(-7)R_2 + R_3 \rightarrow R_3\;]{(2)R_2 + R_1 \;\rightarrow\; R_1} \begin{bmatrix} 1 & 0 & 7 & | & 600 \\ 0 & -1 & 3 & | & 140 \\ 0 & 0 & -22 & | & -1760 \end{bmatrix} \xrightarrow[\;(-1/22)R_3 \rightarrow R_3\;]{(-1)R_2 \;\rightarrow\; R_2} \begin{bmatrix} 1 & 0 & 7 & | & 600 \\ 0 & 1 & -3 & | & -140 \\ 0 & 0 & 1 & | & 80 \end{bmatrix}$

$\xrightarrow[\;(3)R_3 + R_2 \rightarrow R_2\;]{(-7)R_3 + R_1 \;\rightarrow\; R_1} \begin{bmatrix} 1 & 0 & 0 & | & 40 \\ 0 & 1 & 0 & | & 100 \\ 0 & 0 & 1 & | & 80 \end{bmatrix}$

$x = 40,\; y = 100,\; z = 80$

51.
$$\begin{bmatrix} 5 & 2 & 3 & | & 660 \\ 3 & 2 & 4 & | & 480 \\ 1 & 2 & 1 & | & 220 \end{bmatrix} \xrightarrow{R_1 \longleftrightarrow R_3} \begin{bmatrix} 1 & 2 & 1 & | & 220 \\ 3 & 2 & 4 & | & 480 \\ 5 & 2 & 3 & | & 660 \end{bmatrix} \xrightarrow[(-5)R_1 + R_3 \to R_3]{(-3)R_1 + R_2 \to R_2} \begin{bmatrix} 1 & 2 & 1 & | & 220 \\ 0 & -4 & 1 & | & -180 \\ 0 & -8 & -2 & | & -440 \end{bmatrix}$$

$$\xrightarrow{(-1/4)R_2 \to R_2} \begin{bmatrix} 1 & 2 & 1 & | & 220 \\ 0 & 1 & -1/4 & | & 45 \\ 0 & -8 & -2 & | & -440 \end{bmatrix} \xrightarrow[(8)R_2 + R_3 \to R_3]{(-2)R_2 + R_1 \to R_1}$$

$$\begin{bmatrix} 1 & 0 & 3/2 & | & 130 \\ 0 & 1 & -1/4 & | & 45 \\ 0 & 0 & -4 & | & -80 \end{bmatrix} \xrightarrow{(-1/4)R_3 \to R_3} \begin{bmatrix} 1 & 0 & 3/2 & | & 130 \\ 0 & 1 & -1/4 & | & 45 \\ 0 & 0 & 1 & | & 20 \end{bmatrix} \xrightarrow[(1/4)R_3 + R_2 \to R_2]{(-3/2)R_3 + R_1 \to R_1} \begin{bmatrix} 1 & 0 & 0 & | & 100 \\ 0 & 1 & 0 & | & 50 \\ 0 & 0 & 1 & | & 20 \end{bmatrix}$$

$x = 100$, $y = 50$, $z = 20$, where x = no. of conservative sweaters

y = no. of sporty sweaters

z = no. of practical sweaters

53.
$$\begin{bmatrix} 1 & 1 & 0 & 0 & | & 300 \\ 0 & 0 & 1 & 1 & | & 200 \\ 1 & 0 & 1 & 0 & | & 150 \\ 0 & 1 & 0 & 1 & | & 350 \end{bmatrix} \xrightarrow{(-1)R_1 + R_3 \to R_3} \begin{bmatrix} 1 & 1 & 0 & 0 & | & 300 \\ 0 & 0 & 1 & 1 & | & 200 \\ 0 & -1 & 1 & 0 & | & -150 \\ 0 & 1 & 0 & 1 & | & 350 \end{bmatrix} \xrightarrow[R_3 + R_4 \to R_4]{R_3 + R_1 \to R_1} \begin{bmatrix} 1 & 0 & 1 & 0 & | & 150 \\ 0 & 0 & 1 & 1 & | & 200 \\ 0 & -1 & 1 & 0 & | & -150 \\ 0 & 0 & 1 & 1 & | & 350 \end{bmatrix}$$

$$\xrightarrow[\substack{\text{followed by} \\ (-1)R_2 \to R_2}]{R_2 \longleftrightarrow R_3} \begin{bmatrix} 1 & 0 & 1 & 0 & | & 150 \\ 0 & 1 & -1 & 0 & | & 150 \\ 0 & 0 & 1 & 1 & | & 200 \\ 0 & 0 & 1 & 1 & | & 200 \end{bmatrix} \xrightarrow[(-1)R_3 + R_4 \to R_4]{\substack{(-1)R_3 + R_1 \to R_1 \\ R_3 + R_4 \to R_2}} \begin{bmatrix} 1 & 0 & 0 & -1 & | & -50 \\ 0 & 1 & 0 & 1 & | & 350 \\ 0 & 0 & 1 & 1 & | & 200 \\ 0 & 0 & 0 & 0 & | & 0 \end{bmatrix}$$

Infinity of solutions:

$x_1 = -50 + t$, $x_2 = 350 - t$

$x_3 = 200 - t$, $x_4 = t$

Since each $x_i \geq 0$, $50 \leq t \leq 200$

55.
$$\left.\begin{matrix} x_1 + x_2 \qquad\quad = 500 \\ x_3 + x_4 = 300 \end{matrix}\right\} \text{ production capacities}$$

$$\left.\begin{matrix} x_1 \qquad + x_3 \qquad = 600 \\ x_2 \qquad + x_4 = 200 \end{matrix}\right\} \text{ demands}$$

$$\begin{bmatrix} 1 & 1 & 0 & 0 & | & 500 \\ 0 & 0 & 1 & 1 & | & 300 \\ 1 & 0 & 1 & 0 & | & 600 \\ 0 & 1 & 0 & 1 & | & 200 \end{bmatrix} \xrightarrow{(-1)R_1 + R_3 \to R_3} \begin{bmatrix} 1 & 1 & 0 & 0 & | & 500 \\ 0 & 0 & 1 & 1 & | & 300 \\ 0 & -1 & 1 & 0 & | & 100 \\ 0 & 1 & 0 & 1 & | & 200 \end{bmatrix} \xrightarrow[R_3 + R_4 \to R_4]{R_3 + R_1 \to R_1} \begin{bmatrix} 1 & 0 & 1 & 0 & | & 600 \\ 0 & 0 & 1 & 1 & | & 300 \\ 0 & -1 & 1 & 0 & | & 100 \\ 0 & 0 & 1 & 1 & | & 300 \end{bmatrix}$$

$$\xrightarrow[\substack{\text{followed by} \\ (-1)R_2 \to R_2}]{R_2 \longleftrightarrow R_3} \begin{bmatrix} 1 & 0 & 1 & 0 & | & 600 \\ 0 & 1 & -1 & 0 & | & -100 \\ 0 & 0 & 1 & 1 & | & 300 \\ 0 & 0 & 1 & 1 & | & 300 \end{bmatrix} \xrightarrow[(-1)R_3 + R_4 \to R_4]{\substack{(-1)R_3 + R_1 \to R_1 \\ R_3 + R_2 \to R_2}} \begin{bmatrix} 1 & 0 & 0 & -1 & | & 300 \\ 0 & 1 & 0 & 1 & | & 200 \\ 0 & 0 & 1 & 1 & | & 300 \\ 0 & 0 & 0 & 0 & | & 0 \end{bmatrix}$$

Infinity of solutions

$x_1 = 300 + t$, $x_2 = 200 - t$,

$x_3 = 300 - t$, $x_4 = t$

Since each $x_i \geq 0$, $0 \leq t \leq 200$

57. Let x = gallons of A, y = gallons of B, Z = gallons of C

$$x + y + z = 100{,}000 \quad \text{total gallons}$$
$$2x + 4y + 8z = 600{,}000 \quad \text{cost}$$
$$-5.5x \quad + z = 0 \quad z = 5.5x$$

$$\begin{bmatrix} 1 & 1 & 1 & | & 100{,}000 \\ 2 & 4 & 8 & | & 600{,}000 \\ -11 & & 2 & | & 0 \end{bmatrix} \xrightarrow[\;(11)R_1 + R_3 \to R_3\;]{(-2)R_1 + R_2 \to R_2} \begin{bmatrix} 1 & 1 & 1 & | & 100{,}000 \\ 0 & 2 & 6 & | & 400{,}000 \\ 0 & 11 & 13 & | & 1{,}100{,}000 \end{bmatrix}$$

$$\xrightarrow{(1/2)R_2 \to R_2} \begin{bmatrix} 1 & 1 & 1 & | & 100{,}000 \\ 0 & 1 & 3 & | & 200{,}000 \\ 0 & 11 & 13 & | & 1{,}100{,}000 \end{bmatrix} \xrightarrow[\;(-11)R_2 + R_3 \to R_3\;]{(-1)R_2 + R_1 \to R_1} \begin{bmatrix} 1 & 0 & -2 & | & -100{,}000 \\ 0 & 1 & 3 & | & 200{,}000 \\ 0 & 0 & -20 & | & -1{,}100{,}000 \end{bmatrix}$$

$$\xrightarrow{(-1/20)R_3 \to R_3} \begin{bmatrix} 1 & 0 & -2 & | & -100{,}000 \\ 0 & 1 & 3 & | & 200{,}000 \\ 0 & 0 & 1 & | & 55{,}000 \end{bmatrix} \xrightarrow[\;(-3)R_3 + R_2 \to R_2\;]{(2)R_3 + R_1 \to R_1} \begin{bmatrix} 1 & 0 & 0 & | & 10{,}000 \\ 0 & 1 & 0 & | & 35{,}000 \\ 0 & 0 & 1 & | & 55{,}000 \end{bmatrix}$$

$x = 10{,}000$, $y = 35{,}000$, $z = 55{,}000$; solution is unique

59. Let x = ounces of food A, y = ounces of B, z = ounces of C

$$20x + 20y + 40z = 1\,800 \quad \text{calcium requirement}$$
$$10x + 20y + 10z = 800 \quad \text{iron requirement}$$
$$20x + 10y + 10z = 700 \quad \text{vitamin B requirement}$$

Divide all numbers by 10.

$$\begin{bmatrix} 2 & 2 & 4 & | & 180 \\ 1 & 2 & 1 & | & 80 \\ 2 & 1 & 1 & | & 70 \end{bmatrix} \xrightarrow[\;(-2)R_2 + R_3 \to R_3\;]{(-2)R_2 + R_1 \to R_1} \begin{bmatrix} 0 & -2 & 2 & | & 20 \\ 1 & 2 & 1 & | & 80 \\ 0 & -3 & -1 & | & -90 \end{bmatrix} \xrightarrow[\;\substack{\text{followed by} \\ R_1 \longleftrightarrow R_2}\;]{(-1/2)R_1 \to R_1} \begin{bmatrix} 1 & 2 & 1 & | & 80 \\ 0 & 1 & -1 & | & -10 \\ 0 & -3 & -1 & | & -90 \end{bmatrix}$$

$$\xrightarrow[\;(3)R_2 + R_3 \to R_3\;]{(-2)R_2 + R_1 \to R_1} \begin{bmatrix} 1 & 0 & 3 & | & 100 \\ 0 & 1 & -1 & | & -10 \\ 0 & 0 & -4 & | & -120 \end{bmatrix} \xrightarrow{(-1/4)R_3 \to R_3} \begin{bmatrix} 1 & 0 & 3 & | & 100 \\ 0 & 1 & -1 & | & -10 \\ 0 & 0 & 1 & | & 30 \end{bmatrix}$$

$$\xrightarrow[\;R_3 + R_2 \to R_2\;]{(-3)R_3 + R_1 \to R_1} \begin{bmatrix} 1 & 0 & 0 & | & 10 \\ 0 & 1 & 0 & | & 20 \\ 0 & 0 & 1 & | & 30 \end{bmatrix} \quad x = 10,\ y = 20,\ z = 30; \text{ unique solution}$$

5–4 Matrices

1. $\begin{bmatrix} 4 & 3 & -1 \\ 8 & 2 & 6 \end{bmatrix}$

dimension: 2×3

3. $\begin{bmatrix} 8 & 4 & 0 \\ 1 & 1 & 0 \\ 2 & 2 & 0 \end{bmatrix}$

dimension: 3×3

5. $[4 \ -1 \ 6]$

dimension: 1×3

7. $\begin{bmatrix} 5 \\ 0 \\ -1 \\ 4 \end{bmatrix}$

dimension: 4×1

9. $[4 \ 3 \ 0]$
row matrix

11. $\begin{bmatrix} 8 \\ 4 \end{bmatrix}$

column matrix

13. $\begin{bmatrix} 3 & 6 & 1 \\ 8 & 2 & 0 \end{bmatrix}$

not square

15. $\begin{bmatrix} 8 & 1 & 0 \\ 4 & 3 & 0 \\ 8 & 2 & 1 \end{bmatrix}$

square matrix

17. True, since $10/2 = 5$,
$12/4 = 3$, $14/7 = 2$, $9/9 = 1$

19. False

21. $\begin{bmatrix} x \\ y \end{bmatrix} = \begin{bmatrix} 4 \\ -1 \end{bmatrix} \Rightarrow x = 4, y = -1$

23. $\begin{bmatrix} x & y \\ z & w \end{bmatrix} = \begin{bmatrix} 1 & -4 \\ 5 & -7 \end{bmatrix} \Rightarrow$

$x = 1, y = -4, z = 5, w = -7$
In Problems 24 – 47, $A = \begin{bmatrix} 3 & 1 & 2 \\ -1 & 5 & -2 \end{bmatrix}$,

$B = \begin{bmatrix} 0 & 4 & 1 \\ 2 & -5 & 3 \end{bmatrix}, C = \begin{bmatrix} 4 & 3 & 0 \\ -2 & 5 & -1 \end{bmatrix}$

25. $A - B = \begin{bmatrix} 3 & -3 & 1 \\ -3 & 10 & -5 \end{bmatrix}$

27. $A + C = \begin{bmatrix} 7 & 4 & 2 \\ -3 & 10 & -3 \end{bmatrix}$

29. $C - A = \begin{bmatrix} 1 & 2 & -2 \\ -1 & 0 & 1 \end{bmatrix}$

31. $B - C = \begin{bmatrix} -4 & 1 & 1 \\ 4 & -10 & 4 \end{bmatrix}$

33. $A + B + C = \begin{bmatrix} 7 & 8 & 3 \\ -1 & 5 & 0 \end{bmatrix}$

35. $A + B - C = \begin{bmatrix} 7 & 0 & 1 \\ -5 & 15 & -6 \end{bmatrix}$

37. $3B = \begin{bmatrix} 0 & 12 & 3 \\ 6 & -15 & 9 \end{bmatrix}$

39. $-3A = \begin{bmatrix} -9 & -3 & -6 \\ 3 & -15 & 6 \end{bmatrix}$

41. $-2C = \begin{bmatrix} -8 & -6 & 0 \\ 4 & -10 & 2 \end{bmatrix}$

43. $A - 3B = \begin{bmatrix} 3 & -11 & -1 \\ -7 & 20 & -11 \end{bmatrix}$

45. $B - 3A = \begin{bmatrix} -9 & 1 & -5 \\ 5 & -20 & 9 \end{bmatrix}$

47. $A + B - 2C = \begin{bmatrix} -5 & -1 & 3 \\ 5 & -10 & 3 \end{bmatrix}$

In problems 49 – 55, $A = [3\ -4\ 1]$, $B = [2\ 0\ -3]$

49. $A - B = [1\ -4\ 4]$

51. $3A = [9\ -12\ 3]$

53. $A - 2B = [-1\ -4\ 7]$

55. $B - 3A = [-7\ 12\ -6]$

In Problems 57 – 65, $C = \begin{bmatrix} 8 \\ 2 \end{bmatrix}$, $D = \begin{bmatrix} -7 \\ 1 \end{bmatrix}$

57. $C + D$ $\begin{bmatrix} 1 \\ 3 \end{bmatrix}$

59. $D - C = \begin{bmatrix} -15 \\ -1 \end{bmatrix}$

61. $-3D = \begin{bmatrix} 21 \\ -3 \end{bmatrix}$

63. $5C + D = \begin{bmatrix} 33 \\ 11 \end{bmatrix}$

65. $C + 3D = \begin{bmatrix} 13 \\ 5 \end{bmatrix}$

67. $\begin{bmatrix} 1 & -1 \\ 2 & 0 \end{bmatrix} + \begin{bmatrix} -2 & -4 \\ 3 & 9 \end{bmatrix} = \begin{bmatrix} -1 & -5 \\ 5 & 9 \end{bmatrix}$

69. $\begin{bmatrix} -1 & 2 \\ 4 & -3 \\ 0 & -1 \end{bmatrix} - \begin{bmatrix} 4 & -3 \\ 1 & 2 \\ -2 & 6 \end{bmatrix} = \begin{bmatrix} -5 & 5 \\ 3 & -5 \\ 2 & -7 \end{bmatrix}$

71. $\begin{bmatrix} 4 & 2 \\ -1 & 10 \end{bmatrix} - 3\begin{bmatrix} 1 & -2 \\ 2 & -4 \end{bmatrix} = \begin{bmatrix} 1 & 8 \\ -7 & 22 \end{bmatrix}$

73. $2\begin{bmatrix} 10 & 5 \\ 2 & 4 \end{bmatrix} - 3\begin{bmatrix} -1 & -2 \\ 1 & 4 \end{bmatrix}$

$= \begin{bmatrix} 20 & 10 \\ 4 & 8 \end{bmatrix} + \begin{bmatrix} 3 & 6 \\ -3 & -12 \end{bmatrix} = \begin{bmatrix} 23 & 16 \\ 1 & -4 \end{bmatrix}$

75. $5\begin{bmatrix} x_1 \\ x_2 \\ x_3 \end{bmatrix} = \begin{bmatrix} 15 \\ 20 \\ 30 \end{bmatrix} \Rightarrow \begin{matrix} 5x_1 = 15 \\ 5x_2 = 20 \\ 5x_3 = 30 \end{matrix} \Rightarrow \begin{matrix} x_1 = 3 \\ x_2 = 4 \\ x_3 = 6 \end{matrix} \ x = \begin{bmatrix} 3 \\ 4 \\ 6 \end{bmatrix}$

77. $X = \begin{bmatrix} a & b \\ c & d \end{bmatrix}$, $X - X = \begin{bmatrix} a-a & b-b \\ c-c & d-d \end{bmatrix} = \begin{bmatrix} 0 & 0 \\ 0 & 0 \end{bmatrix} = Z$

$X + Z = \begin{bmatrix} a & b \\ c & d \end{bmatrix} + \begin{bmatrix} 0 & 0 \\ 0 & 0 \end{bmatrix} = \begin{bmatrix} a & b \\ c & d \end{bmatrix} = X$

79. $J + A = \begin{bmatrix} 200 & 50 & 70 & 10 & 0 \\ 300 & 50 & 10 & & 0 \end{bmatrix} + \begin{bmatrix} 100 & 30 & 10 & 50 \\ 70 & 40 & 0 & 20 & 0 & 80 \end{bmatrix}$

$= \begin{bmatrix} 300 & 80 & 80 & 150 \\ 370 & 450 & 210 & 80 \end{bmatrix}$

81. $\begin{bmatrix} 30 & 40 & 60 \\ 25 & 40 & 50 \end{bmatrix}$ scarves / mittens

83.
Person	Before	After
1	d350	345
2	249	200
3	260	220
4	195	140
5	275	200
6	295	230

5–5 Multiplying Matrices

1. $[1\ 2] \cdot \begin{bmatrix} -3 \\ 6 \end{bmatrix} = -3 + 12 = 9$

3. $[1\ \ 4\ \ 0\ -3] \cdot \begin{bmatrix} 8 \\ -1 \\ 0 \\ 2 \end{bmatrix} = 8 - 4 + 0 - 6 = -2$

5. $R \cdot C = [1/2\ \ 1/6\ \ 1/3] \cdot \begin{bmatrix} 0.40 \\ 0.60 \\ 0.20 \end{bmatrix}$

$= 0.20 + 0.10 + 0.20/3$

$= 0.37 \text{ (rounded)}$

7. $\underset{3 \times 4}{A} \ \underset{4 \times 5}{B} = \underset{3\ \times 5}{(AB)}$

9. $\underset{4 \times 2}{A} \ \underset{2 \times 4}{B} = \underset{4\ \times 4}{(AB)}$

11. $\underset{2 \times 5}{C} \ \underset{4 \times 2}{D}$ cannot be completed

13. $\underset{2 \times 2}{C} \ \underset{2 \times 2}{D} = \underset{2\ \times 2}{(CD)}$

15. $\underset{4 \times 2}{D} \ \underset{2 \times 5}{C} = \underset{4\ \times 5}{(DC)}$

17. $\underset{2 \times 2}{D} \ \underset{2 \times 2}{C} = \underset{2\ \times 2}{(DC)}$

19. $\begin{bmatrix} 4 & 3 \\ 2 & 1 \end{bmatrix} \begin{bmatrix} -1 & 1 \\ 5 & -2 \end{bmatrix} = \begin{bmatrix} 11 & -2 \\ 3 & 0 \end{bmatrix}$

21. $[-1\ 2] \begin{bmatrix} 8 & 3 \\ 1 & 6 \end{bmatrix} = [-6\ 9]$

23. $\begin{bmatrix} 3 & 4 \\ 2 & 6 \end{bmatrix} \begin{bmatrix} 1 \\ -5 \end{bmatrix} = \begin{bmatrix} -17 \\ -28 \end{bmatrix}$

25. $\begin{bmatrix} 1 & 2 & -3 \\ -1 & 0 & 2 \\ -2 & 1 & -1 \end{bmatrix} \begin{bmatrix} 2 \\ 4 \\ -4 \end{bmatrix} = \begin{bmatrix} 22 \\ -10 \\ 4 \end{bmatrix}$

27. $\begin{bmatrix} 1 & 2 & -3 \\ 0 & -1 & 3 \\ 5 & 0 & -4 \end{bmatrix} \begin{bmatrix} 1 & 3 \\ 2 & 1 \\ -2 & 6 \end{bmatrix} = \begin{bmatrix} 11 & -13 \\ -8 & 17 \\ 13 & -9 \end{bmatrix}$

29. $\begin{bmatrix} 4 & 6 \\ 2 & -1 \\ 8 & 0 \end{bmatrix} \begin{bmatrix} 1 & 3 \\ 2 & 4 \end{bmatrix} = \begin{bmatrix} 16 & 36 \\ 0 & 2 \\ 8 & 24 \end{bmatrix}$

31. $\begin{bmatrix} 1 & -4 & 2 \\ 2 & 0 & -1 \\ 3 & 1 & 1 \end{bmatrix} \begin{bmatrix} -2 & 1 & 3 \\ 1 & 2 & -2 \\ 4 & 0 & -1 \end{bmatrix} = \begin{bmatrix} 2 & -7 & 9 \\ -8 & 2 & 7 \\ -1 & 5 & 6 \end{bmatrix}$

In Problems 33–35, $A = \begin{bmatrix} 4 & 6 \\ -5 & 2 \end{bmatrix}$, $B = \begin{bmatrix} 1 & -2 \\ -3 & 4 \end{bmatrix}$

33. $AB = \begin{bmatrix} -14 & 16 \\ -11 & 18 \end{bmatrix}$

35. $AB \neq BA$

In Problems 37–47, $A = \begin{bmatrix} 1 & 3 & 7 \\ 2 & 4 & 0 \\ -1 & 5 & -2 \end{bmatrix}$, $B = \begin{bmatrix} 1 & 5 \\ 3 & 7 \\ -7 & 2 \end{bmatrix}$,

$C = \begin{bmatrix} 2 & -1 & 0 & 6 \\ -1 & 4 & 3 & 2 \end{bmatrix}$, $D = \begin{bmatrix} 1 & 0 & -2 \\ 3 & -1 & 1 \end{bmatrix}$

37. BA cannot be computed

39. *CB* cannot be computed

41. $DB = \begin{bmatrix} 15 & 1 \\ -7 & 10 \end{bmatrix}$

43. *AD* cannot be computed

45. $(AB)C = \begin{bmatrix} -3 & 9 & 40 \\ 14 & 38 \\ 28 & 26 \end{bmatrix} \begin{bmatrix} 2 & -1 & 0 & 6 \\ -1 & 4 & 3 & 2 \end{bmatrix}$

$= \begin{bmatrix} -118 & 199 & 120 & -154 \\ -10 & 138 & 114 & 160 \\ 30 & 76 & 78 & 220 \end{bmatrix}$

47. $(AB)C = A(BC)$ by inspection.

49. $AB = \begin{bmatrix} 4 & 3 \\ 8 & 0 \end{bmatrix} \begin{bmatrix} 7 & -1 \\ -2 & 4 \end{bmatrix} = \begin{bmatrix} 22 & 8 \\ 56 & -8 \end{bmatrix}$

$BA = \begin{bmatrix} 7 & -1 \\ -2 & 4 \end{bmatrix} \begin{bmatrix} 4 & 3 \\ 8 & 0 \end{bmatrix} = \begin{bmatrix} 20 & 21 \\ 24 & -6 \end{bmatrix} \neq AB$

51. (a) $A^3 = A^2A = \begin{bmatrix} 1 & 18 \\ -6 & 13 \end{bmatrix} \begin{bmatrix} 2 & 3 \\ -1 & 4 \end{bmatrix} = \begin{bmatrix} -16 & 75 \\ -25 & 34 \end{bmatrix}$

(b) $A^3 = A^2A = \begin{bmatrix} 64 & 0 \\ -6 & 4 \end{bmatrix} \begin{bmatrix} -8 & 0 \\ 1 & 2 \end{bmatrix} = \begin{bmatrix} -512 & 0 \\ 52 & 8 \end{bmatrix}$

(c) $A^3 = A^2A = \begin{bmatrix} 11 & 0 & 3 \\ 0 & 9 & 4 \\ 6 & 4 & 6 \end{bmatrix} \begin{bmatrix} 1 & 2 & 1 \\ 4 & -1 & 0 \\ 2 & 0 & 2 \end{bmatrix} = \begin{bmatrix} 17 & 22 & 17 \\ 44 & -9 & 8 \\ 34 & 8 & 18 \end{bmatrix}$

(d) $A^3 = A^2A = \begin{bmatrix} 23 & -18 & -1 \\ -11 & 10 & 13 \\ -24 & 16 & 8 \end{bmatrix} \begin{bmatrix} -1 & 2 & 3 \\ 5 & -2 & 1 \\ 4 & -4 & 0 \end{bmatrix} = \begin{bmatrix} -117 & 86 & 51 \\ 113 & -94 & -23 \\ 136 & -112 & -56 \end{bmatrix}$

53. $A^n = \dfrac{A \cdot A \cdot A \ldots A}{n \text{ times}}$ (*A* must be a square matrix.)

55. $\begin{aligned} x_1 + 5x_2 &= 6 \\ 4x_1 + 8x_2 &= 11 \end{aligned} \rightarrow \begin{bmatrix} 1 & 5 \\ 4 & 8 \end{bmatrix} \begin{bmatrix} x_1 \\ x_2 \end{bmatrix} = \begin{bmatrix} 6 \\ 11 \end{bmatrix}$

57. $\begin{aligned} 2x + 3y + z &= 11 \\ x \quad\quad + 2z &= 9 \\ 4y + 5z &= 17 \end{aligned} \rightarrow \begin{bmatrix} 2 & 3 & 1 \\ 1 & 0 & 2 \\ 0 & 4 & 5 \end{bmatrix} \begin{bmatrix} x \\ y \\ z \end{bmatrix} = \begin{bmatrix} 11 \\ 9 \\ 17 \end{bmatrix}$

59. $\begin{aligned} 2x_1 - x_2 &= 6 \\ 3x_1 + 2x_2 &= 9 \end{aligned} \rightarrow \begin{bmatrix} 2 & -1 \\ 3 & 2 \end{bmatrix} \begin{bmatrix} x_1 \\ x_2 \end{bmatrix} = \begin{bmatrix} 6 \\ 9 \end{bmatrix}$

61. $\begin{aligned} -x_1 + x_2 - 2x_3 &= 10 \\ 6x_1 \quad\quad + 2x_3 &= 5 \\ x_2 - x_3 &= 9 \end{aligned} \rightarrow \begin{bmatrix} -1 & 1 & -2 \\ 6 & 0 & 2 \\ 0 & 1 & -1 \end{bmatrix} \begin{bmatrix} x_1 \\ x_2 \\ x_3 \end{bmatrix} = \begin{bmatrix} 10 \\ 5 \\ 9 \end{bmatrix}$

63. $\begin{bmatrix} 1 & -3 \\ 2 & 4 \end{bmatrix} \begin{bmatrix} x_1 \\ x_2 \end{bmatrix} = \begin{bmatrix} 3 \\ -7 \end{bmatrix} \rightarrow \begin{aligned} x_1 - 3x_2 &= 3 \\ 2x_1 + 4x_2 &= -7 \end{aligned}$

65. $\begin{bmatrix} 4 & 1 & -1 \\ 5 & 0 & 2 \\ -2 & 1 & -2 \end{bmatrix} \begin{bmatrix} x_1 \\ x_2 \\ x_3 \end{bmatrix} = \begin{bmatrix} -4 \\ 1 \\ -1 \end{bmatrix} \rightarrow \begin{aligned} 4x_1 + x_2 - x_3 &= -4 \\ 5x_1 \quad\quad + 2x_3 &= 1 \\ -2x_1 + x_2 - 2x_3 &= -1 \end{aligned}$

67. $\begin{bmatrix} 1 & 4 & -1 \\ 2 & 1 & 0 \\ 4 & 1 & -5 \end{bmatrix} \begin{bmatrix} x_1 \\ x_2 \\ x_3 \end{bmatrix} = \begin{bmatrix} 2 \\ 1 \\ 5 \end{bmatrix} \rightarrow \begin{aligned} x_1 + 4x_2 - x_3 &= 2 \\ 2x_1 + x_2 \quad\quad &= 1 \\ 4x_1 + x_2 - 5x_3 &= 5 \end{aligned}$

69. $AI = \begin{bmatrix} 4 & 3 & 6 \\ 8 & 2 & 7 \\ -1 & 1 & 4 \end{bmatrix} \begin{bmatrix} 1 & 0 & 0 \\ 0 & 1 & 0 \\ 0 & 0 & 1 \end{bmatrix} = \begin{bmatrix} 4 & 3 & 6 \\ 8 & 2 & 7 \\ -1 & 1 & 4 \end{bmatrix} = A$

$IA = \begin{bmatrix} 1 & 0 & 0 \\ 0 & 1 & 0 \\ 0 & 0 & 1 \end{bmatrix} \begin{bmatrix} 4 & 3 & 6 \\ 8 & 2 & 7 \\ -1 & 1 & 4 \end{bmatrix} = \begin{bmatrix} 4 & 3 & 6 \\ 8 & 2 & 7 \\ -1 & 1 & 4 \end{bmatrix} = A$

71. $BI = \begin{bmatrix} 2 & 3 \\ 7 & 4 \\ 5 & 7 \end{bmatrix} \begin{bmatrix} 1 & 0 \\ 0 & 1 \end{bmatrix} = \begin{bmatrix} 2 & 3 \\ 7 & 4 \\ 5 & 7 \end{bmatrix} = B$

73. $\begin{bmatrix} 200 & 500 & 300 \\ 400 & 200 & 600 \\ 600 & 800 & 900 \end{bmatrix} \begin{bmatrix} 30 \\ 20 \\ 50 \end{bmatrix} = \begin{bmatrix} 31,000 \\ 46,000 \\ 79,000 \end{bmatrix}$ April
May
J une

75. $\begin{bmatrix} 0.30 & 0.50 & 0.40 & 0.10 \\ 0.60 & 0.20 & 0.55 & 0.86 \\ 0.10 & 0.30 & 0.05 & 0.04 \end{bmatrix} \begin{bmatrix} 60,000 \\ 100,000 \\ 70,000 \\ 90,000 \end{bmatrix} = \begin{bmatrix} 105,000 \\ 171,900 \\ 43,100 \end{bmatrix}$ Republican
Democrat
Independent

5–6 Inverse of a Square Matrix

1. $\begin{bmatrix} 1 & -3/2 \\ 1 & -2 \end{bmatrix} \begin{bmatrix} 4 & -3 \\ 2 & -2 \end{bmatrix} = \begin{bmatrix} 1 & 0 \\ 0 & 1 \end{bmatrix}$, and $\begin{bmatrix} 4 & -3 \\ 2 & -2 \end{bmatrix} \begin{bmatrix} 1 & -3/2 \\ 1 & -2 \end{bmatrix} = \begin{bmatrix} 1 & 0 \\ 0 & 1 \end{bmatrix}$

The matrices are inverses of each other.

3. $\begin{bmatrix} 7 & -8 \\ 3 & -3 \end{bmatrix} \begin{bmatrix} -1 & 8/3 \\ -1 & 7/3 \end{bmatrix} = \begin{bmatrix} 1 & 0 \\ 0 & 1 \end{bmatrix}$, and $\begin{bmatrix} -1 & 8/3 \\ -1 & 7/3 \end{bmatrix} \begin{bmatrix} 7 & -8 \\ 3 & -3 \end{bmatrix} = \begin{bmatrix} 1 & 0 \\ 0 & 1 \end{bmatrix}$

The matrices are inverses of each other.

5. $\begin{bmatrix} 5 & 6 \\ 3 & 4 \end{bmatrix} \begin{bmatrix} 2 & -3 \\ -3/2 & 5/2 \end{bmatrix} = \begin{bmatrix} 1 & 0 \\ 0 & 1 \end{bmatrix}$, and $\begin{bmatrix} 2 & -3 \\ -3/2 & 5/2 \end{bmatrix} \begin{bmatrix} 5 & 6 \\ 3 & 4 \end{bmatrix} = \begin{bmatrix} 1 & 0 \\ 0 & 1 \end{bmatrix}$

The matrices are inverses of each other.

7. $\begin{bmatrix} 1 & 2 \\ 5 & -1 \end{bmatrix} \begin{bmatrix} 8 & 0 \\ 4 & 1 \end{bmatrix} = \begin{bmatrix} 16 & 2 \\ 36 & -1 \end{bmatrix}$

The matrices are not inverses of each other.

9. (a) $\begin{bmatrix} 1 & 5 & | & 1 & 0 \\ 2 & 11 & | & 0 & 1 \end{bmatrix} \xrightarrow{(-2)R_1 + R_2 \rightarrow R_2} \begin{bmatrix} 1 & 5 & | & 1 & 0 \\ 0 & 1 & | & -2 & 1 \end{bmatrix} \xrightarrow{(-5)R_2 + R_1 \rightarrow R_1} \begin{bmatrix} 1 & 0 & | & 11 & -5 \\ 0 & 1 & | & -2 & 1 \end{bmatrix}$

The inverse matrix is $\begin{bmatrix} 11 & -5 \\ -2 & 1 \end{bmatrix}$.

(b) $AA^{-1} = A^{-1}A = I$

11. $\begin{bmatrix} 5 & -1 & | & 1 & 0 \\ -3 & 7 & | & 0 & 1 \end{bmatrix} \xrightarrow{(-1/5)R_1 \to R_1} \begin{bmatrix} 1 & -1/5 & | & 1/5 & 0 \\ -3 & 7 & | & 0 & 1 \end{bmatrix} \xrightarrow{(3)R_1 + R_2 \to R_2} \begin{bmatrix} 1 & -1/5 & | & 1/5 & 0 \\ 0 & 36/5 & | & 3/5 & 1 \end{bmatrix}$

$\xrightarrow{(5/32)R_2 \to R_2} \begin{bmatrix} 1 & -1/5 & | & 1/5 & 0 \\ 0 & 1 & | & 3/32 & 5/32 \end{bmatrix} \xrightarrow{(1/5)R_2 + R_1 \to R_1} \begin{bmatrix} 1 & 0 & | & 7/32 & 1/32 \\ 0 & 1 & | & 3/32 & 5/32 \end{bmatrix},$

Inverse matrix is $(1/32)\begin{bmatrix} 7 & 1 \\ 3 & 5 \end{bmatrix}$.

13. $\begin{bmatrix} 0 & 1 & | & 1 & 0 \\ 1 & 1 & | & 0 & 1 \end{bmatrix} \xrightarrow{(-1)R_1 + R_2 \to R_2} \begin{bmatrix} 0 & 1 & | & 1 & 0 \\ 1 & 0 & | & -1 & 1 \end{bmatrix} \xleftrightarrow{R_1 \longleftrightarrow R_2} \begin{bmatrix} 1 & 0 & | & -1 & 1 \\ 0 & 1 & | & 1 & 0 \end{bmatrix}$

Inverse matrix is $\begin{bmatrix} -1 & 1 \\ 1 & 0 \end{bmatrix}$.

15. $\begin{bmatrix} 3 & 2 & 1 & | & 1 & 0 & 0 \\ 4 & -3 & 2 & | & 0 & 1 & 0 \\ 2 & 4 & -3 & | & 0 & 0 & 1 \end{bmatrix} \begin{array}{c} \xrightarrow{(-1)R_1 + R_2 \to R_2} \\ \text{followed by} \\ R_1 \longleftrightarrow R_2 \end{array} \begin{bmatrix} 1 & -5 & 1 & | & -1 & 1 & 0 \\ 3 & 2 & 1 & | & 1 & 0 & 0 \\ 2 & 4 & -3 & | & 0 & 0 & 1 \end{bmatrix} \begin{array}{c} \xrightarrow{(-3)R_1 + R_2 \to R_2} \\ \xrightarrow{(-2)R_1 + R3 \to R_3} \end{array}$

$\begin{bmatrix} 1 & -5 & 1 & | & -1 & 1 & 0 \\ 0 & 17 & -2 & | & 4 & -3 & 0 \\ 0 & 14 & -5 & | & 2 & -2 & 1 \end{bmatrix} \xrightarrow{(1/17)R_2 \to R_2} \begin{bmatrix} 1 & -5 & 1 & | & -1 & 1 & 0 \\ 0 & 1 & -2/17 & | & 4/17 & -3/17 & 0 \\ 0 & 14 & -5 & | & 2 & -2 & 1 \end{bmatrix} \begin{array}{c} \xrightarrow{(5)R_2 + R_1 \to R_1} \\ \xrightarrow{(-14)R_2 + R_3 \to R_3} \end{array}$

$\begin{bmatrix} 1 & 0 & 7/17 & | & 3/17 & 2/17 & 0 \\ 0 & 1 & -2/17 & | & 4/17 & -3/17 & 0 \\ 0 & 0 & -57/17 & | & -22/17 & 8/17 & 1 \end{bmatrix} \xrightarrow{(-17/57)R_3 \to R_3} \begin{bmatrix} 1 & 0 & 7/17 & | & 3/17 & 2/17 & 0 \\ 0 & 1 & -2/17 & | & 4/17 & -3/17 & 0 \\ 0 & 0 & 1 & | & 22/57 & -8/57 & -17/57 \end{bmatrix}$

$\begin{array}{c} \xrightarrow{(-7/17)R_3 + R_1 \to R_1} \\ \xrightarrow{(2/17)R_3 + R_2 \to R_2} \end{array} \begin{bmatrix} 1 & 0 & 0 & | & 1/57 & 10/57 & 7/57 \\ 0 & 1 & 0 & | & 16/57 & -11/57 & -2/57 \\ 0 & 0 & 1 & | & 22/57 & -8/57 & -17/57 \end{bmatrix}$

Inverse matrix is $(1/57)\begin{bmatrix} 1 & 10 & 7 \\ 16 & -11 & -2 \\ 22 & -8 & -17 \end{bmatrix}$

17. $\begin{bmatrix} 3 & 4 & | & 1 & 0 \\ 2 & -7 & | & 0 & 1 \end{bmatrix} \begin{array}{c} \xrightarrow{(1/2)R_2 \to R_2} \\ \text{followed by} \\ R_1 \longrightarrow R_2 \end{array} \begin{bmatrix} 1 & -7/2 & | & 0 & 1/2 \\ 3 & 4 & | & 1 & 0 \end{bmatrix} \xrightarrow{(-3)R_1 + R_2 \to R_2} \begin{bmatrix} 1 & -7/2 & | & 0 & 1/2 \\ 0 & 29/2 & | & 1 & -3/2 \end{bmatrix}$

$\xrightarrow{(2/29)R_2 \to R_2} \begin{bmatrix} 1 & -7/2 & | & 0 & 1/2 \\ 0 & 1 & | & 2/29 & -3/29 \end{bmatrix} \xrightarrow{(7/2)R_2 + R_1 \to R_1} \begin{bmatrix} 1 & 0 & | & 7/29 & 4/29 \\ 0 & 1 & | & 2/29 & -3/29 \end{bmatrix}$

Inverse matrix is $(1/29)\begin{bmatrix} 7 & 4 \\ 2 & -3 \end{bmatrix}$

19. $\begin{bmatrix} 2 & 3 \\ 4 & -1 \end{bmatrix} \begin{array}{|cc} 1 & 0 \\ 0 & 1 \end{array}$ $\xrightarrow{(1/2)R_1 \to R_1}$ $\begin{bmatrix} 1 & 3/2 \\ 4 & -1 \end{bmatrix} \begin{array}{|cc} 1/2 & 0 \\ 0 & 1 \end{array}$ $\xrightarrow{(-4)R_1 + R_2 \to R_2}$ $\begin{bmatrix} 1 & 3/2 \\ 0 & -7 \end{bmatrix} \begin{array}{|cc} 1/2 & 0 \\ -2 & 1 \end{array}$

$\xrightarrow{(-1/7)R_2 \to R_2}$ $\begin{bmatrix} 1 & 3/2 \\ 0 & 1 \end{bmatrix} \begin{array}{|cc} 1/2 & 0 \\ 2/7 & -1/7 \end{array}$ $\xrightarrow{(3/2)R_2 + R_1 \to R_1}$ $\begin{bmatrix} 1 & 0 \\ 0 & 1 \end{bmatrix} \begin{array}{|cc} 1/14 & 3/14 \\ 2/7 & -1/7 \end{array}$

Inverse matrix is $(1/14)\begin{bmatrix} 1 & 3 \\ 4 & -2 \end{bmatrix}$

21. $\begin{bmatrix} 1 & 2 & 1 \\ 4 & 1 & 0 \\ 0 & 0 & 1 \end{bmatrix} \begin{array}{|ccc} 1 & 0 & 0 \\ 0 & 1 & 0 \\ 0 & 0 & 1 \end{array}$ $\xrightarrow{(-4)R_1 + R_2 \to R_2}$ $\begin{bmatrix} 1 & 2 & 1 \\ 0 & -7 & -4 \\ 0 & 0 & 1 \end{bmatrix} \begin{array}{|ccc} 1 & 0 & 0 \\ -4 & 1 & 0 \\ 0 & 0 & 1 \end{array}$ $\xrightarrow{(7)R_1 \to R_1}$

$\begin{bmatrix} 7 & 14 & 7 \\ 0 & -7 & -4 \\ 0 & 0 & 1 \end{bmatrix} \begin{array}{|ccc} 7 & 0 & 0 \\ -4 & 1 & 0 \\ 0 & 0 & 1 \end{array}$ $\xrightarrow[\text{followed by } (-1)R_2 \to R_2]{(2)R_2 + R_1 \to R_1}$ $\begin{bmatrix} 7 & 0 & -1 \\ 0 & 7 & 4 \\ 0 & 0 & 1 \end{bmatrix} \begin{array}{|ccc} -1 & 2 & 0 \\ 4 & -1 & 0 \\ 0 & 0 & 1 \end{array}$ $\xrightarrow[(-4)R_3 + R_2 \to R_2]{R_3 + R_1 \to R_1}$

$\begin{bmatrix} 7 & 0 & 0 \\ 0 & 7 & 0 \\ 0 & 0 & 1 \end{bmatrix} \begin{array}{|ccc} -1 & 2 & 1 \\ 4 & -1 & -4 \\ 0 & 0 & 1 \end{array}$ $\xrightarrow{(7)R_3 \to R_3}$ $\begin{bmatrix} 7 & 0 & 0 \\ 0 & 7 & 0 \\ 0 & 0 & 7 \end{bmatrix} \begin{array}{|ccc} -1 & 2 & 1 \\ 4 & -1 & -4 \\ 0 & 0 & 7 \end{array}$ Inverse matrix is $(1/7)\begin{bmatrix} -1 & 2 & 1 \\ 4 & -1 & -4 \\ 0 & 0 & 7 \end{bmatrix}$

23. $\begin{bmatrix} 2 & 3 \\ -10 & -15 \end{bmatrix} \begin{array}{|cc} 1 & 0 \\ 0 & 1 \end{array}$ $\xrightarrow{(5)R_1 + R_2 \to R_2}$ $\begin{bmatrix} 2 & 3 \\ 0 & 0 \end{bmatrix} \begin{array}{|cc} 1 & 0 \\ 5 & 1 \end{array}$ $\Rightarrow K^{-1}$ does not exist.

25. $\begin{bmatrix} -1 & 1 & 1 & 0 \\ 0 & 0 & 0 & 2 \\ 3 & 3 & 0 & 0 \\ 4 & 2 & 2 & 0 \end{bmatrix} \begin{array}{|cccc} 1 & 0 & 0 & 0 \\ 0 & 1 & 0 & 0 \\ 0 & 0 & 1 & 0 \\ 0 & 0 & 0 & 1 \end{array}$ $\xrightarrow[\substack{(4)R_1 + R_4 \to R_4 \\ \text{followed by} \\ (-1)R_1 \to R_1}]{(3)R_1 + R_3 \to R_3}$ $\begin{bmatrix} 1 & -1 & -1 & 0 \\ 0 & 0 & 0 & 2 \\ 0 & 6 & 3 & 0 \\ 0 & 6 & 6 & 0 \end{bmatrix} \begin{array}{|cccc} -1 & 0 & 0 & 0 \\ 0 & 1 & 0 & 0 \\ 3 & 0 & 1 & 0 \\ 4 & 0 & 0 & 1 \end{array}$ $\xrightarrow[\substack{\text{followed by} \\ R_3 + R_1 \to R_1 \\ (-1)R_3 + R_4 \to R_4}]{(6)R_1 \to R_1}$

$\begin{bmatrix} 6 & 0 & -3 & 0 \\ 0 & 0 & 0 & 2 \\ 0 & 6 & 3 & 0 \\ 0 & 0 & 3 & 0 \end{bmatrix} \begin{array}{|cccc} -3 & 0 & 1 & 0 \\ 0 & 1 & 0 & 0 \\ 3 & 0 & 1 & 0 \\ 1 & 0 & -1 & 1 \end{array}$ $\xrightarrow[\substack{(-1)R_4 + R_3 \to R_3}]{R_4 + R_1 \to R_1}$ $\begin{bmatrix} 6 & 0 & 0 & 0 \\ 0 & 0 & 0 & 2 \\ 0 & 6 & 0 & 0 \\ 0 & 0 & 3 & 0 \end{bmatrix} \begin{array}{|cccc} -2 & 0 & 0 & 1 \\ 0 & 1 & 0 & 0 \\ 2 & 0 & 2 & -1 \\ 1 & 0 & -1 & 1 \end{array}$ $\xrightarrow[\substack{\text{followed by} \\ R_3 \longleftrightarrow R_4}]{R_2 \longleftrightarrow R_3}$

$\begin{bmatrix} 6 & 0 & 0 & 0 \\ 0 & 6 & 0 & 0 \\ 0 & 0 & 3 & 0 \\ 0 & 0 & 0 & 2 \end{bmatrix} \begin{array}{|cccc} -2 & 0 & 0 & 1 \\ 2 & 0 & 2 & -1 \\ 1 & 0 & -1 & 1 \\ 0 & 1 & 0 & 0 \end{array}$ $\xrightarrow[(3)R_4 \to R_4]{(2)R_3 \to R_3}$ $\begin{bmatrix} 6 & 0 & 0 & 0 \\ 0 & 6 & 0 & 0 \\ 0 & 0 & 6 & 0 \\ 0 & 0 & 0 & 6 \end{bmatrix} \begin{array}{|cccc} -2 & 0 & 0 & 1 \\ 2 & 0 & 2 & -1 \\ 2 & 0 & -2 & 2 \\ 0 & 3 & 0 & 0 \end{array}$

Inverse matrix is $(1/6)\begin{bmatrix} -2 & 0 & 0 & 1 \\ 2 & 0 & 2 & -1 \\ 2 & 0 & -2 & 2 \\ 0 & 3 & 0 & 0 \end{bmatrix}$

5-7 Solving Square Linear Systems by Matrix Inverses

1. $\begin{bmatrix} 2 & 1 \\ -1 & 3 \end{bmatrix} \begin{bmatrix} 5 \\ 2 \end{bmatrix} = \begin{bmatrix} 12 \\ 1 \end{bmatrix}$ $x_1 = 12, x_2 = 1$

3. $\begin{bmatrix} 4 & -1 \\ -1 & 3 \end{bmatrix} \begin{bmatrix} 5 \\ 2 \end{bmatrix} = \begin{bmatrix} 18 \\ 1 \end{bmatrix}$ $x_1 = 18, x_2 = 1$

5. $\begin{bmatrix} 1 & 2 \\ -1 & 3 \end{bmatrix} \begin{bmatrix} x_1 \\ x_2 \end{bmatrix} = \begin{bmatrix} 9 \\ 1 \end{bmatrix}$

$\left[\begin{array}{cc|cc} 1 & 2 & 1 & 0 \\ -1 & 3 & 0 & 1 \end{array}\right] \xrightarrow{R_1 + R_2 \to R_2} \left[\begin{array}{cc|cc} 1 & 2 & 1 & 0 \\ 0 & 5 & 1 & 1 \end{array}\right] \xrightarrow{(1/5)R_2 \to R_2} \left[\begin{array}{cc|cc} 1 & 2 & 1 & 0 \\ 0 & 1 & 1/5 & 1/5 \end{array}\right]$

$\xrightarrow{(-2)R_2 + R_1 \to R_1} \left[\begin{array}{cc|cc} 1 & 0 & 3/5 & -2/5 \\ 0 & 1 & 1/5 & 1/5 \end{array}\right]$

$\begin{bmatrix} x_1 \\ x_2 \end{bmatrix} = \left[\frac{1}{5}\right] \begin{bmatrix} 3 & -2 \\ 1 & 1 \end{bmatrix} \begin{bmatrix} 9 \\ 1 \end{bmatrix} = \begin{bmatrix} 5 \\ 2 \end{bmatrix}$

In Problems 7–23, details of computing A^{-1} are omitted. In each case A^{-1} can be computed by Gauss–Jordan reduction.

7. $\begin{bmatrix} 1 & 0 & 1 \\ 0 & 1 & 4 \\ 2 & 3 & 1 \end{bmatrix} \begin{bmatrix} x_1 \\ x_2 \\ x_3 \end{bmatrix} = \begin{bmatrix} 11 \\ 39 \\ 22 \end{bmatrix}$

$\begin{bmatrix} x_1 \\ x_2 \\ x_3 \end{bmatrix} = A^{-1}B = \frac{1}{13}\begin{bmatrix} 11 & -3 & 1 \\ -8 & 1 & 4 \\ 2 & 3 & -1 \end{bmatrix} \begin{bmatrix} 11 \\ 39 \\ 22 \end{bmatrix} = \begin{bmatrix} 2 \\ 3 \\ 9 \end{bmatrix}$

9. $\begin{bmatrix} 1 & 1 & 1 \\ 1 & 2 & 3 \\ 0 & 1 & 4 \end{bmatrix} \begin{bmatrix} x_1 \\ x_2 \\ x_3 \end{bmatrix} = \begin{bmatrix} 3 \\ 10 \\ 17 \end{bmatrix}$

$\begin{bmatrix} x_1 \\ x_2 \\ x_3 \end{bmatrix} = A^{-1}B = \frac{1}{2}\begin{bmatrix} 5 & -3 & 1 \\ -4 & 4 & -2 \\ 1 & -1 & 1 \end{bmatrix} \begin{bmatrix} 3 \\ 10 \\ 17 \end{bmatrix} = \begin{bmatrix} 1 \\ -3 \\ 5 \end{bmatrix}$

11. $\begin{bmatrix} 1 & 1 & 0 \\ 0 & 1 & 3 \\ 4 & 6 & 7 \end{bmatrix} \begin{bmatrix} x_1 \\ x_2 \\ x_3 \end{bmatrix} = \begin{bmatrix} 3 \\ -7 \\ -5 \end{bmatrix}$

$\begin{bmatrix} x_1 \\ x_2 \\ x_3 \end{bmatrix} = A^{-1}B = \begin{bmatrix} -11 & -7 & 3 \\ 12 & 7 & -3 \\ -4 & -2 & 1 \end{bmatrix} \begin{bmatrix} 3 \\ -7 \\ -5 \end{bmatrix} = \begin{bmatrix} 1 \\ 2 \\ -3 \end{bmatrix}$

13. $\begin{bmatrix} 2 & -3 \\ 1 & -7 \end{bmatrix} \begin{bmatrix} x \\ y \end{bmatrix} = \begin{bmatrix} 6 \\ 25 \end{bmatrix}$

$\begin{bmatrix} x \\ y \end{bmatrix} = A^{-1}B = -\frac{1}{11} \begin{bmatrix} -7 & 3 \\ -1 & 2 \end{bmatrix} \begin{bmatrix} 6 \\ 25 \end{bmatrix} = \begin{bmatrix} -3 \\ -4 \end{bmatrix}$

15. $\begin{bmatrix} 2 & 3 & -5 \\ 1 & 2 & 3 \\ 3 & -4 & -7 \end{bmatrix} \begin{bmatrix} x \\ y \\ z \end{bmatrix} = \begin{bmatrix} -13 \\ -7 \\ 15 \end{bmatrix}$

$\begin{bmatrix} x \\ y \\ z \end{bmatrix} = A^{-1}B = \frac{1}{94} \begin{bmatrix} -2 & 41 & 19 \\ 16 & 1 & -11 \\ -10 & 17 & 1 \end{bmatrix} \begin{bmatrix} -13 \\ -7 \\ 15 \end{bmatrix} = \frac{1}{94} \begin{bmatrix} 24 \\ -380 \\ 26 \end{bmatrix} = \begin{bmatrix} 12/47 \\ -190/47 \\ 13/47 \end{bmatrix}$

17. $\begin{bmatrix} 5 & 7 & 1 \\ 3 & 2 & 3 \\ 2 & 3 & 5 \end{bmatrix} \begin{bmatrix} x_1 \\ x_2 \\ x_3 \end{bmatrix} = \begin{bmatrix} 1 \\ 8 \\ 19 \end{bmatrix}$

$\begin{bmatrix} x_1 \\ x_2 \\ x_3 \end{bmatrix} = A^{-1}B = \frac{1}{53} \begin{bmatrix} -1 & 32 & -19 \\ 9 & -23 & 12 \\ -5 & 1 & 11 \end{bmatrix} \begin{bmatrix} 1 \\ 8 \\ 19 \end{bmatrix} = \begin{bmatrix} -2 \\ 1 \\ 4 \end{bmatrix}$

19. $\begin{bmatrix} 4 & 1 \\ 6 & -2 \end{bmatrix} \begin{bmatrix} x_1 \\ x_2 \end{bmatrix} = \begin{bmatrix} 8 \\ -9 \end{bmatrix}$

$\begin{bmatrix} x_1 \\ x_2 \end{bmatrix} = A^{-1}B = \frac{1}{14} \begin{bmatrix} 2 & 1 \\ 6 & -4 \end{bmatrix} \begin{bmatrix} 8 \\ -9 \end{bmatrix} = \begin{bmatrix} 1/2 \\ 6 \end{bmatrix}$

21. $\begin{bmatrix} 2 & 1 & 3 \\ 4 & 3 & -2 \\ 6 & 5 & -4 \end{bmatrix} \begin{bmatrix} x \\ y \\ z \end{bmatrix} = \begin{bmatrix} 11 \\ -1 \\ -4 \end{bmatrix}$

$\begin{bmatrix} x \\ y \\ z \end{bmatrix} = A^{-1}B = \frac{1}{6} \begin{bmatrix} -2 & 19 & -11 \\ 4 & -26 & 16 \\ 2 & -4 & 4 \end{bmatrix} \begin{bmatrix} 11 \\ -1 \\ -4 \end{bmatrix} = \begin{bmatrix} 1/2 \\ 1 \\ 3 \end{bmatrix}$

23.
$$\begin{bmatrix} 1 & 0 & 1 & -1 \\ 0 & 1 & -1 & 2 \\ 4 & 0 & 5 & -3 \\ 0 & 0 & 2 & 3 \end{bmatrix} \begin{bmatrix} x_1 \\ x_2 \\ x_3 \\ x_4 \end{bmatrix} = \begin{bmatrix} -2 \\ 12 \\ 2 \\ 27 \end{bmatrix}$$

$$\begin{bmatrix} x_1 \\ x_2 \\ x_3 \\ x_4 \end{bmatrix} = A^{-1}B = \begin{bmatrix} 21 & 0 & -5 & 2 \\ -28 & 1 & 7 & -3 \\ -12 & 0 & 3 & -1 \\ 8 & 0 & -2 & 1 \end{bmatrix} \begin{bmatrix} -2 \\ 12 \\ 2 \\ 27 \end{bmatrix} = \begin{bmatrix} 2 \\ 1 \\ 3 \\ 7 \end{bmatrix}$$

25. (a)
$$\begin{array}{cc} \text{meat} & \text{spinach} \end{array}$$
$$A = \begin{bmatrix} 500 & 200 \\ 100 & 800 \end{bmatrix} \begin{array}{l} \text{protein} \\ \text{iron} \end{array}$$

(b)
$$B = \begin{bmatrix} 1200 \\ 1000 \end{bmatrix} \begin{array}{l} \text{protein} \\ \text{iron} \end{array}$$

(c)
$$\begin{bmatrix} 500 & 200 \\ 100 & 800 \end{bmatrix} \begin{bmatrix} x \\ y \end{bmatrix} = \begin{bmatrix} 1200 \\ 1000 \end{bmatrix}$$

(d)
$$\begin{bmatrix} x \\ y \end{bmatrix} = A^{-1}B = \frac{1}{3800} \begin{bmatrix} 8 & -2 \\ -1 & 5 \end{bmatrix} \begin{bmatrix} 1200 \\ 1000 \end{bmatrix} = \begin{bmatrix} 2 \\ 1 \end{bmatrix}$$

Hence, $x = 2$ lbs meat, $y = $ lb spinach.

(e)
$$\begin{bmatrix} x \\ y \end{bmatrix} = A^{-1}B = \frac{1}{3800} \begin{bmatrix} 8 & -2 \\ -1 & 5 \end{bmatrix} \begin{bmatrix} 2300 \\ 3500 \end{bmatrix} = \begin{bmatrix} 3 \\ 4 \end{bmatrix}$$

27. (a) Let $x = $ units of $XB17$, $y = $ units of $XB18$

$0.01x + 0.03y = 220$

$0.02x + 0.005y = 330$

(b)
$$\begin{bmatrix} 0.01 & 0.03 \\ 0.02 & 0.005 \end{bmatrix} \begin{bmatrix} x \\ y \end{bmatrix} = \begin{bmatrix} 220 \\ 330 \end{bmatrix}$$

(c)
$$A^{-1} = \begin{bmatrix} -100/11 & 600/11 \\ 400/11 & -200/11 \end{bmatrix}$$

(d)
$$X = A^{-1}B = \begin{bmatrix} -100/11 & 600/11 \\ 400/11 & -200/11 \end{bmatrix} \begin{bmatrix} 220 \\ 330 \end{bmatrix} = \begin{bmatrix} 16,000 \\ 2,000 \end{bmatrix} = \begin{bmatrix} x \\ y \end{bmatrix}$$

(e)
$$X = A^{-1}B = \begin{bmatrix} -100/11 & 600/11 \\ 400/11 & -200/11 \end{bmatrix} \begin{bmatrix} 440 \\ 110 \end{bmatrix} = \begin{bmatrix} 2,000 \\ 14,000 \end{bmatrix} = \begin{bmatrix} x \\ y \end{bmatrix}$$

29. (a) $x + y = 100,000$ available funds
$0.07x + 0.09y = 8,500$ yield

(b) $\begin{bmatrix} 1 & 1 \\ 0.07 & 0.09 \end{bmatrix} \begin{bmatrix} x \\ y \end{bmatrix} = \begin{bmatrix} 100{,}000 \\ 8{,}500 \end{bmatrix}$

(c) $A^{-1} = \begin{bmatrix} 4.5 & -50 \\ -3.5 & 50 \end{bmatrix}$

(d) $X = A^{-1}B = \begin{bmatrix} 4.5 & -50 \\ -3.5 & 50 \end{bmatrix} \begin{bmatrix} 100{,}000 \\ 8{,}500 \end{bmatrix} = \begin{bmatrix} 25{,}000 \\ 75{,}000 \end{bmatrix} = \begin{bmatrix} x \\ y \end{bmatrix}$
$x = \$25{,}000$ invested at 7%
$y = \$75{,}000$ invested at 9%

(e) $X = A^{-1}B = \begin{bmatrix} 4.5 & -50 \\ -3.5 & 50 \end{bmatrix} \begin{bmatrix} 200{,}000 \\ 15{,}000 \end{bmatrix} = \begin{bmatrix} 150{,}000 \\ 50{,}000 \end{bmatrix}$
$x = \$150{,}000, \; y = \$50{,}000$

5–8 Leontief's Input–Output Model

1. (a) $A = \begin{bmatrix} 0 & 1/4 \\ 1/2 & 0 \end{bmatrix} \begin{matrix} \text{oil} \\ \text{coal} \end{matrix}$

(b) $X = (I - A)^{-1}D = \left(\begin{bmatrix} 1 & 0 \\ 0 & 1 \end{bmatrix} - \begin{bmatrix} 0 & 1/4 \\ 1/2 & 0 \end{bmatrix} \right)^{-1} \begin{bmatrix} 210 \\ 490 \end{bmatrix} = \begin{bmatrix} 1 & -1/4 \\ -1/2 & 1 \end{bmatrix}^{-1} \begin{bmatrix} 210 \\ 490 \end{bmatrix} = \frac{8}{7} \begin{bmatrix} 1 & 1/4 \\ 1/2 & 1 \end{bmatrix} \begin{bmatrix} 210 \\ 490 \end{bmatrix} = \begin{bmatrix} 380 \\ 680 \end{bmatrix}$

3. $x = (I - A)^{-1}D = \left(\begin{bmatrix} 1 & 0 \\ 0 & 1 \end{bmatrix} - \begin{bmatrix} 0.1 & 0.2 \\ 0.3 & 0.1 \end{bmatrix} \right)^{-1} \begin{bmatrix} 30 \\ 20 \end{bmatrix} = \begin{bmatrix} 0.9 & -0.2 \\ -0.3 & 0.9 \end{bmatrix}^{-1} \begin{bmatrix} 30 \\ 20 \end{bmatrix}$

$= \frac{4}{3} \begin{bmatrix} 0.9 & 0.2 \\ 0.3 & 0.9 \end{bmatrix} \begin{bmatrix} 30 \\ 20 \end{bmatrix} = \frac{4}{3} \begin{bmatrix} 31 \\ 27 \end{bmatrix} = \begin{bmatrix} 41\frac{1}{3} \\ 36 \end{bmatrix}$

5. $X = (I - A)^{-1}D = \left(\begin{bmatrix} 1 & 0 & 0 \\ 0 & 1 & 0 \\ 0 & 0 & 1 \end{bmatrix} - \begin{bmatrix} 0.3 & 0.1 & 0.2 \\ 0.2 & 0.1 & 0.1 \\ 0.2 & 0.1 & 0.1 \end{bmatrix} \right)^{-1} \begin{bmatrix} 50 \\ 30 \\ 90 \end{bmatrix} = \begin{bmatrix} 0.7 & -0.1 & -0.2 \\ -0.2 & 0.9 & -0.1 \\ -0.2 & -0.1 & 0.9 \end{bmatrix}^{-1} \begin{bmatrix} 50 \\ 30 \\ 90 \end{bmatrix}$

$= \frac{1}{50} \begin{bmatrix} 80 & 11 & 19 \\ 20 & 59 & 11 \\ 20 & 9 & 61 \end{bmatrix} \begin{bmatrix} 50 \\ 30 \\ 90 \end{bmatrix} = \frac{1}{5} \begin{bmatrix} 604 \\ 376 \\ 676 \end{bmatrix} = \begin{bmatrix} 1\,8 \\ \,2 \\ 135.2 \end{bmatrix}$

7. $\Delta X = (I - A)^{-1}\Delta D = \begin{bmatrix} 1.4 & 0.8 \\ 1.1 & 0.8 \end{bmatrix} \begin{bmatrix} -2 \\ 6 \end{bmatrix} = \begin{bmatrix} 2 \\ 3.2 \end{bmatrix}$

9.
$$\Delta X = (I - A)^{-1} \Delta D = \begin{bmatrix} 2.3 & 1.5 \\ 0.9 & 1.2 \end{bmatrix} \begin{bmatrix} 4 \\ -1 \end{bmatrix} = \begin{bmatrix} 7.7 \\ 2.4 \end{bmatrix}$$

EXTRA DIVIDENDS: Oil Refinery Scheduling

1.

	Well in Saudia Arabia	Well in Kuwait	Well in Egypt	Demand
Regular	0.2	0.3	0.4	19
Unleaded	0.1	0.2	0.1	10
Kerosene	0.3	0.1	0.1	20

3.
$$\begin{bmatrix} 0.2 & 0.3 & 0.4 \\ 0.1 & 0.2 & 0.1 \\ 0.3 & 0.1 & 0.4 \end{bmatrix} \begin{bmatrix} x \\ y \\ z \end{bmatrix} = \begin{bmatrix} 19 \\ 10 \\ 20 \end{bmatrix}$$

5.
$$X = A^{-1}B = \frac{10}{9} \begin{bmatrix} -7 & 8 & 5 \\ 1 & 4 & -2 \\ 5 & -7 & -1 \end{bmatrix} \begin{bmatrix} 20 \\ 10 \\ 15 \end{bmatrix} = \begin{bmatrix} 150/9 \\ 300/9 \\ 150/9 \end{bmatrix} = \begin{bmatrix} 50/3 \\ 100/3 \\ 50/3 \end{bmatrix}$$

Review Exercises

1. $\begin{aligned} x + 2y &= 2 \\ 3x - 5y &= 17 \end{aligned} \rightarrow \begin{aligned} -3x - 6y &= -6 \\ 3x - 5y &= 17 \end{aligned}$

$$-11y = \implies y = -1$$
$$x = 2 - 2y = 2 + 2 = 4$$
$$(x,y) = (4,-1)$$

3. $\begin{aligned} 3x - 5y &= 8 \\ -6x + 10y &= 22 \end{aligned} \rightarrow \begin{aligned} 6x - 10y &= 16 \\ -6x + 10y &= 12 \end{aligned}$

$$0 = 28 \implies \text{no solution}$$

5. $\dfrac{x}{4} + \dfrac{y}{3} = 6$ $3x + 4y = 72$ $3x + 4y = 72$

$\rightarrow$ $\rightarrow$

$\dfrac{x}{2} - \dfrac{y}{5} = -1$ $5x - 2y = -10$ $10x - 4y = -20$

$\overline{}$

$13x \qquad = 52 \Rightarrow x = 4$

$y = 5\left[\dfrac{x}{2} + 1\right] = 5(3) = 15$

$(x,y) = (4,15)$

7. $3x + y = 48$

$\rightarrow 3x + 5x = 48 \Rightarrow x = 6$

$y = 5x$ $y = 5(6) = 30$

$(x,y) = (6,30)$

9. (a)

(1) $5x - 2y = 20$ $15x - 6y = 60$

$\rightarrow$

(2) $-15x + 6y = -60$ $-15x + 6y = -60$

$\overline{}$

$0 = 0 \Rightarrow$ infinity of solutions

Let $x = t$; then $y = \dfrac{1}{2}(5x - 20) = \dfrac{5}{2}t - 10$

(b) Eq. (1) becomes $y = \dfrac{1}{2}(5x - 20) = \dfrac{5}{2}t - 10$

Eq. (2) becomes $y = \dfrac{1}{6}(-60) + 15x) = \dfrac{5}{2}x - 10$

Equations are the same

(c) Let $t = 0, 1, 2$. Solutions are $(0,-10)$, $\left[1, -\dfrac{15}{2}\right]$, $(2,-5)$

11.

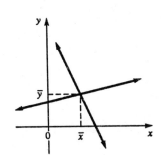

13.

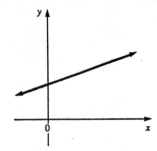

15. $\begin{bmatrix} 2 & 1 & 5 & | & 27 \\ 1 & 0 & 1 & | & 8 \\ -1 & 1 & 0 & | & -7 \end{bmatrix} \xrightarrow[\;R_2 + R_3 \to R_3\;]{(-2)R_2 + R_1 \to R_1} \begin{bmatrix} 0 & 1 & 3 & | & 11 \\ 1 & 0 & 1 & | & 8 \\ 0 & 1 & 1 & | & 1 \end{bmatrix} \xrightarrow{R_1 \longleftrightarrow R_2} \begin{bmatrix} 1 & 0 & 1 & | & 8 \\ 0 & 1 & 3 & | & 11 \\ 0 & 1 & 1 & | & 1 \end{bmatrix} \xrightarrow{(-1)R_2 + R_3 \to R_3}$

$\begin{bmatrix} 1 & 0 & 1 & | & 8 \\ 0 & 1 & 3 & | & 11 \\ 0 & 0 & -2 & | & -10 \end{bmatrix} \xrightarrow{(-1/2)R_3 \to R_3} \begin{bmatrix} 1 & 0 & 1 & | & 8 \\ 0 & 1 & 3 & | & 11 \\ 0 & 0 & 1 & | & 5 \end{bmatrix} \xrightarrow[\;(-3)R_3 + R_2 \to R_2\;]{(-1)R_3 + R_1 \to R_1} \begin{bmatrix} 1 & 0 & 0 & | & 3 \\ 0 & 1 & 0 & | & -4 \\ 0 & 0 & 1 & | & 5 \end{bmatrix}$

$(x,y,z) = (3,-4,5)$

17. $\begin{bmatrix} 2 & 1 & -1 & | & 8 \\ -6 & -3 & 3 & | & 15 \end{bmatrix} \xrightarrow{(3)R_1 + R_2 \to R_2} \begin{bmatrix} 2 & 1 & -1 & | & 8 \\ 0 & 0 & 0 & | & 39 \end{bmatrix} \Rightarrow$ no solution

19. $\begin{bmatrix} 1 & -1 & | & 7 \\ 2 & -3 & | & 13 \\ 1 & 1 & | & 6 \end{bmatrix} \xrightarrow[\;(-1)R_1 + R_3 \to R_3\;]{(-2)R_1 + R_2 \to R_2} \begin{bmatrix} 1 & -1 & | & 7 \\ 0 & -1 & | & -1 \\ 0 & 2 & | & -1 \end{bmatrix} \xrightarrow[\;(2)R_2 + R_3 \to R_3\;]{(-1)R_2 + R_1 \to R_1} \begin{bmatrix} 1 & 0 & | & 8 \\ 0 & -1 & | & -1 \\ 0 & 0 & | & -3 \end{bmatrix} \Rightarrow$ no solution

21. $\begin{bmatrix} 1 & -1 & 2 & | & 5 \\ -2 & 3 & 0 & | & -2 \\ -1 & 0 & 1 & | & 4 \end{bmatrix} \xrightarrow[\;R_1 + R_3 \to R_3\;]{(2)R_1 + R_2 \to R_2} \begin{bmatrix} 1 & -1 & 2 & | & 5 \\ 0 & 1 & 4 & | & 8 \\ 0 & -1 & 3 & | & 9 \end{bmatrix} \xrightarrow[\;R_2 + R_3 \to R_3\;]{R_2 + R_1 \to R_1} \begin{bmatrix} 1 & 0 & 6 & | & 13 \\ 0 & 1 & 4 & | & 8 \\ 0 & 0 & 7 & | & 17 \end{bmatrix}$

$\xrightarrow[\;(7)R_2 \to R_2\;]{(7)R_1 \to R_1} \begin{bmatrix} 7 & 0 & 42 & | & 91 \\ 0 & 7 & 28 & | & 56 \\ 0 & 0 & 7 & | & 17 \end{bmatrix} \xrightarrow[\;(-4)R_3 + R_2 \to R_2\;]{(-6)R_3 + R_1 \to R_1} \begin{bmatrix} 7 & 0 & 0 & | & -11 \\ 0 & 7 & 0 & | & -12 \\ 0 & 0 & 7 & | & 17 \end{bmatrix} \rightarrow \begin{bmatrix} 1 & 0 & 0 & | & -11/7 \\ 0 & 1 & 0 & | & -12/7 \\ 0 & 0 & 1 & | & 17/7 \end{bmatrix}$

$(x_1, x_2, x_3) = (-11/7, -12/7, 17/7)$

23. $\begin{bmatrix} 1 & 0 & 4 & | & 3 \\ 0 & 1 & 0 & | & 5 \\ 0 & 0 & 0 & | & 4 \end{bmatrix} \Rightarrow$ no solution

25. $\begin{bmatrix} 1 & 0 & | & -3 \\ 0 & 1 & | & 4 \\ 0 & 0 & | & 6 \end{bmatrix} \Rightarrow$ no solution

27. $\begin{bmatrix} 1 & 0 & 0 & | & 9 \\ 0 & 1 & 0 & | & -3 \\ 0 & 0 & 0 & | & 0 \end{bmatrix} \Rightarrow$ infinity of solutions

$x_1 = 9$, $x_2 = -3$, $x_3 = t$,
a real number
Note that x_1 and x_2 are unique.

In Problems 29–35,

$A = \begin{bmatrix} 3 & 4 \\ 1 & -2 \\ 0 & 1 \end{bmatrix}$, $B = \begin{bmatrix} -2 & 3 \\ 1 & -1 \\ 1 & 2 \end{bmatrix}$, $C = \begin{bmatrix} 1 & 2 & 1 \\ 2 & 1 & 0 \end{bmatrix}$

29.
$A - B = \begin{bmatrix} 5 & 1 \\ 0 & -1 \\ -1 & -1 \end{bmatrix}$

31.
$A - 3B = \begin{bmatrix} 9 & -5 \\ -2 & 1 \\ -3 & -5 \end{bmatrix}$

33. $CA = \begin{bmatrix} 5 & 1 \\ 7 & 6 \end{bmatrix}$

35.
$(BC)A = \begin{bmatrix} 11 & 16 \\ -2 & -5 \\ 19 & 13 \end{bmatrix}$

37. (a) $\begin{bmatrix} 1 & 2 & 2 & | & 1 & 0 & 0 \\ 3 & 7 & 1 & | & 0 & 1 & 0 \\ 0 & 1 & 0 & | & 0 & 0 & 1 \end{bmatrix} \xrightarrow{(-3)R_1 + R_2 \rightarrow R_2} \begin{bmatrix} 1 & 2 & 2 & | & 1 & 0 & 0 \\ 0 & 1 & -5 & | & -3 & 1 & 0 \\ 0 & 1 & 0 & | & 0 & 0 & 1 \end{bmatrix} \xrightarrow[\;(-1)R_2 + R_3 \rightarrow R_3\;]{(-2)R_2 + R_1 \rightarrow R_1}$

$\begin{bmatrix} 1 & 0 & 12 & | & 7 & -2 & 0 \\ 0 & 1 & -5 & | & -3 & 1 & 0 \\ 0 & 0 & 5 & | & 3 & -1 & 1 \end{bmatrix} \xrightarrow[\;(5)R_2 \rightarrow R_2\;]{(5)R_1 \rightarrow R_1} \begin{bmatrix} 5 & 0 & 60 & | & 35 & -10 & 0 \\ 0 & 5 & -25 & | & -15 & 5 & 0 \\ 0 & 0 & 5 & | & 3 & -1 & 1 \end{bmatrix} \xrightarrow[\;(5)R_3 + R_2 \rightarrow R_2\;]{(-7)R_3 + R_1 \rightarrow R_1}$

$\begin{bmatrix} 5 & 0 & 0 & | & -1 & 2 & -12 \\ 0 & 5 & 0 & | & 0 & 0 & 5 \\ 0 & 0 & 5 & | & 3 & -1 & 1 \end{bmatrix} \quad A^{-1} = \begin{bmatrix} -1/5 & 2/5 & -12/5 \\ 0 & 0 & 1 \\ 3/5 & -1/5 & 1/5 \end{bmatrix}$

(b) $AA^{-1} = A^{-1}A = I$
First product shown:

$$\begin{bmatrix} -1/5 & 2/5 & -12/5 \\ 0 & 0 & 1 \\ 3/5 & -1/5 & 1/5 \end{bmatrix} \begin{bmatrix} 1 & 2 & 2 \\ 3 & 7 & 1 \\ 0 & 1 & 0 \end{bmatrix} = \begin{bmatrix} 1 & 0 & 0 \\ 0 & 1 & 0 \\ 0 & 0 & 1 \end{bmatrix}$$

39. $\begin{bmatrix} 2 & 3 & | & 18 \\ 6 & 1 & | & 22 \end{bmatrix} \xrightarrow{(-3)R_1 + R_2 \rightarrow R_2} \begin{bmatrix} 2 & 3 & | & 18 \\ 0 & -8 & | & -32 \end{bmatrix} \xrightarrow{(-1/8)R_2 \rightarrow R_2} \begin{bmatrix} 2 & 3 & | & 18 \\ 0 & 1 & | & 4 \end{bmatrix}$

$\xrightarrow{(-3)R_2 + R_1 \rightarrow R_1} \begin{bmatrix} 2 & 0 & | & 6 \\ 0 & 1 & | & 4 \end{bmatrix} \xrightarrow{(1/2)R_1 \rightarrow R_1} \begin{bmatrix} 1 & 0 & | & 3 \\ 0 & 1 & | & 4 \end{bmatrix}$

41. $\begin{bmatrix} 2 & 1 & 5 & | & 27 \\ 1 & 0 & 1 & | & 8 \\ -1 & 1 & 0 & | & -7 \end{bmatrix} \xrightarrow[\;R_2 + R_3 \rightarrow R_3\;]{(-2)R_2 + R_1 \rightarrow R_1} \begin{bmatrix} 0 & 1 & 3 & | & 11 \\ 1 & 0 & 1 & | & 8 \\ 0 & 1 & 1 & | & 1 \end{bmatrix} \xrightarrow{R_1 \longleftrightarrow R_2} \begin{bmatrix} 1 & 0 & 1 & | & 8 \\ 0 & 1 & 3 & | & 11 \\ 0 & 1 & 1 & | & 1 \end{bmatrix} \xrightarrow{(-1)R_2 + R_3 \rightarrow R_3}$

$\begin{bmatrix} 1 & 0 & 1 & | & 0 \\ 0 & 1 & 3 & | & 11 \\ 0 & 0 & -2 & | & -10 \end{bmatrix} \xrightarrow{(-1/2)R_3 \rightarrow R_3} \begin{bmatrix} 1 & 0 & 1 & | & 8 \\ 0 & 1 & 3 & | & 11 \\ 0 & 0 & 1 & | & 5 \end{bmatrix} \xrightarrow[\;(-3)R_3 + R_2 \rightarrow R_2\;]{(-1)R_3 + R_1 \rightarrow R_1} \begin{bmatrix} 1 & 0 & 0 & | & 3 \\ 0 & 1 & 0 & | & -4 \\ 0 & 0 & 1 & | & 5 \end{bmatrix}$

$(x_1, x_2, x_3) = (3, -4, 5)$

43. Let x_1 = lbs alloy containing 30% silver, x_2 = lbs alloy containing 80% silver.

$x_1 + x_2 = 100$ total amounts
$0.3x_1 + 0.8x_2 = 0.7(100) = 70$ silver content

$\begin{bmatrix} 1 & 1 & | & 100 \\ 3 & 8 & | & 700 \end{bmatrix} \rightarrow \begin{bmatrix} 1 & 1 & | & 100 \\ 0 & 5 & | & 400 \end{bmatrix} \rightarrow \begin{bmatrix} 1 & 1 & | & 100 \\ 0 & 1 & | & 80 \end{bmatrix} \rightarrow \begin{bmatrix} 1 & 0 & | & 20 \\ 0 & 1 & | & 80 \end{bmatrix} \quad (x_1, x_2) = (20, 80)$

45.

$$\left.\begin{array}{l} x_1 + x_2 \qquad\quad = 800 \\ \qquad x_3 + x_4 = 500 \end{array}\right\} \text{ capacities}$$

$$\left.\begin{array}{l} x_1 \qquad + x_3 \qquad = 700 \\ \quad x_2 \qquad + x_4 = 600 \end{array}\right\} \text{ demands}$$

$$\begin{bmatrix} 1 & 1 & 0 & 0 & | & 800 \\ 0 & 0 & 1 & 1 & | & 500 \\ 1 & 0 & 1 & 0 & | & 700 \\ 0 & 1 & 0 & 1 & | & 600 \end{bmatrix} \xrightarrow{(-1)R_1 + R_3 \to R_3} \begin{bmatrix} 1 & 1 & 0 & 0 & | & 800 \\ 0 & 0 & 1 & 1 & | & 500 \\ 0 & -1 & 1 & 0 & | & -100 \\ 0 & 1 & 0 & 1 & | & 600 \end{bmatrix} \begin{array}{l} \xrightarrow{(-1)R_3 + R_1 \to R_1} \\ \hline R_3 + R_4 \to R_4 \end{array}$$

$$\begin{bmatrix} 1 & 0 & 1 & 0 & | & 700 \\ 0 & 0 & 1 & 1 & | & 500 \\ 0 & -1 & 1 & 0 & | & -100 \\ 0 & 0 & 1 & 1 & | & 500 \end{bmatrix} \begin{array}{l} \xrightarrow{(-1)R_2 + R_1 \to R_1} \\ \xrightarrow{(-1)R_2 + R_3 \to R_3} \\ (-1)R_2 + R_4 \to R_4 \end{array} \begin{bmatrix} 1 & 0 & 0 & -1 & | & 200 \\ 0 & 0 & 1 & 1 & | & 500 \\ 0 & -1 & 0 & -1 & | & -600 \\ 0 & 0 & 0 & 0 & | & 0 \end{bmatrix} \begin{array}{l} \xrightarrow{(-1)R_3 \to R_3} \\ \text{followed by} \\ \hline R_3 \longleftrightarrow R_3 \end{array}$$

$$\begin{bmatrix} 1 & 0 & 0 & -1 & | & 200 \\ 0 & 1 & 0 & 1 & | & 600 \\ 0 & 0 & 1 & 1 & | & 500 \\ 0 & 0 & 0 & 0 & | & 0 \end{bmatrix} \Rightarrow \text{infinitely many solutions}$$

$$x_1 = 200 + t, \quad x_2 = 600 - t, \quad x_3 = 500 - t, \quad x_4 = t. \quad \text{Since each } x_i \geq 0,\ 0 \leq t \leq 500$$

47. (a)

$$A = \begin{bmatrix} 0.4 & 0.2 & 0.3 \\ 0.2 & 0.4 & 0.3 \\ 0.3 & 0.3 & 0.4 \end{bmatrix}$$

a_{ij} = amount of item i consumed in making one unit of item j.

(b)

$$D = \begin{bmatrix} 200 \\ 500 \\ 400 \end{bmatrix}$$

d_i = units of item i for external demand

(c)

$$X = (I - A)^{-1}D = \left[\begin{bmatrix} 1 & 0 & 0 \\ 0 & 1 & 0 \\ 0 & 0 & 1 \end{bmatrix} - \begin{bmatrix} 0.4 & 0.2 & 0.3 \\ 0.2 & 0.4 & 0.3 \\ 0.3 & 0.3 & 0.4 \end{bmatrix} \right]^{-1} \begin{bmatrix} 200 \\ 500 \\ 400 \end{bmatrix}$$

$$= \begin{bmatrix} 0.6 & -0.2 & -0.3 \\ -0.2 & 0.6 & -0.3 \\ -0.3 & -0.3 & 0.6 \end{bmatrix}^{-1} \begin{bmatrix} 200 \\ 500 \\ 400 \end{bmatrix} = \frac{1}{24} \begin{bmatrix} 135 & 105 & 120 \\ 105 & 135 & 120 \\ 120 & 120 & 160 \end{bmatrix} \begin{bmatrix} 200 \\ 500 \\ 400 \end{bmatrix} = \begin{bmatrix} 5312\frac{1}{2} \\ 5687\frac{1}{2} \\ 6166\frac{1}{3} \end{bmatrix}$$

x_i = total output of item i.

CHAPTER
6
LINEAR PROGRAMMING: THE GRAPHICAL METHOD

6–1 Linear Inequalities in Two Variables

The solutions to Problems 1–23 are graphs, shown in the accompanying figures. Shaded areas include points that do **not** satisfy the given inequalities.

1.

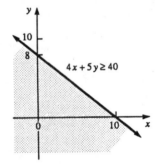

3.

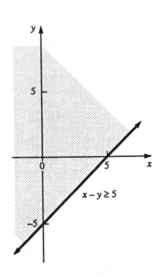

5.

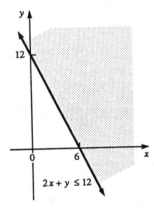

7.

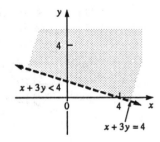

9.

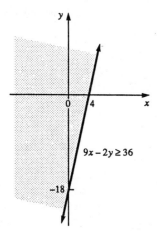

$9x - 2y \geq 36$

11.

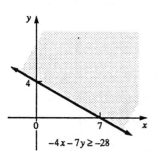

$-4x - 7y \geq -28$

13.

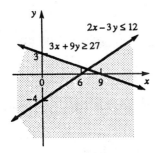

$2x - 3y \leq 12$

$3x + 9y \geq 27$

15.

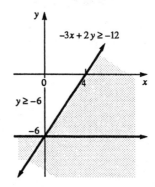

$-3x + 2y \geq -12$

$y \geq -6$

17.

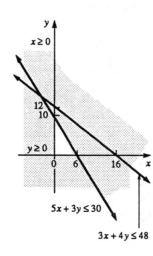

$x \geq 0$

$y \geq 0$

$5x + 3y \leq 30$

$3x + 4y \leq 48$

19.

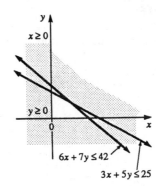

$x \geq 0$

$y \geq 0$

$6x + 7y \leq 42$

$3x + 5y \leq 25$

21.

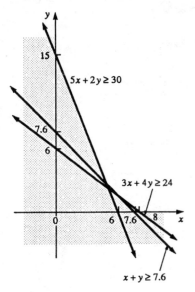

5x + 2y ≥ 30

3x + 4y ≥ 24

x + y ≥ 7.6

23.

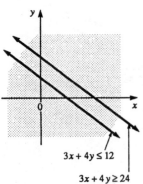

3x + 4y ≤ 12

3x + 4y ≥ 24

25. Let x and y represent the number of units of models x and y, respectively.
$$5x + 7y \le 105$$

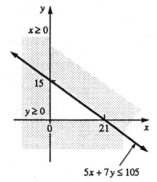

5x + 7y ≤ 105

27. Let g and w represent the number of gadgets and widgets, respectively.
$$4g + 5w \le 80$$

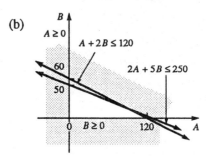

4g + 5w ≤ 80

29. (a) Let A and B represent number of lots of type A tires and type B tires, respectively.

$$2A + 5B \le 250 \quad \text{(department 1)}$$
$$A + 2B \le 120 \quad \text{(department 2)}$$

(b)

B
A ≥ 0
A + 2B ≤ 120
2A + 5B ≤ 250
60
50
0   B ≥ 0   120   A

(c) Vertex points are (0,0), (0,50), (100,10), and (120,0).

31. Let x = number of families in city 1, y = number of families in city 2.

(a)
$$30x + 40y \le 12,000 \quad \text{(cost)}$$
$$x \ge 100$$
$$\ge 120$$

(continued)

31. (*continued*)

(b)

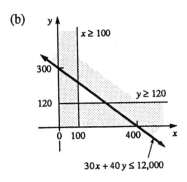

(c) Vertex points: $x = 100 \Rightarrow y = 9000/40 = 225$
$y = 120 \Rightarrow x = 7200/30 = 240$
Hence, (100, 120), (100, 225), (240, 120)

33. (a) Let x = acres to be zoned A–2, y = acres to be zoned A–3.

$x + y \leq 1{,}000{,}000$ (acres)

$\frac{x}{2} + \frac{y}{3} \geq 480{,}000$ (lots)

$x \geq 200{,}000$

$y \geq 60{,}000$

(b)

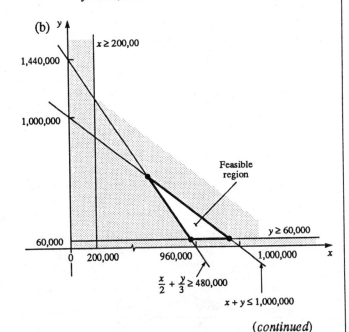

(*continued*)

33. (*continued*)

(c) Vertex points: Solve $x + y = 1{,}000{,}000$
and $\frac{x}{2} + \frac{y}{3} = 480{,}000$
simultaneously
to get $x = 880{,}000$, $y = 120{,}000$
Solve $y = 60{,}000$ and $x + y = 1{,}000{,}000$ to get $x = 940{,}000$
Solve $y = 60{,}000$ and $\frac{x}{2} + \frac{y}{3} = 480{,}000$ to get $x = 920{,}000$
Hence, (880,000, 120,000), (920,000, 60,000), (940,000, 60,000)

6–2 Linear Programming

1. Maximize $z = 4x + 7y$
(See Figure 6–30).
Evaluate z at the vertex points of the feasible region:

Vertex Point	z	
(0,0)	0	
(0,8)	56	
(6,7)	73	(maximum)
(10,3)	61	
(11,0)	44	

3. Minimize $z = 0.70x + 0.40y$
(See Figure 6–32).
Evaluate z at the vertex points of the feasible region:

Vertex Point	z	
(0,14)	5.6	
(2,8)	4.6	(minimum)
(8,2)	6.4	
(13,0)	9.1	

5. Maximize $z = 0.80x + 0.30y$
(See Figure 6–34).
Evaluate z at the vertex points of the feasible region:

Vertex Point	z	
(3,9)	5.1	
(7,3)	6.5	
(10,16)	12.8	(maximum)
(12,6)	11.4	

7. Maximize $P = 5x + 4y$

subject to $3x + 4y \leq 24$
$\qquad\quad x + y \leq 14$
$\qquad\quad x, y \geq 0$

Vertex points are (0,0), (0,6), (7,0), and the intersection of $3x + 4y = 24$ and $2x + y = 14$, which is (6.4,1.2)

Vertex Point	P	
(0,0)	0	
(0,6)	24	
(6.4,1.2)	36.8	(maximum)
(7,0)	35	

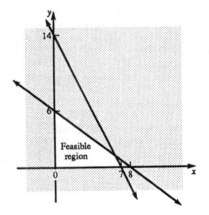

9. Minimize $C = 4x + 9y$
subject to $\quad x + y \geq 12$
$\qquad\qquad 3x + 2y \geq 30$
$\qquad\qquad x, y \geq 0$

Vertex points: (0,15), (12,0), and the intersection of $x + y = 12$ and $3x + 2y = 30$, which is (6,6).

Vertex Point	P	
(0,15)	135	
(6,6)	78	
(12,0)	48	(minimum)

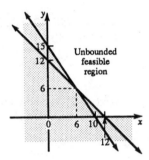

11. Maximize $z = 10x + 30y$

subject to $3x + 5y \leq 30$
$\qquad\quad 2x + 4y \leq 22$
$\qquad\quad x, y \geq 0$

Vertex points are (0,0), (0,5.5), (5,3), (10,0)

Vertex Point	z	
(0,0)	0	
(0,5.5)	165	(maximum)
(5,3)	140	
(10,0)	100	

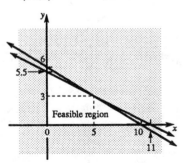

13. Minimize $z = 20x + 40y$
subject to $x + y \geq 10$
$\quad\quad\quad 3x + 4y \geq 36$
$\quad\quad\quad\quad x + y \leq 40$
$\quad\quad\quad\quad\quad x, y \geq 0$

Vertex points: $(0,10)$, $(0,40)$, $(12,0)$, $(40,0)$ and the intersection of $x + y = 10$ and $3x + 4y = 36$, which is $(4,6)$.

Vertex Point	z	
(0,10)	400	
(0,40)	1600	
(4,6)	320	
(12,0)	240	(minimum)
(40,0)	800	

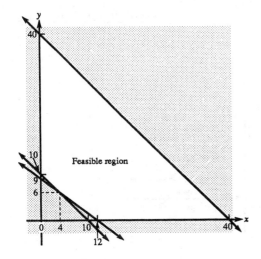

15. Maximize $z = 3x + 5y$

subject to $2x + y \leq 30$
$\quad\quad\quad x - 2y \leq 10$
$\quad\quad\quad\quad x \geq 4$
$\quad\quad\quad\quad y \geq 1$

Vertex points are $(4,1)$, $(4,22)$, $(12,1)$, $(14,2)$

Vertex Point	z	
(4,1)	17	
(4,22)	122	(maximum)
(12,1)	41	
(14,2)	52	

(continued)

15. *(continued)*

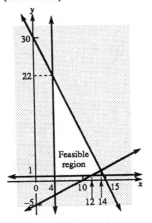

17. Maximize $z = 20x + 30y$

subject to $x + 4y \leq 20$
$\quad\quad\quad 2x - y \leq 30$
$\quad\quad\quad -x - y \leq 4$
$\quad\quad\quad\quad x, y \geq 0$

Intersections:

$x + 4y = 20$ and $-x + y = 4$: $\left[\dfrac{4}{5}, \dfrac{24}{5}\right]$

$x + 4y = 20$ and $2x - y = 30$: $\left[\dfrac{140}{9}, \dfrac{10}{9}\right]$

Vertex Point	z	
(0,0)	0	
(0,4)	120	
(4/5,24/5)	160	
(15,0)	300	
(140/9,10/9)	$344\dfrac{4}{9}$	(maximum)

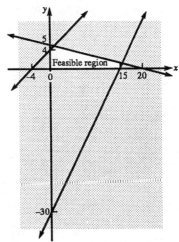

19. (a) x = number of casual jackets
     y = number of formal jackets
     Objective function: Maximize $P = 30x + 50y$

	Casual	Formal
Unit profits	30	50
Cutting dept hrs/jacket	2	3 960 hrs at most
Sewing dept hrs/jacket	1	1 400 hrs at most

At least 150 casual jackets must be produced.
At least 50 formal jackets must be produced.

Constraints:
Cutting dept: $2x + 3y \le 960$
Sewing dept: $x + y \le 400$
Demand: $x \ge 150$
               $y \ge 50$

Intersections: $2x + 3y = 960$ and $x + y = 400$:
                (240,160)
                $2x + 3y = 960$ and $x = 150$: (150,220)

Vertex Point	P
(150,50)	$ 7,000
(150,220)	$ 15,500
(240,160)	$ 15,200
(350,50)	$ 13,000

Cutting dept., slack = 0
Sewing dept., slack = 30
Constraint: $x \ge 150$, surplus = 0
Constraint: $y \ge 50$, surplus = 170

(b) Person–hours used in the cutting dept. = 960

(c) Person–hours used in the sewing dept. = 370

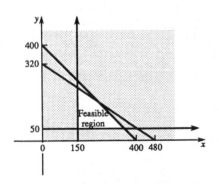

21. Maximize $R = 0.18x + 0.22y$

subject to $x + y \le 20,000$
                  $-x + 3y \le 0$
                  $x \ge 6000$
                  $y \le 4000$

Vertex Point	R
(6000, 0)	$1080
(6000, 2000)	$1520
(12,000, 4000)	$3040
(16,000, 4000)	$3760 (maximum)
(20,000, 0)	$3600

Constraint 1: Total amount invested = 20,000
               Slack = 0

Constraint 2: Average risk level for the optimal
               solution

$$= \frac{2(16,000) + 6(4000)}{20,000} = 2.8$$

               Slack = 0.2

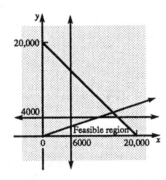

23. Minimize $C = 1200x + 2000y$ cost function

subject to $x + y \ge 1200$

$1000x + 1500y \ge 1,500,000$ } constraints

The vertex points are the same as in Exercise 22.

(continued)

23. *(continued)*

Vertex Point	C
(0, 1200)	$2,400,000
(600, 600)	$1,920,000
(1500, 0)	$1,800,000 (minimum)

(a) 1500 radio ads and no TV ads

(b) $1,800,000

(c) Constraint 1: surplus = 300 ads
Constraint 2: surplus = 0 families

(d) 1,500,000 families

(e) $x + y = 1500$ ads

25. Let x = lbs meat, y = lbs spinach

Minimize $C = 3x + 1.5y$ cost function
subject to $500x + 200y \geq 1200$ protein
$100x + 800y \geq 1000$ iron
$x, y \geq 0$

Vertex Point	C
(0,6)	$9.00
(2,1)	$7.50 (minimum)
(10,0)	$30.00

The minimum cost of $7.50 occurs at $x = 2$, $y = 1$.

Constraint 1: surplus = 0
Constraint 2: surplus = 0

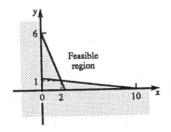

27. Let x = acres planted with crop A, y = acres planted with crop B.

Maximize $P = 400x + 500y$ profit function
subject to $x + y \leq 100$ acres available
$60x + 80y \leq 6600$ person–hours available

$x, y \geq 0$

Intersection of $x + y = 100$ and $60x + 80y = 6600$: (70,30)

Vertex Point	P
(0,0)	$0
(0,82.5)	$41,250
(70,30)	$43,000 (maximum)
(100,0)	$40,000

(a) 70 acres of A and 30 acres of B

(b) $43,000

(c) Constraint 1: slack = 0 acres
Constraint 2: slack = 0 person–hours

(d) 100

(e) 6600

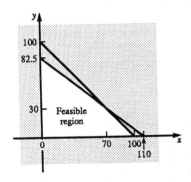

29. Let x = gallons of F10 fuel, y = gallons of F20 fuel.

Minimize $C = 1.5x + 1.1y$ cost function
subject to $x + y \geq 3800$ demand
    $0.02x + 0.05y \geq 120$ ash
    $0.06x + 0.01y \geq 136$ soot
        $x, y \geq 0$

Intersection of $x + y = 3800$ and
      $0.06x + 0.01y = 136$: (1960,1840)
Intersection of $x + y = 3800$ and
      $0.02x + 0.05y = 120$:

$$\left[2333\tfrac{1}{3},\ 1466\tfrac{2}{3}\right]$$

Vertex Point	P
(0, 13,600)	$14,960
(1960,1840)	$4,964 (minimum)
$\left[2333\tfrac{1}{3},\ 1466\tfrac{2}{3}\right]$	$5,113.33
(6000,0)	$9,000

(a) 1960 gallons of F10, 1840 gallons of F20

(b) $4964

(c) Constraint 1: surplus = 0
Constraint 2: surplus = 131.2 − 120 = 11.2 lbs ash
Constraint 3: surplus = 0

(d) 3800

(e) 131.2

(f) 136

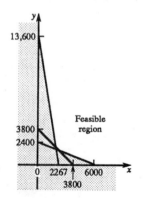

31. Objective function: $P = 4x + 2y$

Constraints: $2x + y \leq 420$ machining
          $2x + 2y \leq 500$ assembling
          $2x + 3y \leq 600$ finishing
             $x \geq 50$ demand
             $y \geq 50$ demand

Intersections: $x = 50$ and $2x + 3y = 600$:

$$\left[50,\ 166\tfrac{2}{3}\right]$$

$2x + 3y = 600$ and $2x + 2y = 500$: (150,100)
$2x + 2y = 500$ and $2x + y = 420$: (170,80)
$2x + y = 420$ and $y = 50$: (185,50)

Vertex Point	P
(50,50)	$300
$\left[50,\ 166\tfrac{2}{3}\right]$	$533.33
(150,100)	$800
(170,80)	$840
(185,50)	$840

(a) All points on the line connecting (170,80) and (185,50) are equally profitable.

(b) $840

(c) At (170,80),   Constraint 1: slack = 0
               Constraint 2: slack = 0
               Constraint 3: slack = 20
               Constraint 4: surplus = 120
               Constraint 5: surplus = 30

(d) 420

(e) 500

(f) 580

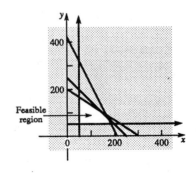

33. Let x = units sold to wholesale outlets, y = units sold to retail outlets.

Maximize $P = 90x + 120y$
subject to $2.4x + 3.6y \leq 17,280$ hours available
$x + y \leq 5600$ demand
$x, y \geq 0$

Vertex Point	P	
(0,0)	\$0	
(0,4800)	\$576,000	
(2400,3200)	\$600,000	(maximum)
(5600,0)	\$504,000	

(a) 2400 units to wholesale outlets, 3200 units to retail outlets

(b) \$600,000

(c) Constraint 1: slack = 0
Constraint 2: slack = 0

(d) 5600

(e) 17,280

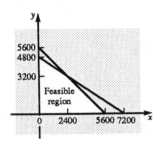

35. Maximize $P = 90x + 120y$

subject to $2.4x + 3.6y \leq 17,280$ hours available
$x + y \leq 5600$ demand
$x \geq 4y$, or $x - 4y \geq 0$

Intersection of $x + y = 5600$ and $x = 4y$ is (4480,1120)

Vertex Point	P	
(0,0)	\$0	
(4480,1120)	\$537,600	(maximum)
(5600,0)	\$504,000	

(a) 4480 units to wholesale outlets, 1120 units to retail outlets

(b) \$537,600

(c) Constraint 1: slack = 2496 hours
Constraint 2: slack = 0
Constraint 3: slack = 0

(d) 5600

(e) 14,784

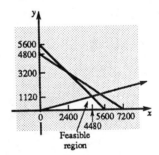

6–3 Fundamental Theorem of Linear Programming

1. The optimal solution is (2,9).

3. The optimal solution is (9,0).

5. The optimal solution occurs along the boundary from (0,8) to (7,5).

7. The optimal solution occurs at (2,4).

9. The optimal solution occurs at (9,4).

11. The optimal solution occurs at (9,4).

13. The optimal solution occurs at (1,2.5).

15. Maximize $z = 3x + 5y$

subject to $\quad x + y \leq 5$
$\qquad\qquad x + y \geq 7$
$\qquad\qquad\quad x \geq 0$
$\qquad\qquad\quad y \geq 0$

There is no feasible region, since $x + y$ cannot be both ≤ 5 and ≥ 7.

17. Minimize $z = 12x + 15y$

subject to $3x + 2y \leq 48$
$\qquad\qquad\quad x \geq 4$
$\qquad\quad 0 \leq y \leq 20$

Intersection of $3x + 2y = 48$ and $x = 4$: (4,18)

Vertex Point	z	
(4,0)	48	(minimum)
(4,18)	318	
(16,0)	192	

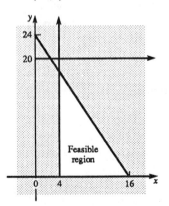

19. Maximize $z = 10x + 20y$

subject to $3x + 4y \geq 60$
$\qquad\qquad\quad x \geq 4$
$\qquad\qquad\quad y \geq 3$

Vertex Point	z
(4,12)	280
(16,3)	220

Unbounded solution: An optimal solution does not exist because there is always a better solution.

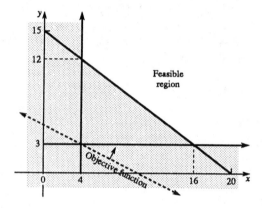

Review Exercises

The solutions of Exercises 1–8 are graphs, shown in the accompanying figures. Shaded areas include points that do **not** satisfy the given inequalities.

1.

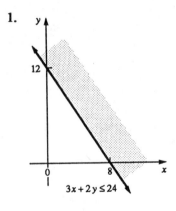

3.

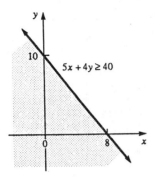

5.

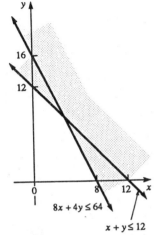

7.

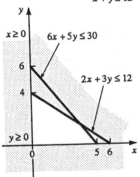

9. Maximize $z = 2x + 5y$

subject to $2x + y \leq 40$
$3x + y \leq 48$
$x, y \geq 0$

Vertex Point	z	
(0,0)	0	
(0,40)	200	(maximum)
(8,24)	136	
(16,0)	32	

(*continued*)

9. (*continued*)

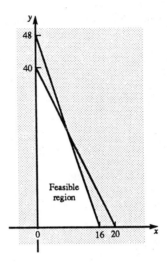

11. Minimize $z = 5x + 2y$

subject to $x + y \geq 20$
$x + 3y \geq 30$
$x, y \geq 0$

Vertex Point	z	
(0,20)	40	(minimum)
(15,5)	85	
(30,0)	150	

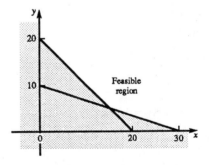

13. (a) Let x = number of type A branches,
y = number of type B branches.
Maximize $P = 46{,}000x + 18{,}000y$ profit
function
subject to
$650{,}000x + 335{,}000y \le 5{,}200{,}000$ costs
$10x + 5y \le 100$ personnel
$x, y \ge 0$

(b)
Vertex Point	P
(0,0)	$0
(0,15.52)	$279,403
(8,0)	$368,000 (maximum)

(c) At the optimal solution, type A branches = 8,
type B branches = 0.
Constraint 1: slack = 0
Constraint 2: slack = 100 − 80 = 20 people

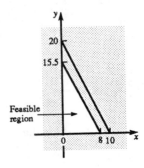

15. Let x = units of product A, y = units of product B
Maximize $P = 10x + 4y$

subject to $3x + 4y \le 300$ Raw material 1
$\qquad\quad 7x + 2y \le 400$ Raw material 2
$\qquad\qquad\; x, y \ge 0$

Intersections: $3x + 4y = 300$ and $7x + 2y = 400$:
$$\left[\frac{500}{11}, \frac{450}{11}\right]$$
$7x + 2y = 400$ and $y = 0$: $\left[\frac{400}{7}, 0\right]$

15. (continued)

Vertex Point	P
(0,0)	$0
(0,75)	$300
(500/11,450/11)	$618.18 (maximum)
(400/7,0)	$571.43

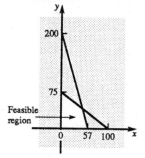

17.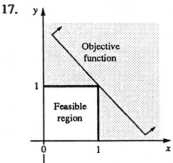

19.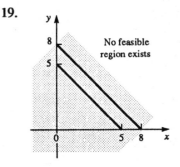

(continued)

CHAPTER

7

LINEAR PROGRAMMING: THE SIMPLEX METHOD

7–1 A Geometric Introduction and the Simplex Tableau

1. Maximize $Z = 120x_1 + 60x_2$
subject to $2x_1 + 3x_2 \leq 90$
$\qquad\qquad 5x_1 + x_2 \leq 160$

(a) $\quad 2x_1 + 3x_2 + s_1 \qquad\qquad = 90$
$\qquad\; 5x_1 + x_2 \qquad + s_2 \qquad = 160$
$\;\; -120x_1 - 60x_2 \qquad\qquad + Z = 0$
$\qquad\qquad\quad x_1, x_2 \geq 0$

(b) Simplex tableau:

x_1	x_2	s_1	s_2	Z	
2	3	1	0	0	90
5	1	0	1	0	160
−120	−60	0	0	1	0

(c) Basic feasible solution: $x_1 = 0$, $x_2 = 0$,
$S_1 = 90$, $S_2 = 160$
Current value of objective function: $z = 0$

3. Maximize $Z = 400x_1 + 500x_2$
subject to $\quad x_1 + x_2 \leq 100$
$\qquad\qquad 60x_1 + 80x_2 \leq 6600$
where x_1 and $x_2 \geq 0$

(a) $x_1 + x_2 + s_1 \qquad\qquad = 100$
$\quad 60x_1 + 80x_2 \qquad + s_2 \qquad = 6600$
$-400x_1 - 500x_2 \qquad\qquad + Z = 0$
$\qquad\qquad\quad x_1, x_2 \geq 0$

(b) Simplex tableau:

x_1	x_2	s_1	s_2	Z	
1	1	1	0	0	100
60	80	0	1	0	6600
−400	−500	0	0	1	0

(c) $x_1 = 0$, $x_2 = 0$, $\quad s_1 = 100$, $s_2 = 6600$ $\quad Z = 0$

$\qquad\quad$ Nonbasic variables $\qquad$ Basis

5. Maximize $Z = 20x_1 + 42x_2 + 56x_3$
    subject to $2x_1 + 3x_2 + x_3 \leq 6$
    $\qquad\qquad 4x_1 + 2x_2 + 3x_3 \leq 12$
    $\qquad\qquad 4x_1 + 2x_2 + x_3 \leq 8$
    where x_1, x_2 and $x_3 \geq 0$

    (a) $2x_1 + 3x_2 + x_3 + s_1 \qquad\qquad = 6$
    $\qquad 4x_1 + 2x_2 + 3x_3 \qquad + s_2 \qquad = 12$
    $\qquad 4x_1 + 2x_2 + x_3 \qquad\qquad + s_3 \quad = 8$
    $\qquad -20x_1 - 42x_2 - 56x_3 \qquad\qquad + Z = 0$
    $\qquad\qquad x_1, x_2, x_3 \geq 0$

    (b) Simplex tableau:

x_1	x_2	x_3	s_1	s_2	s_3	Z	
2	3	1	1	0	0	0	6
4	2	3	0	1	0	0	12
4	2	1	0	0	1	0	8
−20	−42	−56	0	0	0	1	0

    (c) $x_1 = 0, x_2 = 0, x_3 = 0 \qquad s_1 = 6, s_2 = 12, s_3 = 8$

    $\underbrace{\qquad\qquad\qquad\qquad}$ $\underbrace{\qquad\qquad\qquad\qquad}$
    Nonbasic variables $\qquad\qquad$ Basis
    $Z = 0$

7. Basic feasible solution: $x_1 = 3, x_2 = 0, x_3 = 6$,
    $\qquad\qquad\qquad s_1 = 0, s_2 = 5, s_3 = 0$
    Current value of objective function: $Z = 5670$

9. Basic feasible solution: $x_1 = 75, x_2 = 0, x_3 = 60$,
    $\qquad\qquad\qquad x_4 = 0, s_1 = 0, s_2 = 0, s_3 = 0$
    Current value of objective function: $Z = 50$

7–2 Simplex Method: Maximization

1. Substitution of the quotient values for x_2 into the
    expressions for s_1, s_2, and s_3 gives:

Quotient Value	s_1	s_2	s_3
425	0	150	−225
500	−150	0	−900
400	50	200	0

    Only the quotient value 400 yields nonnegative
    values for s_1, s_2, and s_3.

3. (a)

	Equations	Quotients
$2x_1 + s_1 = 240$	$s_1 = 240 - 2x_1$	120
$-4x_1 + s_2 = 400$	$s_2 = 400 + 4x_1$	
$6x_1 + s_3 = 600$	$s_3 = 600 - 6x_1$	100

    Hence, x_1 can increase 120 units before s_1
    becomes negative and x_1 can increase 100 units
    before s_3 becomes negative. x_2 is not limited
    and can increase indefinitely.
    Therefore, x_1 cannot be increased by more than
    the minimum quotient. x_1 can increase by
    100 units.

    (b) Quotient Value

Quotient Value	s_1	s_2	s_3
120	0	880	−120
100	40	800	0

    An increase of 100 units in x_1 is the maximum
    that retains nonnegative values of s_1, s_2, and s_3.

5. $\qquad 3x_1 + 6x_2 + s_1 \qquad\qquad = 60$
    $\qquad 4x_1 + 4x_2 \qquad + s_2 \qquad = 60$
    $\qquad -10x_1 - 30x_2 \qquad\qquad + Z = 0$
    $\qquad\qquad x_1, x_2 \geq 0$

    Simplex tableau:

x_1	x_2	s_1	s_2	Z		Quotients (x_2)
3	⑥	1	0	0	60	10
4	4	0	1	0	60	15
−10	−30	0	0	1	0	

    $(1/6)R_1 \to R_1; \; (-4)R_1 + R_2 \to R_2; \; (30)R_1 + R_3 \to R_3$

1/2	1	1/6	0	0	10
2	0	−2/3	1	0	20
5	0	5	0	1	300

    Solution: $x_1 = 0, x_2 = 10, s_1 = 0, s_2 = 20, Z = 300$

7.
$$2x_1 + 3x_2 + 5x_3 + s_1 \qquad\qquad = 30$$
$$6x_1 + 2x_2 + 3x_3 \qquad + s_2 \qquad = 14$$
$$x_1 + 3x_2 + 4x_3 \qquad\qquad + s_3 \qquad = 24$$
$$-5x_1 - 6x_2 - x_3 \qquad\qquad\qquad + Z = 0$$
$$x_1, x_2, x_3 \geq 0$$

Simplex tableau:

x_1	x_2	x_3	s_1	s_2	s_3	Z		Quotients (x_2)
2	③	5	1	0	0	0	30	10
6	②	3	0	1	0	0	14	7
1	3	4	0	0	1	0	24	8
-5	-6	-1	0	0	0	1	0	

$(1/2)R_2 \to R_2$; $(-3)R_2 + R_1 \to R_1$; $(-3)R_2 + R_3 \to R_3$;
$(6)R_3 + R_4 \to R_4$

-7	0	1/2	1	-3/2	0	0	9
3	1	3/2	0	1/2	0	0	7
-8	0	-1/2	0	-3/2	1	0	3
13	0	8	0	3	0	1	42

Solution: $x_1 = 0$, $x_2 = 7$, $x_3 = 0$, $s_1 = 9$, $s_2 = 0$, $s_3 = 3$, $Z = 42$

9.
$$x_1 + x_2 + x_3 + x_4 + s_1 \qquad\qquad = 40$$
$$2x_1 + x_2 + 4x_3 + x_4 \qquad + s_2 \qquad = 90$$
$$-80x_1 - 10x_2 - 16x_3 - 12x_4 \qquad\qquad + Z = 0$$
$$x_1, x_2, x_3, x_4 \geq 0$$

Simplex tableau:

x_1	x_2	x_3	x_4	s_1	s_2	Z		Quotients (x_1)
①	1	1	1	1	0	0	40	40
2	1	4	1	0	1	0	90	45
-80	-10	-16	-12	0	0	1	0	

$(-2)R_1 + R_2 \to R_2$; $(80)R_1 + R_3 \to R_3$

1	1	1	1	1	0	0	40
0	-1	2	-1	-2	1	0	10
0	70	64	68	80	0	1	3200

Solution: $x_1 = 40$, $x_2 = 0$, $x_3 = 0$, $x_4 = 0$, $s_1 = 0$, $s_2 = 10$, $Z = 3200$

11.
$$4x_1 + 2x_2 + 6x_3 + s_1 \qquad\qquad = 60$$
$$2x_1 + 2x_2 + 4x_3 \qquad + s_2 \qquad = 120$$
$$x_1 + 2x_2 + 3x_3 \qquad\qquad + s_3 \qquad = 90$$
$$-5x_1 - 8x_2 + 9x_3 \qquad\qquad\qquad + Z = 0$$
$$x_1, x_2, x_3 \geq 0$$

Simplex tableau:

x_1	x_2	x_3	s_1	s_2	s_3	Z		Quotients (x_2)
4	②	6	1	0	0	0	60	30
2	2	4	0	1	0	0	120	60
1	2	3	0	0	1	0	90	45
-5	-8	9	0	0	0	1	0	

$(1/2)R_1 \to R_1$; $(-2)R_1 + R_2 \to R_2$; $(-2)R_1 + R_3 \to R_3$;
$(8)R_1 + R_4 \to R_4$

2	1	3	1/2	0	0	0	30
-2	0	-2	-1	1	0	0	60
-3	0	-3	-1	0	1	0	30
11	0	33	4	0	0	1	240

Solution: $x_1 = 0$, $x_2 = 30$, $x_3 = 0$, $s_1 = 0$, $s_2 = 60$, $s_3 = 30$, $Z = 240$

13.
$$x_1 + 2x_2 + x_3 + s_1 \qquad\qquad = 50$$
$$2x_1 + 4x_2 + 2x_3 \qquad + s_2 \qquad = 120$$
$$x_1 \qquad\qquad\qquad + s_3 \qquad = 20$$
$$-8x_1 - 5x_2 - 2x_3 \qquad\qquad\qquad + Z = 0$$
$$x_1, x_2, x_3 \geq 0$$

Simplex tableau:

x_1	x_2	x_3	s_1	s_2	s_3	Z		Quotients (x_1)
1	2	1	1	0	0	0	50	50
2	4	2	0	1	0	0	120	60
①	0	0	0	0	1	0	20	20
-8	-5	-2	0	0	0	1	0	

$(-1)R_3 + R_1 \to R_1$; $(-2)R_3 + R_2 \to R_2$; $(8)R_3 + R_4 \to R_4$;

								Quotients (x_2)
0	②	1	1	0	-1	0	30	15
0	4	2	0	1	-2	0	80	20
1	0	0	0	0	1	0	20	
0	-5	-2	0	0	8	1	160	

(continued)

13. *(continued)*

$(1/2)R_1 \to R_1;\ (-4)R_1 + R_2 \to R_2;\ (5)R_1 + R_4 \to R_4$

0	1	1/2	1/2	0	−1/2	0	15
0	0	0	−2	1	0	0	20
1	0	0	0		1	0	20
0	0	1/2	5/2	0	11/2	1	235

Solution: $x_1 = 20$, $x_2 = 15$, $x_3 = 0$, $s_1 = 0$, $s_2 = 20$, $s_3 = 0$, $Z = 235$

15.
$$
\begin{aligned}
x_1 + x_2 + x_3 + s_1 &= 150 \\
x_1 \qquad\qquad + s_2 &= 30 \\
x_2 \qquad\quad + s_3 &= 40 \\
x_3 \quad + s_4 &= 60 \\
-5x_1 - 3x_2 - x_3 \qquad\qquad + Z &= 0 \\
x_1, x_2, x_3 &\geq 0
\end{aligned}
$$

Simplex tableau:

x_1	x_2	x_3	s_1	s_2	s_3	s_4	Z		Quotients (x_1)
1	1	1	1	0	0	0	0	150	150
①	0	0	0	1	0	0	0	30	30
0	1	0	0	0	1	0	0	40	
0	0	1	0	0	0	1	0	60	
−5	−3	−1	0	0	0	0	1	0	

$(-1)R_2 + R_1 \to R_1;\ (5)R_2 + R_5 \to R_5$

									Quotients (x_2)
0	1	1	1	−1	0	0	0	120	120
1	0	0	0	1	0	0	0	30	
0	1	0	0	0	1	0	0	40	40
0	0	1	0	0	0	1	0	60	
0	−3	−1	0	5	0	0	1	150	

$(-1)R_3 + R_1 \to R_1;\ (3)R_3 + R_5 \to R_5$

									Quotients (x_3)
0	0	1	1	−1	−1	0	0	80	80
1	0	0	0	1	0	0	0	30	
0	①	0	0	0	1	0	0	40	
0	0	1	0	0	0	1	0	60	60
0	0	−1	0	5	3	1	1	270	

(continued)

15. *(continued)*

$(-1)R_4 + R_1 \to R_1;\ R_4 + R_5 \to R_5$

0	0	0	1	−1	−1	−1	0	20
1	0	0	0	1	0	0	0	30
0	1	0	0	0	1	0	0	40
0	0	①	0	0	0	1	0	60
0	0	0	0	5	3	2	1	330

Solution: $x_1 = 30$, $x_2 = 40$, $x_3 = 60$, $s_1 = 20$, $s_2 = 0$, $s_3 = 0$, $s_4 = 0$, $Z = 330$

17.
$$
\begin{aligned}
x_1 + x_2 + x_3 + s_1 &= 100 \\
5x_1 + 8x_2 \qquad\quad + s_2 &= 180 \\
x_1 \quad + 10x_3 \qquad\quad + s_3 &= 50 \\
-3x_1 - 4x_2 - 5x_3 \qquad\qquad + Z &= 0 \\
x_1, x_2, x_3 &\geq 0
\end{aligned}
$$

(a)　Simplex tableau:

x_1	x_2	x_3	s_1	s_2	s_3	Z		Quotients (x_3)
1	1	1	1	0	0	0	100	100
5	8	0	0	1	0	0	180	
1	0	⑩	0	0	1	0	50	5
−3	−4	−5	0	0	0	1	0	

$(1/10)R_3 \to R_3;\ (-1)R_3 + R_1 \to R_1;\ (5)R_3 + R_4 \to R_4$

								Quotients (x_2)
9/10	1	0	1	0	−1/10	0	95	95
5	⑧	0	0	1	0	0	180	22.5
1/10	0	1	0	0	1/10	0	5	
−5/2	−4	0	0	0	1/2	1	25	

$(1/8)R_2 \to R_2;\ (-1)R_2 + R_1 \to R_1;\ (4)R_2 + R_4 \to R_4$

11/40	0	0	1	−1/8	−1/10	0	72.5
5/8	1	0	0	1/8	0	0	22.5
1/10	0	1	0	0	1/10	0	5
0	0	0	0	1/2	1/2	1	115

Solution: $x_1 = 0$, $x_2 = 22.5$, $x_3 = 5$, $s_1 = 72.5$, $s_2 = 0$, $s_3 = 0$, $Z = 115$

(continued)

17. *(continued)*

(b) The nonbasic variable with a zero indicator is x_1.

(c) Using row 2 as the pivot now gives

$$
\begin{array}{cccccccc}
0 & -11/25 & 0 & 1 & -9/50 & -1/10 & 0 & 62.6 \\
1 & 8/5 & 0 & 0 & 1/5 & 0 & 0 & 36 \\
0 & -4/25 & 1 & 0 & -1/50 & 1/10 & 0 & 7/5 \\
0 & 0 & 0 & 0 & 1/2 & 1/2 & 1 & 115
\end{array}
$$

Solution: $\underbrace{x_1 = 36,\ x_3 = 1.4,\ s_1 = 62.6}_{\text{Basis}}$

$\underbrace{x_2 = 0,\ s_2 = 0,\ s_3 = 0}_{\text{Nonbasic variables}}$

$Z = 115$, as in the previous solution

(d) x_1 can be varied from 0 to 36 without changing the optimal value of the objective function. Note that for the solution to part (c) x_1 is nonzero and $x_2 = 0$ whereas for the solution to part (a) the opposite is true. Also note the change in the value of s_1 when comparing both solutions.

19. See Figure below for the graphical solution.

$$
\begin{array}{rcl}
2x_1 + 4x_2 + s_1 &=& 100 \\
x_1 + 3x_2 \quad + s_2 &=& 90 \\
x_1 \qquad\quad + s_3 &=& 80 \\
x_2 \qquad\qquad + s_4 &=& 40 \\
-10x_1 - 15x_2 \qquad\qquad + Z &=& 0 \\
x_1, x_2 \geq 0
\end{array}
$$

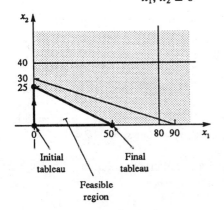

Initial tableau
Final tableau
Feasible region

(continued)

19. *(continued)*

Simplex tableau:

x_1	x_2	s_1	s_2	s_3	s_4	Z		Quotients (x_2)
2	④	1	0	0	0	0	100	25
1	3	0	1	0	0	0	90	30
1	0	0	0	1	0	0	80	
0	1	0	0	0	1	0	40	40
−10	−15	0	0	0	0	1	0	

Vertex Point (0,0)

$(1/4)R_1 \to R_1$; $(-3)R_1 + R_2 \to R_2$; $(-1)R_1 + R_4 \to R_4$; $(15)R_1 + R_5 \to R_5$

								Quotients (x_1)
① /2	1	1/4	0	0	0	0	25	50
−1/2	0	−3/4	1	0	0	0	15	
1	0	0	0	1	0	0	80	80
−1/2	0	−1/4	0	0	1	0	15	
−5/2	0	15/4	0	0	0	1	375	

Vertex Point (0,25)

$(2)R_1 \to R_1$; $(1/2)R_1 + R_2 \to R_2$; $(-1)R_1 + R_3 \to R_3$; $(1/2)R_1 + R_4 \to R_4$; $(5/2)R_1 + R_5 \to R_5$

1	2	1/2	0	0	0	0	50
0	1	−1/2	1	0	0	0	40
0	−2	−1/2	0	1	0	0	30
0	1	0	0	0	1	0	40
0	5	5	0	0	0	1	500

Vertex Point (50,0)

Solution: $x_1 = 50$, $x_2 = 0$, $s_1 = 0$, $s_2 = 40$, $s_3 = 30$, $s_4 = 40$, $Z = 500$

21. Let x_1 = number of units of product A, x_2 = number of units of product B, x_3 = number of units of product C.

Maximize $P = 50x_1 + 40x_2 + 60x_3$
subject to $2x_1 + 4x_2 + 2x_3 \leq 1500$
$\qquad\qquad 3x_1 + 2x_2 + x_3 \leq 1200$
$\qquad\qquad x_1, x_2 \geq 0$

(continued)

21. *(continued)*

Simplex tableau:

x_1	x_2	x_3	s_1	s_2	P		Quotients (x_3)
2	4	②	1	0	0	1500	750
3	2	1	0	1	0	1200	1200
−50	−40	−60	0	0	1	0	

$(1/2)R_1 \to R_1$; $(-1)R_1 + R_2 \to R_2$; $(60)R_1 + R_3 \to R_3$

1	2	1	1/2	0	0	750
2	0	0	−1/2	1	0	450
10	80	0	30	0	1	45,000

Solution: $x_1 = 0$, $x_2 = 0$, $x_3 = 750$, $s_1 = 0$, $s_2 = 450$, $P = 45,000$

23. Let x_1 = dollars invested in stocks, x_2 = dollars invested in bonds, x_3 = dollars invested in money market funds.

Maximize $R = 0.10x_1 + 0.08x_2 + 0.06x_3$
subject to
$$x_1 + x_2 \quad\ \le 1,000,000$$
$$x_2 + x_3 \le 1,200,000$$
$$x_1 + x_2 + x_3 \le 2,000,000$$
$$x_1,\ x_2,\ x_3 \ge 0$$

Simplex tableau:

x_1	x_2	x_3	s_1	s_2	s_3	R		Quotients (x_1)
①	1	0	1	0	0	0	1,000 000	1,000,000
0	1	1	0	1	0	0	1,200 000	
1	1	1	0	0	1	0	2,000 000	2,000,000
−0.10	−0.08	−0.06	0	0	0	1	0	

$(-1)R_1 + R_3 \to R_3$; $(0.10)R_1 + R_4 \to R_4$

x_1	x_2	x_3	s_1	s_2	s_3	R		Quotients (x_3)
1	1	0	1	0	0	0	1,000,000	
0	1	1	0	1	0	0	1,200,000	1,200,000
0	0	①	−1	0	1	0	1,000,000	1,000,000
0	0.02	−0.06	0.1	0	0	1	100,000	

$(-1)R_3 + R_2 \to R_2$; $(0.06)R_3 + R_4 \to R_4$

1	1	0	1	0	0	0	1,000,000
0	1	0	1	1	−1	0	200,000
0	0	1	−1	0	1	0	1,000,000
0	0.02	0	0.04	0	0.06	1	160,000

Solution: $x_1 = 1,000,000$, $x_2 = 0$, $x_3 = 1,000,000$, $s_1 = 0$, $s_2 = 200,000$, $s_3 = 0$, $R = 160,000$

25. Maximize total profit:
$$P = (15 - 10)x_1 + (18 - 10)x_2 + (15 - 12)x_3$$
$$+ (18 - 12)x_4$$
$$= 5x_1 + 8x_2 + 3x_3 + 6x_4$$

subject to
$$\left.\begin{array}{l} x_1 + x_2 \le 1000 \\ x_3 + x_4 \le 1800 \end{array}\right\} \text{ shipping constraints}$$

$$\left.\begin{array}{l} x_1 + x_3 \le 1200 \\ x_2 + x_4 \le 2000 \end{array}\right\} \text{ demand constraints}$$

$$x_1,\ x_2,\ x_3,\ x_4 \ge 0$$

Simplex tableau:

x_1	x_2	x_3	x_4	s_1	s_2	s_3	s_4	P	
1	①	0	0	1	0	0	0	0	1000
0	0	1	1	0	1	0	0	0	1800
1	0	1	0	0	0	1	0	0	1200
0	1	0	1	0	0	0	1	0	2000
−5	−8	−3	−6	0	0	0	0	1	0

$(-1)R_1 + R_4 \to R_4$; $(8)R_1 + R_5 \to R_5$

1	1	0	0	1	0	0	0	0	1000
0	0	1	1	0	1	0	0	0	1800
1	0	1	0	0	0	1	0	0	1200
−1	0	0	①	−1	0	0	1	0	1000
3	0	−3	−6	8	0	0	0	1	8000

$(-1)R_4 + R_2 \to R_2$; $(6)R_4 + R_5 \to R_5$

1	1	0	0	1	0	0	0	0	1000
①	0	1	0	1	1	0	−1	0	800
1	0	1	0	0	0	1	0	0	1200
−1	0	0	1	−1	0	0	1	0	1000
−3	0	−3	0	2	0	0	6	1	14,000

$(-1)R_2 + R_1 \to R_1$; $(-1)R_2 + R_3 \to R_3$; $R_2 + R_4 \to R_4$; $(3)R_2 + R_5 \to R_5$

0	1	−1	0	0	−1	0	1	0	200
1	0	1	0	1	1	0	−1	0	800
0	0	0	0	−1	−1	1	1	0	400
0	0	1	1	0	1	0	0	0	1800
0	0	0	0	5	3	0	3	1	16,400

Solution: $x_1 = 800$, $x_2 = 200$, $x_3 = 0$, $x_4 = 1800$, $s_1 = 0$, $s_2 = 0$, $s_3 = 400$, $s_4 = 0$, $P = 16,400$

Note: There are alternate solutions, since x_3 can be brought into the basis.

27. Additional constraints: $x_1 \leq 500$, $x_3 \leq 300$

Simplex tableau:

x_1	x_2	x_3	s_1	s_2	s_3	s_4	P	
3	5	4	1	0	0	0	0	6000
1	1	1	0	1	0	0	0	1000
①	0	0	0	0	1	0	0	500
0	0	1	0	0	0	1	0	300
−600	−300	−500	0	0	0	0	1	0

$(-3)R_3 + R_1 \to R_1$; $(-1)R_3 + R_2 \to R_2$;
$(600)R_3 + R_5 \to R_5$

0	5	4	1	0	−3	0	0	4500
0	1	1	0	1	−1	0	0	500
1	0	0	0	0	1	0	0	500
0	0	①	0	0	0	1	0	300
0	−300	−500	0	0	600	0	1	300,000

$(-4)R_4 + R_1 \to R_1$; $(-1)R_4 + R_2 \to R_2$;
$(500)R_4 + R_5 \to R_5$

0	5	0	1	0	−3	−4	0	3300
0	①	0	0	1	−1	−1	0	200
1	0	0	0	0	1	0	0	500
0	0	1	0	0	0	1	0	300
0	−300	0	0	0	600	500	1	450,000

$(-5)R_2 + R_1 \to R_1$; $(300)R_2 + R_5 \to R_5$

0	0	0	1	−5	2	1	0	2300
0	1	0	0	1	−1	−1	0	200
1	0	0	0	0	1	0	0	500
0	0	1	0	0	0	1	0	300
0	0	0	0	300	300	200	1	510,000

Solution: $x_1 = 500$, $x_2 = 200$, $x_3 = 300$, $s_1 = 2300$, $s_2 = 0$, $s_3 = 0$, $s_4 = 0$, $P = 510,000$

29. Let x_1 = units marketed through Retail Outlet 1,
x_2 = units marketed through Retail Outlet 2,
x_3 = units marketed through the Wholesale Outlet.

Maximize $P = 400x_1 + 350x_2 + 300x_3$
subject to $1.5x_1 + x_2 + 0.5x_3 \leq 8000$
$\qquad\qquad x_1 + x_2 + x_3 \leq 10,000$
$\qquad\qquad x_1, x_2, x_3 \geq 0$

Simplex tableau:

x_1	x_2	x_3	s_1	s_2	P	
⑴.5	1	0.5	1	0	0	8000
1	1	1	0	1	0	10,000
−400	−350	−300	0	0	1	0

$(2/3)R_1 \to R_1$, $(-1)R_1 + R_2 \to R_2$; $(400)R_1 + R_3 \to R_3$

1	2/3	1/3	2/3	0	0	16,000/3
0	1/3	⑵/3	−2/3	1	0	14,000/3
0	−250/3	−500/3	800/3	0	1	6,400,000/3

$(3/2)R_2 \to R_2$; $(-1/3)R_2 + R_1 \to R_1$; $(500/3)R_2 + R_3 \to R_3$

1	1/2	0	1	−1/2	0	3000
0	1/2	1	−1	3/2	0	7000
0	0	0	100	250	1	3,300,000

Solution: $x_1 = 3000$, $x_2 = 0$, $x_3 = 7000$, $s_1 = 0$, $s_2 = 0$, $P = 3,300,000$

31. Let x_1 = number of professional pollsters at location 1,
$\qquad x_2$ = number of semiprofessional pollsters at location 1
$\qquad x_3$ = number of professional pollsters at location 2
$\qquad x_4$ = number of semiprofessional pollsters at location 2

Maximize $V = 100x_1 + 60x_2 + 100x_3 + 60x_4$
subject to
$500x_1 + 200x_2 \qquad\qquad\qquad \leq 3000$ Location 1
$\qquad\qquad 500x_3 + 200x_4 \leq 5000$ Location 2
$\qquad\qquad x_1, x_2, x_3, x^4 \geq 0$

(continued)

31. *(continued)*

Simplex tableau:

x_1	x_2	x_3	x_4	s_1	s_2	V	
(500)	200	0	1	1	0	0	3000
0	0	500	200	0	1	0	5000
−100	−60	−100	−60	0	0	1	0

$(1/500)R_1 \to R_1$, $(100)R_1 + R_3 \to R_3$

1	2/5	0	0	1/500	0	0	6
0	0	(500)	200	0	1	0	5000
0	−20	−100	−60	1/5	0	1	600

$(1/500)R_2 \to R_2$; $(100)R_2 + R_3 \to R_3$

1	(2/5)	0	0	1/500	0		0	6
0	0	1	2/5	0		1/500	0	10
0	−20	0	−20	1/5	100/500	1		1600

$(5/2)R_1 \to R_1$; $(20)R_1 + R_3 \to R_3$

5/2	1	0	0	1/200	0	0	15
0	0	1	(2/5)	0	1/500	0	10
50	0	0	−20	3/10	1/5	1	1900

$(5/2)R_2 \to R_2$; $(20)R_2 + R_3 \to R_3$

5/2	1	0	0	1/200	0	0	15
0	0	5/2	1	0	1/200	0	25
50	0	50	0	3/10	3/10	1	2400

Solution: $x_1 = 0$, $x_2 = 15$, $x_3 = 0$, $x_4 = 25$, $s_1 = 0$, $s_2 = 0$, $V = 2400$
Note that no professional pollsters are used in the optimal solution.

7–3 Minimization; The Dual; Shadow Prices

1. (a) Primal:
Minimize $Z = 40x_1 + 60x_2$
subject to $2x_1 + x_2 \geq 6$
$4x_1 + 6x_2 \geq 8$
$x_1, x_2 \geq 0$

(continued)

1. *(continued)*

Dual:
Maximize $Z = 6y_1 + 8y_2$
subject to $2y_1 + 4y_2 \leq 40$
$y_1 + 6y_2 \leq 60$
$y_1, y_2 \geq 0$

(b) Simplex tableau for dual:

y_1	y_2	s_1	s_2	Z		
2	(4)	1	0	0	40	40/4 = 10
1	6	0	1	0	60	60/6 = 10
−6	−8	0	0	1	0	

$(1/4)R_1 \to R_1$; $(−6)R_1 + R_2 \to R_2$; $(8)R + R_3 \to R_3$

(1/2)	1	1/4	0	0	10
−2	0	−3/2	1	0	0
−2	0	2	0	1	80

$(2)R_1 \to R_1$; $(2)R_1 + R_2 \to R_2$; $(2)R_1 + R_3 \to R_3$

1	2	1/2	0	0	20
0	4	−1/2	1	0	40
0	4	3	0	1	120

Solution: $y_1 = 20$, $y_2 = 0$, $s_1 = 0$, $s_2 = 40$, $z = 120$

(c) From the indicators corresponding to s_1 and s_2, $x_1 = 3$, $x_2 = 0$. Also, $Z = 120$.

(d) The surplus in $2x_1 + x_2 \geq 6$ is 0. The surplus in $4x_1 + 6x_2 \geq 8$ is 4.

(e) Shadow prices: $y_1 = 20$, $y_2 = 0$.
$y_2 = 20$ is the change that will occur in the optimal value of the objective function if the right–hand side of the first constraint in the primal is increased by 1 unit, that is, to $2x_1 + x_2 \geq 7$. Similarly, $y_2 = 0$ is the change that will occur in the optimal value of the objective function if the right–hand side of the first constraint in the primal is increased by 1 unit, that is, to $4x_1 + 6x_2 \geq 9$. Note: Each shadow price is valid only for a specific interval within which its corresponding right–hand–side value can vary.

3. (a) Primal:
       Minimize $Z = 300x_1 + 150x_2$
       subject to $4x_1 + 2x_2 \geq 120$
       $x_1 + 8x_2 \geq 100$
       $x_1, x_2 \geq 0$

   Dual:
   Maximize $Z = 120y_1 + 100y_2$
   subject to $4y_1 + y_2 \leq 300$
   $2y_1 + 8y_2 \leq 150$
   $y_1, y_2 \geq 0$

   (b) Simplex tableau for dual:

y_1	y_2	s_1	s_2	Z		
4	1	1	0	0	300	300/4 = 75
②	8	0	1	0	150	150/2 = 75
–120	–100	0	0	1	0	

   $(1/2)R_2 \to R_2$; $(-4)R_2 + R_1 \to R_1$;
   $(120)R_2 + R_3 \to R_3$

0	–15	1	–2	0	0
1	4	0	1/2	0	75
0	380	0	60	1	9000

   Solution: $y_1 = 75$, $y = 0$, $s_1 = 0$, $s_2 = 0$,
   $x = 9000$

   (c) From the indicators corresponding to s_1 and s_2, $x_1 = 0$, $x_2 = 60$. Also, $Z = 9000$.

   (d) The surplus amounts in the primal constraints are 0 and 380.

   (e) Shadow prices: $y_1 = 75$, $y_2 = 0$. An increase of 1 unit in the right–hand side of the first primal constraint will increase the optimal value of the objective function by 75 units.

   Note: The solution for (x_1, x_2) is not unique.

   Any point between $(0,60)$ and $\left[\dfrac{380}{15}, \dfrac{140}{15}\right]$ will satisfy the constraints and yield the same minimum.

5. (a) Primal:
       Minimize $Z = 200x_1 + 240x_2$
       subject to $x_1 + x_2 \geq 6$
       $x_1 + 6x_2 \geq 8$
       $x_1 + 4x_2 \geq 4$
       $x_1, x_2 \geq 0$

   Dual:
   Maximize $Z = 6y_1 + 8y_2 + 4y_3$
   subject to $y_1 + y_2 + y_3 \leq 200$
   $2y_1 + 6y_2 + 4y_3 \leq 240$
   $y_1, y_2 \geq 0$

   (b) Simplex tableau for dual:

y_1	y_2	y_3	s_1	s_2	Z		
1	1	1	1	0	0	200	200/1 = 200
2	⑥	4	0	1	0	240	240/6 = 40
–6	–8	–4	0	0	1	0	

   $(1/6)R_2 \to R_2$; $R_1 - R_2 \to R_1$; $R_3 + 8R_2 \to R_3$

2/3	0	1/3	1	–1/6	0	160
①/3	1	2/3	0	1/6	0	40
–10/3	0	4/3	0	4/3	1	320

   $(3)R_2 \to R_2$; $R_1 - (2/3)R_2 \to R_1$; $R_3 + (10/3)R_2 \to R_3$

0	–2	–1	1	–1/2	0	80
1	3	2	0	1/2	0	120
0	10	8	0	3	1	720

   Solution: $y_1 = 120$, $y_2 = 0$, $y_3 = 0$, $s_1 = 80$, $s_2 = 0$, $Z = 720$

   (c) From the indicators corresponding to s_1 and s_2, $x_1 = 0$, $x_2 = 3$. Also, $Z = 720$.

   (d) The surplus amounts in the primal constraints are 0, 10, and 8.

   (e) Shadow prices: $y_1 = 120$, $y_2 = 0$; $y_3 = 0$. An increase of 1 unit in the right–hand side of the first primal constraint will increase the optimal value of the objective function by 120 units; changes in the right–hand–side values of the remaining constraints will not result in changes in the value of the objective function because the shadow prices are 0.

7. (a) Primal:

$$\text{Minimize } Z = 80x_1 + 100x_2 + 10x_3$$
$$\text{subject to } 2x_1 + x_2 + x_3 \geq 6$$
$$4x_1 + 2x_2 \qquad \geq 9$$
$$2x_2 \qquad \geq 5$$
$$x_1, x_2, x_3 \geq 0$$

Dual:

$$\text{Maximize } Z = 6y_1 + 9y_2 + 5y_3$$
$$\text{subject to } 2y_1 + 4y_2 \qquad \leq 80$$
$$y_1 + 2y_2 + 2y_3 \leq 100$$
$$y_1 \qquad \leq 10$$
$$y_1, y_2, y_3 \geq 0$$

(b) Simplex tableau for dual:

y_1	y_2	y_3	s_1	s_2	s_3	Z	
2	④	0	1	0	0	0	80
1	2	2	0	1	0	0	100
1	0	0	0	0	1	0	10
−6	−9	−5	0	0	0	1	0

$(1/4)R_1 \to R_1$; $(-2)R_1 + R_2 \to R_2$; $(9)R_1 + R_4 \to R_4$

1/2	1	0	1/4	0	0	0	20
0	0	②	−1/2	1	0	0	60
1	0	0	0	0	1	0	10
−3/2	0	−5	9/4	0	0	1	180

$(1/2)R_2 \to R_2$; $(5)R_2 + R_4 \to R_4$

1/2	1	0	1/4	0	0	0	20
0	0	1	−1/4	1/2	0	0	30
①	0	0	0	0	1	0	10
−3/2	0	0	1	5/2	0	1	330

$(-1/2)R_3 + R_1 \to R_1$; $(3/2)R_3 + R_4 \to R_4$

0	1	0	1/4	0	−1/2	0	15
0	0	1	−1/4	1/2	0	0	30
1	0	0	0	0	1	0	10
0	0	0	1	5/2	3/2	1	345

Solution: $y_1 = 10$, $y_2 = 15$, $y_3 = 30$, $s_1 = 0$, $s_2 = 0$, $s_3 = 0$, $Z = 345$

(c) From the indicators corresponding to s_1, s_2, and s_3, $x_1 = 1$, $x_2 = 5/2$, $x_3 = 3/2$. Also, $Z = 345$.

(d) The surplus amounts in the primal constraints are all 0.

(e) Shadow prices: $y_1 = 10$, $y_2 = 15$; $y_3 = 30$. An increase of 1 unit in the right–hand side of the i–th primal constraint ($i = 1, 2, 3$) will increase the optimal value of the objective function by y_i units.

9. (a) Primal:

$$\text{Minimize } Z = 600x_1 + 460x_2 + 600x_3 + 800x_4$$
$$\text{subject to } 2x_1 + x_2 + x_3 \qquad \geq 40$$
$$x_1 + 2x_2 + x_3 + x_4 \geq 60$$
$$2x_1 + 2x_2 \qquad + x_4 \geq 50$$
$$x_1, x_2, x_3, x_4 \geq 0$$

Dual:

$$\text{Maximize } Z = 40y_1 + 60y_2 + 50y_3$$
$$\text{subject to } 2y_1 + y_2 + 2y_3 \leq 600$$
$$y_1 + 2y_2 + 2y_3 \leq 460$$
$$y_1 + y_2 \qquad \leq 600$$
$$y_2 + y_3 \leq 800$$
$$y_1, y_2, y_3 \geq 0$$

(b) Simplex tableau for dual:

y_1	y_2	y_3	s_1	s_2	s_3	s_4	Z	
2	1	2	1	0	0	0	0	600
1	②	2	0	1	0	0	0	460
1	1	0	0	0	1	0	0	600
0	1	1	0	0	0	1	0	800
−40	−60	−50	0	0	0	0	1	0

$(1/2)R_2 \to R_1$; $(-1)R_2 + R_1 \to R_1$;
$(-1)R_2 + R_3 \to R_3$; $(-1)R_2 + R_4 \to R_4$;
$(60)R_2 + R_5 \to R_5$

③/②	0	1	1	−1/2	0	0	0	370
1/2	1	1	0	1/2	0	0	0	230
1/2	0	−1	0	−1/2	1	0	0	370
−1/2	0	0	0	−1/2	0	1	0	570
−10	0	10	0	30	0	0	1	13,800

$(2/3)R_1 \to R_1$; $(-1/2)R_1 + R_2 \to R_2$;
$(-1/2)R_1 + R_3 \to R_3$; $(1/2)R_1 + R_4 \to R_4$;
$(10)R_1 + R_5 \to R_5$

1	0	2/3	2/3	−1/3	0	0	0	740/3
0	1	2/3	−1/3	2/3	0	0	0	320/3
0	0	−4/3	−1/3	−1/3	1	0	0	740/3
0	0	1/3	1/3	−2/3	0	1	0	2080/3
0	0	50/3	20/3	80/3	0	0	1	48,800/3

Solution: $y_1 = 740/3$, $y_2 = 320/3$, $y_3 = 0$, $s_1 = 0$, $s_2 = 0$, $s_3 = 740/3$, $s_4 = 2080/3$, $Z = 48,800/3$

(continued)

9. *(continued)*

(c) From the indicators corresponding to s_1, s_2, s_3, and s_4, $x_1 = 20/3$, $x_2 = 80/3$, $x_3 = 0$, $x_4 = 0$. Also, $Z = 48,800/3$.

(d) The surplus amounts in the primal constraints are 0, 0, and 50/3.

(e) Shadow prices: $y_1 = 740/3$, $y_2 = 320/3$; $y_3 = 0$. An increase of 1 unit in the right–hand side of the i–th primal constraint ($i = 1, 2, 3$) will increase the optimum value of the objective function by y_i units.

11. (a) Primal:
Minimize $Z = 90x_1 + 100x_2 + 60x_3 + 80x_4$
subject to
$$
\begin{aligned}
x_1 \quad\quad + x_3 \quad\quad &\geq 7 \\
x_1 \quad\quad\quad\quad + x_4 &\geq 8 \\
x_2 \quad\quad + x_4 &\geq 9 \\
x_2 + x_3 \quad\quad &\geq 10 \\
x_1, x_2, x_3, x_4 &\geq 0
\end{aligned}
$$

Dual:
Maximize $Z = 7y_1 + 8y_2 + 9y_3 + 10y_4$
subject to
$$
\begin{aligned}
y_1 + y_2 \quad\quad\quad &\leq 90 \\
y_3 + y_4 &\leq 100 \\
y_1 \quad\quad + y_4 &\leq 60 \\
y_2 + y_3 \quad\quad &\leq 80
\end{aligned}
$$

(b) Simplex tableau for dual:

y_1	y_2	y_3	y_4	s_1	s_2	s_3	s_4	Z	
1	1	0	0	1	0	0	0	0	90
0	0	1	1	0	1	0	0	0	100
1	0	0	①	0	0	1	0	0	60
0	1	1	0	0	0	0	1	0	80
-7	-8	-9	-10	0	0	0	0	1	0

$(-1)R_3 + R_2 \to R_2$; $(10)R_3 + R_5 \to R_5$

1	1	0	0	1	0	0	0	0	90
-1	0	①	0	0	1	-1	0	0	40
1	0	0	1	0	0	1	0	0	60
0	1	1	0	0	0	0	1	0	80
3	-8	-9	0	0	0	10	0	1	600

(continued)

11. *(continued)*

$R_4 - R_2 \to R_4$; $R_5 + (9)R_2 \to R_5$

1	1	0	0	1	0	0	0	0	90
-1	0	1	0	0	1	-1	0	0	40
1	0	0	1	0	0	1	0	0	60
1	①	0	0	0	-1	1	1	0	40
-6	-8	0	0	0	9	1	0	1	960

$R_1 - R_4 \to R_1$; $R_5 + (8)R_4 \to R_5$

0	0	0	0	1	1	-1	-1	0	50
-1	0	1	0	0	1	-1	0	0	40
1	0	0	1	0	0	1	0	0	60
1	1	0	0	0	-1	1	1	0	40
2	0	0	0	0	1	9	8	1	1280

Solution: $y_1 = 0$, $y_2 = 40$, $y_3 = 40$, $y_4 = 60$, $s_1 = 50$, $s_2 = 0$, $s_3 = 0$, $s_4 = 0$, $Z = 1280$

(c) From the indicators corresponding to s_1, s_2, s_3, and s_4, $x_1 = 0$, $x_2 = 1$, $x_3 = 9$, $x_4 = 8$. Also, $Z = 1280$.

(d) The surplus amounts in the primal constraints are 2, 0, 0, and 0.

(e) Shadow prices: $y_1 = 0$, $y_2 = 40$; $y_3 = 40$, $y_4 = 60$. An increase of 1 unit in the right–hand side of the i–th primal constraint ($i = 1, 2, 3, 4$) will increase the optimum value of the objective function by y_i units.

13. (a) Minimize $C = 6x_1 + 8x_2 + 10x_3 + 9x_4$
subject to
$$
\left.
\begin{aligned}
x_1 + x_2 \quad\quad\quad &\leq 500 \\
x_3 + x_4 &\leq 600
\end{aligned}
\right\} \text{capacity}
$$
$$
\left.
\begin{aligned}
x_1 \quad\quad + x_3 \quad\quad &\geq 400 \\
x_2 \quad\quad + x_4 &\geq 650
\end{aligned}
\right\} \text{demand}
$$

(b) Minimize $C = 6x_1 + 8x_2 + 10x_3 + 9x_4$
subject to
$$
\begin{aligned}
-x_1 - x_2 \quad\quad\quad &\geq -500 \\
-x_3 - x_4 &\geq -600 \\
x_1 + \quad\quad x_3 \quad\quad &\geq 400 \\
x_2 \quad\quad + x_4 &\geq 650 \\
x_1, x_2, x_3, x_4 &\geq 0
\end{aligned}
$$

(continued)

13. *(continued)*

Dual:

Maximize $C = -500y_1 - 600y_2 + 400y_3 + 650y_4$

subject to
$$
\begin{aligned}
-y_1 \quad\quad + y_3 \quad\quad &\le 6 \\
-y_1 \quad\quad\quad\quad + y_4 &\le 8 \\
-y_2 + y_3 \quad\quad &\le 10 \\
-y_2 \quad\quad + y_4 &\le 9 \\
y_1,\ y_2,\ y_3,\ y_4 &\ge 0
\end{aligned}
$$

(c) Simplex tableau for dual:

y_1	y_2	y_3	y_4	s_1	s_2	s_3	s_4	C	
-1	0	1	0	1	0	0	0	0	6
-1	0	0	①	0	1	0	0	0	8
0	-1	1	0	0	0	1	0	0	10
0	-1	0	1	0	0	0	1	0	9
500	600	-400	-650	0	0	0	0	1	0

$(-1)R_2 + R_4 \rightarrow R_4;\ (650)R_2 + R_5 \rightarrow R_5$

-1	0	①	0	1	0	0	0	0	6
-1	0	0	1	0	1	0	0	0	8
0	-1	1	0	0	0	1	0	0	10
1	-1	0	0	0	-1	0	1	0	1
-150	600	-400	0	0	650	0	0	1	5200

$R_3 - R_1 \rightarrow R_3;\ R_5 + (400)R_1 \rightarrow R_5$

-1	0	1	0	1	0	0	0	0	6
-1	0	0	1	0	1	0	0	0	8
1	-1	0	0	-1	0	1	0	0	4
①	-1	0	0	0	-1	0	1	0	1
-550	600	0	0	400	650	0	0	1	7600

$R_1 + R_4 \rightarrow R_1;\ R_2 + R_4 \rightarrow R_2;\ R_3 - R_4 \rightarrow R_3;$
$R_5 + (550)R_4 \rightarrow R_5$

0	-1	1	0	1	-1	0	1	0	7
0	-1	0	1	0	0	0	1	0	9
0	0	0	0	-1	1	1	-1	0	3
1	-1	0	0	0	-1	0	1	0	1
0	50	0	0	400	100	0	550	1	8150

Solution: $y_1 = 1,\ y_2 = 0,\ y_3 = 7,\ y_4 = 9,\ s_1 = 0,$
$s_2 = 0,\ s_3 = 3,\ s_4 = 0,\ C = 8150$

(continued)

13. *(continued)*

(d) From the indicators corresponding to $s_1,\ s_2,\ s_3,$ and $s_4,\ x_1 = 400,\ x_2 = 100,\ x_3 = 0,\ x_4 = 550.$ Also, $C = 8150.$

(e) The slack or surplus amounts in the primal constraints are 0, 50(slack), 0, and 0.

(f) Shadow prices: $y_1 = 0,\ y_2 = 0;\ y_3 = 7,\ y_4 = 9.$ An increase of 1 unit in the right–hand side of the i–th primal constraint $(i = 1, 2, 3, 4)$ will increase the optimum value of the objective function by y_i units.

15. (a) Minimize $C = 8x_1 + 10x_2 + 15x_3 + 9x_4$

subject to
$$
\left.\begin{aligned}
x_1 + x_2 \quad\quad &\le 450 \\
x_3 + x_4 &\le 600
\end{aligned}\right\}\ \text{capacities}
$$
$$
\left.\begin{aligned}
x_1 \quad + x_3 \quad\quad &\ge 400 \\
x_2 \quad\quad + x_4 &\ge 500
\end{aligned}\right\}\ \text{demands}
$$

(b) Minimize $C = 8x_1 + 10x_2 + 15x_3 + 9x_4$

subject to
$$
\begin{aligned}
-x_1 - x_2 \quad\quad &\ge -450 \\
-x_3 - x_4 &\ge -600 \\
x_1 \quad + x_3 \quad\quad &\ge 400 \\
x_2 \quad\quad + x_4 &\ge 500 \\
x_1,\ x_2,\ x_3,\ x_4 &\ge 0
\end{aligned}
$$

Dual:

Maximize $C = -450y_1 - 600y_2 + 400y_3 + 500y_4$

subject to
$$
\begin{aligned}
-y_1 \quad\quad + y_3 \quad\quad &\le 8 \\
-y_1 \quad\quad\quad\quad + y_4 &\le 10 \\
-y_2 + y_3 \quad\quad &\le 15 \\
-y_2 \quad\quad + y_4 &\le 9 \\
y_1,\ y_2,\ y_3,\ y_4 &\ge 0
\end{aligned}
$$

(c) Simplex tableau for dual:

y_1	y_2	y_3	y_4	s_1	s_2	s_3	s_4	C	
-1	0	1	0	1	0	0	0	0	8
-1	0	0	1	0	1	0	0	0	10
0	-1	1	0	0	0	1	0	0	15
0	-1	0	①	0	0	0	1	0	9
450	600	-400	-500	0	0	0	0	1	0

(continued)

15. *(continued)*

$$(-1)R_4 + R_2 \to R_2; \ (500)R_4 + R_5 \to R_5$$

−1	0	①	0	1	0	0	0	0	8
−1	1	0	0	0	1	0	−1	0	1
0	−1	1	0	0	0	1	0	0	15
0	−1	0	1	0	0	0	1	0	9
450	100	−400	0	0	0	0	500	1	4500

$$(-1)R_1 + R_3 \to R_3; \ (400)R_1 + R_5 \to R_5$$

−1	0	1	0	1	0	0	0	0	8
−1	1	0	0	0	1	0	−1	0	1
1	−1	0	0	−1	0	1	0	0	7
0	−1	0	1	0	0	0	1	0	9
50	100	0	0	400	0	0	500	1	7700

Solution: $y_1 = 0$, $y_2 = 0$, $y_3 = 8$, $y_4 = 9$, $s_1 = 0$, $s_2 = 1$, $s_3 = 7$, $s_4 = 0$, $C = 7700$

(d) From the indicators corresponding to s_1, s_2, s_3, and s_4, $x_1 = 400$, $x_2 = 0$, $x_3 = 0$, $x_4 = 500$. Also, $C = 7700$.

(e) The slack or surplus amounts in the primal constraints are 50(slack), 100(slack), 0, and 0.

(f) Shadow prices: $y_1 = 0$, $y_2 = 0$; $y_3 = 8$, $y_4 = 9$. An increase of 1 unit in the right–hand side of the i–th primal constraint ($i = 1, 2, 3, 4$) will increase the optimum value of the objective function by y_i units.

17. (a) Let x_1 = units of food A, x_2 = units of food B, x_3 = units of food C

Minimize $C = 800x_1 + 500x_2 + 600x_3$ calories
subject to $5x_1 + 4x_2 + 2x_3 \geq 25$ protein
$\qquad\qquad 10x_1 + 5x_2 + 8x_3 \geq 60$ calcium
$\qquad\qquad 2x_1 + \ x_2 + 3x_3 \geq 10$ iron
$\qquad\qquad\qquad x_1, x_2, x_3 \geq 0$

(continued)

17. *(continued)*

(b) Dual:
Maximize $C = 25y_1 + 60y_2 + 10y_3$
subject to $5y_1 + 10y_2 + 2y_3 \leq 800$
$\qquad\qquad 4y_1 + \ 5y_2 + \ y_3 \leq 500$
$\qquad\qquad 2y_1 + \ 8y_2 + 3y_3 \leq 600$
$\qquad\qquad\qquad y_1, y_2, y_3 \geq 0$

(c) Simplex tableau for dual:

y_1	y_2	y_3	s_1	s_2	s_3	C	
5	10	2	1	0	0	0	800
4	5	1	0	1	0	0	500
2	⑧	3	0	0	1	0	600
−25	−60	−10	0	0	0	1	0

$$(1/8)R_3 \to R_3; \ (-10)R_3 + R_1 \to R_1;$$
$$(-5)R_3 + R_2 \to R_2; \ (60)R_3 + R_4 \to R_4$$

⑤/2	0	−7/4	1	0	−5/4	0	50
11/4	0	−7/8	0	1	−5/8	0	125
1/4	1	3/8	0	0	1/8	0	75
−10	0	25/2	0	0	15/2	1	4500

$$(2/5)R_1 \to R_1; \ (-11/4)R_1 + R_2 \to R_2;$$
$$(-1/4)R_1 + R_3 \to R_3; \ (10)R_1 + R_4 \to R_4;$$

1	0	−7/10	2/5	0	−1/2	0	20
0	0	21/20	−11/10	1	3/4	0	70
0	1	11/20	−1/10	0	1/4	0	70
0	0	11/2	4	0	5/2	1	4700

Solution: $y_1 = 20$, $y_2 = 70$, $y_3 = 0$, $s_1 = 0$, $s_2 = 70$, $s_3 = 0$, $C = 4700$

(d) From the indicators corresponding to s_1, s_2, and s_3, $x_1 = 4$, $x_2 = 0$, $x_3 = 5/2$. Also, $C = 4700$.

(e) The surplus amounts in the primal constraints are 0, 0, and 11/2.

(f) Shadow prices: $y_1 = 20$, $y_2 = 70$, $y_3 = 0$. An increase of 1 unit in the right–hand side of the i–th primal constraint ($i = 1, 2, 3, 4$) will increase the optimum value of the objective function by y_i units.

19. Let x_1 = number of professional pollsters at location 1,
x_2 = number of semiprofessional pollsters at location 1
x_3 = number of professional pollsters at location 2,
x_4 = number of semiprofessional pollsters at location 2

$$\begin{aligned} &\text{Minimize } C = 400x_1 + 300x_2 + 400x_3 + 300x_4 \\ &\text{subject to } 80x_1 + 50x_2 \geq 440 \\ &\qquad 80x_3 + 50x_4 \geq 800 \\ &\qquad 50x_2 + 50x_4 \geq 200 \\ &\qquad x_1, x_2, x_3, x_4 \geq 0 \end{aligned}$$

(b) Dual:

$$\begin{aligned} &\text{Maximize } C = 440y_1 + 800y_2 + 200y_3 \\ &\text{subject to } 80y_1 \leq 400 \\ &\qquad 50y_1 + 50y_3 \leq 300 \\ &\qquad 80y_2 \leq 400 \\ &\qquad 50y_2 + 50y_3 \leq 300 \\ &\qquad y_1, y_2, y_3, y_4 \geq 0 \end{aligned}$$

(c) Simplex tableau for dual:

y_1	y_2	y_3	s_1	s_2	s_3	s_4	C	
80	0	0	1	0	0	0	0	400
50	0	50	0	1	0	0	0	300
0	(80)	0	0	0	1	0	0	400
0	50	50	0	0	0	1	0	300
–440	–800	–200	0	0	0	0	1	0

$(1/80)R_3 \to R_3$; $(-50)R_3 + R_4 \to R_4$;
$(800)R_3 + R_5 \to R_5$

(80)	0	0	1	0	0	0	0	400
50	0	50	0	1	0	0	0	300
0	1	0	0	0	1/80	0	0	5
0	0	50	0	0	–5/8	1	0	50
–440	0	–200	0	0	10	0	1	4000

$(1/80)R_1 \to R_1$; $(-50)R_1 + R_2 \to R_2$;
$(440)R_1 + R_5 \to R_5$

1	0	0	1/80	0	0	0	0	5
0	0	50	–5/8	1	0	0	0	50
0	1	0	0	0	1/80	0	0	5
0	0	(50)	0	0	–5/8	1	0	50
0	0	–200	11/2	0	10	0	1	6200

(continued)

19. *(continued)*

$(1/50)R_4 \to R_4$; $(-50)R_4 + R_2 \to R_2$;
$(200)R_4 + R_5 \to R_5$

1	0	0	1/80	0	0	0	0	5
0	0	0	–5/8	1	5/8	–1	0	0
0	1	0	0	0	1/80	0	0	5
0	0	1	0	0	–1/80	1/50	0	1
0	0	0	11/2	0	15/2	4	1	6400

Solution: $y_1 = 5$, $y_2 = 5$, $y_3 = 1$, $s_1 = 0$, $s_2 = 0$, $s_3 = 0$, $s_4 = 0$, $C = 6400$

(d) From the indicators corresponding to s_1, s_2, s_3, and s_4, $x_1 = 3$, $x_2 = 4$, $x_3 = 10$, $x_4 = 0$. Also, $C = 6400$. The constraints are still satisfied, but the cost is \$6600.

(e) Using $x_1 = 5$, $x_2 = 1$, $x_3 = 7$, and $x_4 = 5$ gives surplus amounts in the primal constraints of 0 for all constraints.

(f) Shadow prices: $y_1 = 5$, $y_2 = 5$, $y_3 = 1$. An increase of 1 unit in the right–hand side of the i–th primal constraint (i = 1, 2, 3) will increase the optimum value of the objective function by y_i units.

7–4 Mixed Constraints; Shadow Prices (Primal); Computer Solution

1. (a) Maximize
$x = 5x_1 + 8x_2 + 6x_3 + 0s_1 + 0s_2 - Ma_1 - Ma_2$
subject to

$$\begin{aligned} 2x_1 + x_2 + 4x_3 + s_1 &= 190 \\ x_1 + 2x_2 + 4x_3 - s_2 + a_1 &= 20 \\ x_1 + x_2 + x_3 + a_2 &= 160 \\ x_1, x_2, x_3, s_1, s_2, a_1, a_2 &\geq 0 \end{aligned}$$

(continued)

1. *(continued)*

(b) Optimal solution:
$$x_1 = 0, \quad x_3 = 0$$
$\underbrace{\hspace{3cm}}$
Nonbasic variables

$$x_2 = 160, \ s_1 \ = \ 30, \ s_2 = 300$$
$\underbrace{\hspace{4cm}}$
Basis

$Z = 1280$
Value of objective function

3. (a) Optimal solution:
$$x_2 \ = \ 0, \ x_3 \ = \ 0, \ s_1 = 0$$
$\underbrace{\hspace{3cm}}$
Nonbasic variables

$$x_1 = 33.333, \ s_2 = 83.333, \ s_3 = 126.667$$
$\underbrace{\hspace{5cm}}$
Basis

$Z = 333.333$
Value of objective function

(b) Only the first constraint is binding.

(c) The shadow prices are 3.333, 0, and 0. An increase of 1 unit in the right–hand side of the first constraint increases the optimal value of the objective function by 3.333 units. The other constraints have no effect.

5. (a) Optimal solution:
$$x_1 = 0, \ s_1 = 0, \ s_2 = 0 \quad x_2 = 2.778, \ x_3 = 39.444$$
$\underbrace{\hspace{3.5cm}}$ $\underbrace{\hspace{3.5cm}}$
Nonbasic variables   Basis

$Z = 290$
Value of objective function

(b) Both constraints are binding.

(c) The shadow prices are 0.5 and 1. An increase of 1 unit in the right–hand side of the first constraint will increase the optimum value of the objective function by 0.5 unit. For the second constraint, an increase of 1 unit will increase the value of the objective function by 1 unit.

7. Optimal solution:
$$x_3 = 0, \ s_1 = 0, \ s_3 = 0$$
$\underbrace{\hspace{3cm}}$
Nonbasic variables

$$x_1 = 146.67, \ x_2 = 53.33, \ s_2 = 606.67$$
$\underbrace{\hspace{5cm}}$
Basis

$Z = 15,066.67$
Value of objective function

(b) The first and third constraints are binding.

(c) The shadow prices are given as the dual prices: 6.67, 0, and 56.67. An increase of 1 unit in the right–hand side of the first constraint increases the optimum value of the objective function by 6.67 units (the first dual price). For the third constraint an increase of 1 unit increases the value of the objective function by 56.67 units.

9. (a) Optimal solution:
$$x_3 = 0, \ s_1 = 0, \ s_3 = 0 \quad x_1 = 90, \ x_2 = 10, \ s_2 = 60$$
$\underbrace{\hspace{3.5cm}}$ $\underbrace{\hspace{3.5cm}}$
Nonbasic variables   Basis

$Z = 1100$
Value of objective function

(b) The first and third constraints are binding.

(c) The shadow prices are –5, 0, and –5. An increase of 1 unit in the right–hand side of the first or third constraint will not lead to an improvement in the value of the objective function but will increase that value by 5 units.

11. (a) Optimal solution:
$$x_2 = 0, \ s_2 = 0, \ s_3 = 0$$
$\underbrace{\hspace{3cm}}$
Nonbasic variables

$$x_1 = 24, \ x_3 = 14.5, \ s_1 = 108$$
$\underbrace{\hspace{4cm}}$
Basis

$Z = 216.5$, value of objective function

(continued)

11. *(continued)*

(b) The surplus in the first constraint is 108; the remaining constraints are binding.

(c) The shadow prices are 0, 0.45, and –2.15. An increase of 1 unit in the right–hand side of the second constraint will improve the value of the objective function by 0.45 (Z will decrease). For the third constraint an increase in the right–hand side will not improve the value of the objective function (Z will increase by 2.15).

13. (a) Let x_1 = number of chairs, x_2 = number of desks, x_3 = number of tables.
Maximize $P = 30x_1 + 100x_2 + 160x_3$ profit function

subject to
$$30x_1 + 20x_2 + 10x_3 \leq 46{,}000 \text{ cutting}$$
$$20x_1 + 60x_2 + 20x_3 \leq 32{,}000 \text{ assembly}$$
$$10x_1 + 40x_2 + 90x_3 \leq 26{,}000 \text{ finishing}$$
$$x_1, x_2, x_3 \geq 0$$

(b) Optimal solution:
$$\underbrace{x_2 = 0,\ s_2 = 0,\ s_3 = 0}_{\text{Nonbasic variables}}$$

$$\underbrace{x_1 = 1475,\ x_3 = 125,\ s_1 = 500}_{\text{Basis}}$$

$Z = 64{,}250$
Value of objective function

(c) The second and third constraints are binding. The slack in the first constraint is 500 hours.

(d) Since there is already slack time in the cutting department, additional hours are worth no money.

(e) From the shadow price of 0.6875, one additional hour in the assembly department will increase the profit by 0.6875. Thus, the maximum amount that should be paid is $0.69.

(f) From the shadow price of 1.625, one additional hour in the finishing department will increase the profit by 1.625. Thus, the maximum amount that should be paid is $1.625.

15. (a) Let x_1 = dollars invested in treasury bills,
x_2 = dollars invested in municipal bonds,
x_3 = dollars invested in real estate,
x_4 = dollars invested in mutual fund,
x_5 = dollars invested in energy stocks

Maximize
$0.083x_1 + 0.098x_2 + 0.159x_3 + 0.163x_4 + 0.184x_5$
subject to
$$x_1 + x_2 + x_3 + x_4 + x_5 = 800{,}000$$
$$0x_1 + 1x_2 + 3x_3 + 4x_4 + 6x_5 \leq 5.7(800{,}000)$$
$$\text{or } 4{,}560{,}000$$
$$x_5 \leq 200{,}000$$
$$x_1, x_2, x_3, x_4, x_5 \geq 0$$

(b) Optimal solution:
$$\underbrace{x_1 = 0,\ x_2 = 0,\ x_3 = 0,\ s_1 = 0,\ s_3 = 0}_{\text{Nonbasic variables}}$$

$$\underbrace{x_4 = 600{,}000,\ x_5 = 200{,}000,\ s_2 = 960{,}000}_{\text{Basis}}$$

$Z = 134{,}600$
Value of objective function

(c) Constraints 1 and 3 are binding.

(d) If the amount invested is increased by $1, the annual return is increased by $0.163.

(e) If the restriction that no more than $200,000 be invested in energy stocks is lifted, then an additional dollar invested in such stocks will increase the annual return by $0.021.

17. (a) Let x_1 = truckloads from plant 1 shipped to distribution center 1,

x_2 = truckloads from plant 1 shipped to distribution center 2,

x_3 = truckloads from plant 1 shipped to distribution center 3,

x_4 = truckloads from plant 2 shipped to distribution center 1,

x_5 = truckloads from plant 2 shipped to distribution center 2,

x_6 = truckloads from plant 2 shipped to distribution center 3.

Minimize $400x_1 + 800x_2 + 600x_3 + 500x_4 + 700x_5 + 900 x_6$

subject to

$$x_1 + x_2 + x_3 \leq 50 \text{ plant 1 capacity}$$
$$x_4 + x_5 + x_6 \leq 40 \text{ plant 2 capacity}$$
$$x_1 + x_4 \geq 20 \text{ center 1 demand}$$
$$x_2 + x_5 \geq 50 \text{ center 2 demand}$$
$$x_3 + x_6 \geq 15 \text{ center 3 demand}$$
$$x_1, \, x_2, \, x_3, \, x_4, \, x_5, \, x_6 \geq 0$$

(b) Optimal solution:

$x_4 = 0, \, x_6 = 0, \, s_2 = 0, \, s_3 = 0, \, s_4 = 0, \, s_5 = 0$

$\underbrace{}$

Nonbasic variables

$x_1 = 20, \, x_2 = 10, \, x_3 = 15, \, x_5 = 40, \, s_1 = 5$

$\underbrace{}$

Basis

Z = 53,000
Value of objective function

(c) Constraints 2, 3, 4 and 5 are binding.

(d) Plant 1 should not increase its capacity, since the slack variable is not zero for constraint 1.

(e) Plant 2 should increase its capacity, since constraint 2 is binding and the shadow price is positive. An increase of 1 truckload in the capacity will improve the optimal value of the objective function by 100 units, i.e., decrease the total transportation cost.

(continued)

17. *(continued)*

(f) If distribution center 1 increases its capacity by 1 truckload, the total transportation cost will not improve (shadow price = –400) but will increase by 400.

(g) If distribution center 2 increases its capacity by 1 truckload, the total transportation cost will not improve (shadow price = –800) but will increase by 800.

(h) If distribution center 3 increases its capacity by 1 truckload, the total transportation cost will not improve (shadow price = –600) but will increase by 600.

(i) distribution center 2

19. (a) Minimize $x_1 + x_2 + x_3 + x_4 + x_5 + x_6$ total manpower

subject to
$$x_1 + x_6 \geq 3 \quad \text{midnight to 4 A.M.}$$
$$x_1 + x_2 \geq 5 \quad \text{4 A.M. to 8 A.M.}$$
$$x_2 + x_3 \geq 14 \quad \text{8 A.M. to noon}$$
$$x_3 + x_4 \geq 16 \quad \text{noon to 4 P.M.}$$
$$x_4 + x_5 \geq 14 \quad \text{4 P.M. to 8 P.M.}$$
$$x_5 + x_6 \geq 10 \quad \text{8 P.M. to midnight}$$
$$x_1, x_2, x_3 \, x_4, x_5, x_6 \geq 0$$

(b) Optimal solution:

$x_6 = 0, s_1 = 0, s_2 = 0, s_3 = 0, s_4 = 0, s_5 = 0, s_6 = 0$

$\underbrace{}$

Nonbasic variables

$x_1 = 3, \, x_2 = 2, \, x_3 = 12, \, x_4 = 4 \, x_5 = 10$

$\underbrace{}$

Basis

Z = 31
Value of objective function

(c) The total number of officers needed to satisfy minimal personnel requirements is 31.

(d) All constraints are binding. There are no surplus amounts.

(continued)

19. *(continued)*

(e) If the right–hand side of constraint 2 is increased by 1 unit, the total number of officers needed to satisfy minimal personnel requirements will be increased by 1 (shadow price = –1).

(f) If the right–hand side of constraint 3 is increased by 1 unit, no additional officers will be needed, provided the right–hand side can vary in this way. In fact, it cannot do so without violating other constraints. Several solutions to the new problem, with $x_2 + x_3 \geq 15$, are possible but the total number of officers is at least 32. Two solutions are $x_1 = 3$, $x_2 = 3$, $x_3 = 12$, $x_4 = 4$, $x_5 = 10$, $x_6 = 0$ and $x_1 = 3$, $x_2 = 2$, $x_3 = 13$, $x_4 = 4$, $x_5 = 10$, $x_6 = 0$.

7–5 Sensitivity Analysis

1. (a) Optimal solution:
$$x_1 = 0, \ s_1 = 0, \ s_3 = 0$$
$$\underbrace{}$$
Nonbasic variables

$$x_1 = 28.2609, \ x_3 = 8.6957, \ s_2 = 30$$
$$\underbrace{}$$
Basis

$Z = 182.6087$
Value of objective function

(b) The slack for constraint 2 is 30 units. Constraints 1 and 3 are binding.

(c) The dual prices corresponding to constraints 1 and 3 are 1.56522 and 0.1739. An increase of 1 unit in the right–hand side of constraint 1 or 3 will increase the value of the optimal solution by 1.56522 or 0.1739, respectively.

1. *(continued)*

(d) The coefficient of x_1 can vary from 3.2 to 5.5 without changing the optimum values of x_1, x_3, or s_2. The coefficient of x_2 has a maximum value of 8.3478 and no lower limit. The coefficient of x_3 has a range of 6.7143 to 10, within which the optimum values of the basic variables (x_1, x_3, and s_2) will not change.

(e) The range for the right–hand side of constraint 1 is 60 – 750, for constraint 2 is 150 up to no upper limit, and for constraint 3 is 20 – 180. Within each range, taken individually, the optimal basis still contains x_1, x_3, and s_2, and the interpretation of the dual prices remains valid. The values of the basic variables may change.

3. (a) Optimal solution:
$$x_2 = 0, \ s_2 = 0, \ s_3 = 0$$
$$\underbrace{}$$
Nonbasic variables

$$x_1 = 8.6557, \ x_3 = 7.82609, \ s_1 = 36.52174$$
$$\underbrace{}$$
Basis

$Z = 140$
Value of objective function

(b) Constraints 2 and 3 are binding. The surplus in constraint 1 is 36.52174.

(c) The dual prices for constraints 2 and 3 are –1 and –2, indicating that an increase of 1 unit in the right–hand side of either constraint will not improve the optimal value of Z but will lead to increases of 1 and 2 units, respectively.

(d) The coefficients of x_1, x_2, x_3 can vary from 2.25 – 14.3889, 0 to no upper limit, and –1.8125 – 32, respectively. Within these individual ranges the optimal values of the basic variables will not change.

(continued)

(continued)

3. *(continued)*

(e) The ranges for the right–hand sides of constraints 1, 2, and 3 are no lower limit to 56.52174, 10 – 240, and 10 – 240, respectively. Within each range, taken individually, the optimal basis still contains x_1, x_3, and s_1, and the interpretation of the dual prices remains valid. The values of the basic variables may change.

5. (a) Optimal solution:
$$s_1 = 0, \ s_2 = 0, \ s_3 = 0$$
$$\underbrace{}$$
Nonbasic variables

$$x_1 = 10.7143, \ x_2 = 16.0714, \ x_3 = 5$$
$$\underbrace{}$$
Basis

$Z = 255$
Value of objective function

(b) All constraints are binding.

(c) The dual prices for constraints 1 and 2 are positive, indicating that an increase of 1 unit in the right–hand side either constraint will improve (increase) the optimum value of Z by 1 unit (the dual price). The dual price for constraint 3 is negative, indicating that an increase of 1 unit in the right–hand side will not improve the optimum value of Z but will lead to a decrease of 1 unit in Z.

(d) The coefficients of x_1, x_2, and x_3 can vary from 6.6667 – 9.6364, 7.6216 – 10.8, and no lower limit to 7, respectively. Within each range, taken individually, the optimal values of the basic variables will not change.

(e) The ranges for the right–hand sides of the constraints are 45 – 115, 133.75 – 265 and 0 – 11.0811, respectively. Within each range, taken individually, the optimal basis still contains x_1, x_2, and x_3, and the interpretation of the dual prices remains valid. The values of the basic variables may change.

7. (a) Let x_1 = units of product A, x_2 = units of product B, x_3 = units of product C
Maximize $50x_1 + 40x_2 + 60x_3$ profit function
subject to
$$2x_1 + 3x_2 + \ x_3 \le 760 \ \text{capacity, department 1}$$
$$x_1 + 2x_2 + 4x_3 \le 600 \ \text{capacity, department 2}$$
$$2x_1 + \ x_2 + 2x_3 \le 660 \ \text{capacity, department 3}$$
$$x_1, x_2, x_3 \ge \ 0$$

(b) Optimal solution:
$$s_1 = 0, \ s_2 = 0, \ s_3 = 0$$
$$\underbrace{}$$
Nonbasic variables

$$x_1 = 240, \ x_2 = 76, \ x_3 = 52$$
$$\underbrace{}$$
Basis

$Z = 18,160$
Value of objective function

(c) All constraints are binding.

(d) If additional hours are made available in department 1, one hour is worth $4 (the shadow price for constraint 1).

(e) The range within which the dual price for department 1 is valid is from 570 – 1020 (right–hand side). Since the current value is 760, up to 260 hours can be added within this range.

(f) Department 3 has the greatest potential increase in profit per unit increase in labor capacity. An increase of one person–hour in capacity yields $18.67 in profit (dual price 3). The upper limit on labor capacity within which this dual price holds is 822.857 person–hours.

(g) The unit profit on product B can vary from 30 to 80 (objective coefficient range for x_2) without changing the optimal basis.

9. (a) Let x_1 = units of food A, x_2 = units of food B,
        x_3 = units of food C.
    Minimize
    $$30x_1 + 20x_2 + 35x_3 \quad \text{cost function}$$
    $$3x_1 + 5x_2 + 6x_3 \geq 30 \text{ protein}$$
    $$6x_1 + 12x_2 + 5x_3 \geq 72 \text{ calcium}$$
    $$4x_1 + 2x_2 + x_3 \geq 15 \text{ iron}$$
    $$x_1, x_2, x_3 \geq 0$$

    (b) Optimal solution:
        $$\underbrace{x_3 = 0, s_2 = 0, s_3 = 0}_{\text{Nonbasic variables}}$$

        $$\underbrace{x_1 = 1, x_2 = 5.5, s_1 = 0.5}_{\text{Basis}}$$

        $Z = 140$
        Value of objective function

    (c) Constraints 2 and 3 are binding. The surplus
        for constraint 1 is 0.5 milligrams of protein.

    (d) If the minimal protein requirement is decreased
        by 1 milligram, there will be no effect on the
        optimal diet cost, since constraint 1 is not
        binding.

    (e) A one–milligram decrease in the requirement
        for iron yields the longest decrease in diet cost
        (a decrease of 6.67 cost units). The minimal
        requirement for iron can go down to 12
        milligrams with the answer still being valid.

    (f) The unit cost of food B can vary from 15 to 60
        without changing the optimal basis.

    (g) The unit cost of food C can vary from 9.44 to
        no upper limit without changing the optimal
        basis.

11. (a) Let x_1 = units of product A produced at plant 1,
        x_2 = units of product B produced at plant 1,
        x_3 = units of product C produced at plant 1,
        x_4 = units of product A produced at plant 2,
        x_5 = units of product B produced at plant 2,
        x_6 = units of product C produced at plant 2.

    Minimize
    $$80x_1 + 100x_2 + 150x_3 + 90x_4 + 70x_5 + 60x_6 \text{ cost function}$$

    subject to
    $$\left.\begin{array}{l} x_1 + x_2 + x_3 \leq 900 \\ x_4 + x_5 + x_6 \leq 1000 \end{array}\right\} \text{capacities}$$

    $$\left.\begin{array}{l} x_1 + x_4 \geq 800 \\ x_2 + x_5 \geq 600 \\ x_3 + x_6 \geq 400 \end{array}\right\} \text{demands}$$

    $$x_1, x_2, x_3, x_4, x_5, x_6 \geq 0$$

    (b) Optimal solution:
        $$\underbrace{x_2 = 0, x_3 = 0, x_4 = 0, s_2 = 0,}_{} $$
        $$\underbrace{s_3 = 0, s_4 = 0, s_5 = 0}_{\text{Nonbasic variables}}$$

        $$\underbrace{x_1 = 800, x_5 = 600, x_6 = 400, s_1 = 100}_{\text{Basis}}$$

        $Z = 130{,}000$
        Value of objective function

    (c) Constraints 2, 3, 4, and 5 are binding. The
        slack in constraint 1 is 100 units.

    (d) The dual price for constraint 2 is zero,
        indicating that there is no advantage in
        increasing the capacity of plant 2. Plant 1
        already has excess capacity.

    (e) The dual price for constraint 5 is –60,
        indicating that an increase in the demand for
        product C will not improve the optimal cost but
        will cause it to increase by $60.

    (f) Optimal cost will improve (decrease) by $70.

    (g) Product A. Decrease of $80. Lower limit is 0.

    (h) $80 to no upper limit.

    (i) $0 to $150.

13. (a) Minimize $50x_1 + 70x_2 + 60x_3 + 90x_4$ cost function

subject to

$$x_1 + x_2 \geq 150 \}\, \text{minimum number to}$$
$$x_3 + x_4 \geq 120 \}\, \text{be interviewed by type}$$
$$x_1 + x_3 \leq 200 \}\, \text{maximum number to}$$
$$x_2 + x_4 \leq 250 \}\, \text{be interviewed by city}$$
$$x_4 \geq 20 \quad \text{minimum number of specialists in city 2}$$

$$x_1, x_2, x_3, x_4 \geq 0$$

(b) Optimal solution:

$$s_1 = 0,\ s_2 = 0,\ s_3 = 0,\ s_5 = 0$$

Nonbasic variables

$$x_1 = 100,\ x_2 = 50,\ x_3 = 100,\ x_4 = 20,\ s_4 = 180$$

Basis

$Z = \$16,300$
Value of objective function

(c) Constraints 1, 2, 3, and 5 are binding. The slack in constraint 4 is 180 interviewees.

(d) If the right–hand side of constraint 3 is increased by 1, the total interview cost will be improved (decreased) by $20. This interpretation is valid for the right–hand side between 100 and 250.

(e) If the right–hand side of constraint 2 is increased by 1, the total interview cost will be increased by $80. This interpretation is valid for the right–hand side between 70 and 220.

(f) The coefficient of x_2 can vary from $50 to $80 without changing the optimal basis.

15. (a) Let x_1 = number of dollars invested in money market funds,

x_2 = number of dollars invested in mutual funds,

x_3 = number of dollars invested in growth and income stocks,

x_4 = number of dollars invested in aggressive growth stocks.

Maximize
$0.07x_1 + 0.15x_2 + 0.18x_3 + 0.25x_4$ total return
subject to
$$x_1 + x_2 + x_3 + x_4 = 5,000,000 \quad \text{total funds}$$
$$0x_1 + 3x_2 + 5x_3 + 9x_4 \leq 4.7(5,000,000)$$
$$= 23,500,000$$
$$x_1, x_2, x_3, x_4 \geq 0$$

(b) Optimal solution:

$$x_1 = 0,\ x_3 = 0,\ s_1 = 0,\ s_2 = 0$$

Nonbasic variables

$$x_2 = 3,583,333.80,\ x_4 = 1,416,666.50$$

Basis

$Z = \$891,666.69$
Value of objective function

(c) Both constraints are binding.

(d) If the total amount invested is increased by $1, the optimum projected annual return will increase by $0.10 (dual price for constraint 1). The answer is valid for the total amount invested lying between $2,611,111 and $7,833,333.

(e) The projected rate of return for aggressive growth stocks can vary from 0.24 to 0.31 without changing the optimal basis.

17. (a) Let x_1 = number of units of product 1,
x_2 = number of units of product 2,
x_3 = number of units of product 3.

Minimize $30x_1 + 50x_2 + 50x_3$ total variable cost
subject to
$$50x_1 + 70x_2 + 40x_3 = 63,000 \text{ break--even point}$$
$$x_1 \qquad\qquad\qquad \geq \quad 200 \text{ demand}$$
$$x_2 \qquad\qquad \leq \quad 300 \text{ production limit}$$
$$x_3 \geq \quad 400 \text{ demand}$$
$$x_1, x_2, x_3 \geq \quad 0$$

(b) Optimal solution:
$$\underbrace{x_2 = 0, s_1 = 0, s_4 = 0}_{\text{Nonbasic variables}}$$

$$\underbrace{x_1 = 940, x_3 = 400, s_2 = 740, s_3 = 300}_{\text{Basis}}$$

Z = $48,200
Value of objective function

(c) Constraints 1 and 4 are binding. The surplus in constraint 2 is 740 units. The slack in constraint 3 is 300 units.

(d) If the total fixed cost is increased by $1, the optimum total variable cost will be increased by $0.60 (dual cost for constraint 1).

(e) If the right–hand side of constraint 4 is decreased by 1 unit, there will be a decrease of $26.00 in the optimal total variable cost. The answer is valid within the range of 0 to 1325 for the right–hand side.

(f) The variable cost for product 1 can vary from no lower limit to $35.71 without changing the optimal basis.

(g) The variable cost for product 3 can vary from $24 to no upper limit without changing the optimal basis.

19. (a) Let x_1 = number of units of product A,
x_2 = number of units of product B,
x_3 = number of untis of product C.

Maximize $90x_1 + 60x_2 + 70x_3$ profit function
subject to
$$3x_1 + 2x_2 + 5x_3 \leq 30,000 \quad \text{raw material 1}$$
$$2x_1 + 4x_2 + 6x_3 \leq 33,000 \quad \text{raw material 2}$$
$$4x_1 + 2x_2 + 7x_3 \leq 37,000 \quad \text{raw material 3}$$
$$x_1 \qquad\qquad\qquad \geq \quad 800$$
$$x_2 \qquad\qquad \geq \quad 400 \Big\} \text{ demand}$$
$$x_3 \geq \quad 500$$
$$x_1, x_2, x_3 \geq \quad 0$$

(b) On your computer.

(c) Optimal solution:
$$\underbrace{s_2 = 0, s_3 = 0, s_6 = 0}_{\text{Nonbasic variables}}$$

$$\underbrace{\begin{array}{l} x_1 = 6166.67, x_2 = 4416.67, x_3 = 500, \\ s_1 = 166.67, s_4 = 5366.67, s_5 = 4016.67 \end{array}}_{\text{Basis}}$$

Z = $855,000
Objective function value

(d) Constraints 2, 3, and 6 are binding. The slack in constraint 1 is 166.67 units. The surplus amounts in constraints 4 and 5 are 5366.67 and 4016.67, respectively.

(e) If additional pounds of raw material 1 become available, the maximum amount to be paid per pound is 0, since there is already a surplus.

(f) If additional pounds of raw material 2 become available, the maximum amount to be paid per pound is $5, the dual price for constraint 2. The answer is valid for up to 1000 additional pounds.

(g) Raw material 3 would yield the greatest increase in the total profit ($20 per additional pound). The limit is 250 additional pounds.

(h) The unit profit from product B can vary from $45 to $180 without changing the optimal basis.

Review Exercises

1. (a) $\underbrace{x_2 = 4, x_3 = 7, s_3 = 9}_{\text{Basic feasible solution}}$ $Z = 856$, current value of objective function

    (b) The solution is not optional, since a negative value is in the indicator row.

3. Maximize $Z = 5x_1 + 8x_2 + 6x_3$
    subject to $x_1 + x_2 + x_3 \le 30$
    $\qquad\qquad 2x_1 + x_2 + 2x_3 \le 40$
    $\qquad\qquad\quad x_1, x_2, x_3 \ge 0$

Simplex tableau:

x_1	x_2	x_3	s_1	s_2	Z	
1	①	1	1	0	0	30
2	1	2	0	1	0	40
-5	-8	-6	0	0	1	0

$(-1)R_1 + R_2 \rightarrow R_2$; $(8)R_1 + R_3 \rightarrow R_3$

1	1	1	1	0	0	30
1	0	1	-1	1	0	10
3	0	2	8	0	1	240

Solution: $\underbrace{x_2 = 30, s_2 = 10}_{\text{Basis}}$ $\underbrace{x_1 = 0, x_3 = 0, s_1 = 0}_{\text{Nonbasic variables}}$

$Z = 240$
Value of objective function

5. Maximize $Z = 6x_1 + 8x_2 + 4x_3$
    subject to $x_1 + 4x_2 + 8x_3 \le 40$
    $\qquad\qquad 2x_1 + x_2 + 6x_3 \le 30$
    $\qquad\qquad\ x_1 + x_2 + x_3 \le 16$
    $\qquad\qquad\quad x_1, x_2, x_3 \ge 0$

Simplex tableau:

x_1	x_2	x_3	s_1	s_2	s_3	Z	
1	④	8	1	0	0	0	40
2	1	6	0	1	0	0	30
1	1	1	0	0	1	0	16
-6	-8	-4	0	0	0	1	0

(continued)

5. *(continued)*

$(1/4)R_1 \rightarrow R_1$; $(-1)R_1 + R_2 \rightarrow R_2$; $(-1)R_1 + R_3 \rightarrow R_3$;
$(8)R_1 + R_4 \rightarrow R_4$

1/4	1	2	1/4	0	0	0	10
7/4	0	4	-1/4	1	0	0	20
③/④	0	-1	-1/4	0	1	0	6
-4	0	12	2	0	0	1	80

$(4/3)R_3 \rightarrow R_3$; $(-1/4)R_3 + R_1 \rightarrow R_1$;
$(-7/4)R_3 + R_2 \rightarrow R_2$; $(4)R_3 + R_4 \rightarrow R_4$

0	1	7/3	1/3	0	-1/3	0	8
0	0	19/3	1/3	1	-7/3	0	6
1	0	-4/3	-1/3	0	4/3	0	8
0	0	20/3	2/3	0	16/3	1	112

Solution: $\underbrace{x_1 = 8, x_2 = 8, s_2 = 6}_{\text{Basis}}$ $\underbrace{x_3 = 0, s_1 = 0, s_3 = 0}_{\text{Nonbasic variables}}$

$Z = 112$
Value of objective function

7. (a) Maximize $Z = 8x_1 + 4x_2$
    subject to $20x_1 + 10x_2 \le 220$
    $\qquad\qquad 10x_1 + 12x_2 \le 194$
    $\qquad\qquad\quad x_1, x_2 \ge 0$
    Simplex tableau:

x_1	x_2	s_1	s_2	Z	
⑳	10	1	0	0	220
10	12	0	1	0	194
-8	-4	0	0	1	0

$(1/20)R_1 \rightarrow R_1$; $(-10)R_1 + R_2 \rightarrow R_2$;
$(8)R_1 + R_3 \rightarrow R_3$

1	1/2	1/20	0	0	11
0	7	-1/2	1	0	84
0	7	2/5	1	1	88

Solution: $\underbrace{x_1 = 11, s_2 = 84}_{\text{Basis}}$ $\underbrace{x_2 = 0, s_1 = 0}_{\text{Nonbasic variables}}$

$Z = 88$
Value of objective function

(continued)

7. *(continued)*

(b) x_2 has a zero indicator in the tableau.

(c) To bring x_2 into th basis, use row 2 as the pivot row:

$(1/7)R_2 \to R_2;\ (-1/2)R_2 + R_1 \to R_1$

1	0	3/35	-1/14	0	5
0	1	-1/14	1/7	0	12
0	0	2/5	0	1	88

Alternate optimal solution:
$\underbrace{x_1 = 5,\ x_2 = 12}_{\text{Basis}}\ \underbrace{s_1 = 0,\ s_2 = 0}_{\text{Nonbasic variables}}$

$Z = 88$
Value of objective function

(d) Note that for the solution in part (a), constraint 2 has 84 units of slack whereas for the solution in part (c), constraint 2 has no slack. Thus, one optimal solution uses up all resource capacities and another does not.

9. Maximize $Z = 6x_1 + 4x_2$
subject to
$$\begin{aligned} x_1 + \ \ x_2 &\le 30 \\ 20x_1 + 40x_2 &\le 800 \\ 60x_1 + 15x_2 &\le 1800 \\ x_1,\ x_2 &\ge \ \ 0 \end{aligned}$$

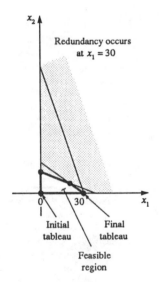

Redundancy occurs at $x_1 = 30$

0   30 x_1

Initial tableau   Final tableau

Feasible region

(continued)

9. *(continued)*

Simplex tableau:

x_1	x_2	s_1	s_2	s_3	Z	
①	1	1	0	0	0	30
20	40	0	1	0	0	800
60	15	0	0	1	0	1800
-6	-4	0	0	0	1	0

$(-20)R_1 + R_2 \to R_2$
$(-20)R_1 + R_2 \to R_2$
$\overline{(-60)R_1 + R_3 \to R_3}$
$(6)R_1 + R_4 \to R_4$

1	1	1	0	0	0	30
0	20	-20	1	0	0	200
0	-45	-60	0	1	0	0
0	2	6	0	0	1	180

$\underbrace{x_1 = 30,\ s_2 = 200,\ s_3 = 0}_{\text{Basis}}\quad \underbrace{x_2 = 0,\ s_1 = 0}_{\text{Nonbasic variables}}$

$Z = 180$
Value of objective function.

There is a degeneracy because s_3, a basic variable, is zero. The solution is an optimal degenerate solution. The feasible region is shown in Figure __.

11. (a) Minimize $Z = 5x_1 + 8x_2 + 4x_3$
subject to
$$\begin{aligned} x_1 + \ \ x_2 + \ \ x_3 &\ge 10 \\ 4x_1 + \ \ x_2 + 2x_3 &\ge 12 \\ x_1 + 2x_2 + 2x_3 &\ge 8 \\ x_1,\ x_2,\ x_3 &\ge \ \ 0 \end{aligned}$$

Dual: Maximize $Z = 10y_1 + 12y_2 + 8y_3$
subject to
$$\begin{aligned} y_1 + 4y_2 + \ \ y_3 &\le 4 \\ y_1 + \ \ y_2 + 2y_3 &\le 8 \\ y_1 + 2y_2 + 2y_3 &\le 2 \\ y_1,\ y_2,\ y_3 &\ge 0 \end{aligned}$$

(b) Simplex tableau for dual:

y_1	y_2	y_3	s_1	s_2	s_3	Z	
1	4	1	1	0	0	0	4
1	1	2	0	1	0	0	8
1	②	2	0	0	1	0	2
-10	-12	-8	0	0	0	1	0

(continued)

11. *(continued)*

$(1/2)R_3 \to R_3$; $(-4)R_3 + R_1 \to R_1$;
$(-1)R_3 + R_2 \to R_2$; $(12)R_3 + R_4 \to R_4$

-1	0	-3	1	0	-2	0	0
1/2	0	1	0	1	-1/2	0	7
⓵⁄₂	1	1	0	0	1/2	0	1
-4	0	4	0	0	6	1	12

$(2)R_3 \to R_3$; $R_3 + R_1 \to R_1$; $(-1/2)R_3 + R_2 \to R_2$;
$(4)R_3 + R_4 \to R_4$

0	2	-1	1	0	-1	0	2
0	-1	0	0	1	-1	0	6
1	2	2	0	0	1	0	2
0	8	12	0	0	10	1	20

Solution: $y_1 = 2, s_1 = 2, s_2 = 6$ $y_2 = 0, y_3 = 0, s_3 = 0$

$\underbrace{\qquad\qquad}_{\text{Basis}}$ $\underbrace{\qquad\qquad}_{\text{Nonbasic variables}}$

$Z = 20$
Value of objective function

(c) Solution of primal problem: From the indicators below s_1, s_2, and s_3, $x_1 = 0$, $x_2 = 0$, and $x_3 = 10$. Also, $Z = 20$..

(d) The surplus amounts in the primal constraints are $s_1 = 0$, $s_2 = 8$, and $s_3 = 12$.

(e) The shadow prices are the y values: 2, 0, and 0. ($i = 1, 2, 3$) If the right–hand side of constraint 1 is increased by 1 unit, then the optimal value of the objective function will be increased by 2 units.

13. (a) Minimize $Z = 96x_1 + 64x_2 + 48x_3$
  subject to $2x_1 + x_2 + 2x_3 \geq 5$
  $6x_1 + 4x_2 + x_3 \geq 7$
  $4x_1 + 2x_2 + x_3 \geq 6$
  $x_1, x_2, x_3 \geq 0$

Dual: Maximize $Z = 5y_1 + 7y_2 + 6y_3$
  subject to $2y_1 + 6y_2 + 4y_3 \leq 96$
  $y_1 + 4y_2 + 2y_3 \leq 64$
  $2y_1 + y_2 + y_3 \leq 48$
  $y_1, y_2, y_3 \geq 0$

(continued)

13. *(continued)*

(b) Simplex tableau for dual:

y_1	y_2	y_3	s_1	s_2	s_3	Z	
2	6	4	1	0	0	0	96
1	④	2	0	1	0	0	64
2	1	1	0	0	1	0	48
-5	-7	-6	0	0	0	1	0

$(1/4)R_2 \to R_2$; $R_1 - (6)R_2 \to R_1$; $R_3 - R_2 \to R_3$;
$R_4 + (7)R_2 \to R_4$

⓵⁄₂	0	1	1	-3/2	0	0	0
1/4	1	1/2	0	1/4	0	0	16
7/4	0	1/2	0	-1/4	1	0	32
-13/4	0	-5/2	0	7/4	0	1	112

$(2)R_1 \to R_1$; $R_2 - (1/4)R_1 \to R_2$;
$R_3 - (7/4)R_1 \to R_3$; $R_4 + (13/4)R_1 \to R_4$

1	0	2	2	-3	0	0	0
0	1	0	-1/2	1	0	0	16
0	0	-3	-7/2	⑤	1	0	32
0	0	4	13/2	-8	0	1	112

$(1/3)R_3 \to R_3$; $R_1 + 3(R_3) \to R_1$; $R_2 - R_3 \to R_2$;
$R_4 + 8R_3 \to R_4$

1	0	1/5	-1/10	0	3/5	0	96/5
0	1	③⁄₅	1/5	0	-1/5	0	48/5
0	0	-3/5	-7/10	1	1/5	0	32/5
0	0	-4/5	9/10	0	8/5	1	816/5

$(5/3)R_2 \to R_2$; $R_1 - (1/5)R_2 \to R_1$;
$R_3 + (3/5)R_2 \to R_3$; $R_4 + (4/5)R_2 \to R_4$

1	-1/3	0	-1/6	0	4/5	0	16
0	5/3	1	1/3	0	-1/3	0	16
0	1	0	-1/2	1	0	0	16
0	4/3	0	7/6	0	4/3	1	176

Solution: $y_1 = 16$, $y_3 = 16$, $s_2 = 16$

$\underbrace{\qquad\qquad}_{\text{Basis}}$

$y_2 = 0, s_1 = 0, s_3 = 0$

$\underbrace{\qquad\qquad}_{\text{Nonbasic variables}}$

$Z = 176$, value of objective function

(continued)

13. *(continued)*

(c) Solution of primal problem: From the indicators below s_1, s_2, and s_3, $x_1 = 7/6$, $x_2 = 0$, and $x_3 = 4/3$. Also, $Z = 176$.

(d) The surplus amounts in the primal constraints are $s_1 = 0$, $s_2 = 4/3$, and $s_3 = 0$.

(e) The shadow prices are the y values: 16, 0, and 16. If the right–hand side of either constraint 1 or 3 is increased by 1 unit, then the optimal value of the objective function will be increased by 16 units. For the second constraint, there is no effect.

15. (a) From the indicators under s_1, s_2, and s_3, $x_1 = 0$, $x_2 = 1$, and $x_3 = 3$ is the solution to the primal problem. $Z = 51$ is the value of the objective function.

(b) The surplus amounts in the primal constraints are 0, 0, and 4, respectively. These values are given in the indicator row under y_1, y_2, and y_3.

(c) The shadow prices are the y values: 3, 6, and 0. If the right–hand side of the first primal constraint is increased by 1 unit, the optimal value of the objective function will be increased by 3 units. For the second constraint a change of 1 unit will lead to a change of 6 units in the value of the objective function. For the third constraint there is no effect.

17. (a) The optimal solution is $x_1 = 3$, $x_2 = 6$, and $x_3 = 0$. $Z = 48$ is the value of the objective function.

(b) The values of the slack variables are $s_1 = 1$, $s_2 = 0$, and $s_3 = 0$. The second and third constraints are binding.

(c) The shadow prices are given in the indicator row under s_1, s_2, and s_3. They are 0, 1.333, and 2.333. An increase of 1 unit in the right–hand side of the second constraint, for example, will lead to an increase of 1.333 units in the optimal value of the objective function.

19. (a) The optimal solution is $x_1 = 16.364$, $x_2 = 14.545$, and $x_3 = 0$. $Z = 1072.72$ is the value of the objective function.

(b) The values of the slack variables are $s_1 = 22.727$, $s_2 = 0$, and $s_3 = 0$. The second and third constraints are binding.

(c) The shadow prices are given in the indicator row under s_1, s_2, and s_3. They are 0, 11.818, and 4.545. An increase of 1 unit in the right–hand side of the second constraint, for example, will lead to an increase of 11.818 units in the optimal value of the objective function.

21. Maximize
$$Z = 19x_1 + 17x_2 + 15x_3 + 0s_1 + 0s_2 - Ma_1 - Ma_2$$
subject to
$$5x_1 + x_2 + 5x_3 - s_1 + a_1 \qquad\qquad = 15$$
$$3x_1 + 6x_2 + 8x_3 \qquad\quad + s_2 \qquad = 90$$
$$2x_1 + 5x_2 + 8x_3 \qquad\qquad\quad + a_2 = 180$$
$$x_1,\ x_2,\ x_3,\ s_1,\ s_2,\ a_1,\ a_2 \geq 0$$

23. Minimize
$$Z = 5x_1 + 9x_2 + 8x_3 + 0s_1 + 0s_2 + Ma_1 + Ma_2$$
subject to
$$x_1 + 5x_2 + 8x_3 + s_1 \qquad\qquad = 100$$
$$x_1 + 15x_2 + 8x_3 \qquad + a_1 \qquad = 160$$
$$2x_1 + 3x_2 + 9x_3 \quad - s_2 \quad + a_2 = 50$$
$$x_1,\ x_2,\ x_3,\ s_1,\ s_2,\ a_1,\ a_2 \geq 0$$

25. Minimize $Z = 8x_1 + 4x_2 + 6x_3$
subject to
$$3x_1 + x_2 + 4x_3 \geq 20$$
$$x_1 + 4x_2 + x_3 = 70$$
$$2x_1 + x_2 + 8x_3 \leq 160$$
$$x_1, x_2, x_3 \geq 0$$

(a) Optimal solution: $x_1 = 0$, $x_2 = 17.333$, $x_3 = 0.667$.

$Z = 73.333$
Value of objective function

(continued)

25. *(continued)*

(b) The surplus amount for constraint 1 is 0. The constraint is binding. The slack for constraint 3 is 13.7333 units.

(c) The dual prices are −1.333, −0.667, and 0. An increase of 1 unit in the right–hand side of either constraint 1 or constraint 2 will not improve the optimal value of the objective function but will lead to an increase of 1.333 units or 0.667 units, respectively. Constraint 3 is not binding.

27. Maximize $Z = 7x_1 + 5x_2 + 9x_3$
subject to $9x_1 + 5x_2 + x_3 \le 50$
$3x_1 + x_2 + 2x_3 \ge 20$
$x_1 + 3x_2 + 4x_3 = 60$
$x_1, x_2, x_3 \ge 0$

(a) Optimal solution: $x_1 = 4$, $x_2 = 0$, $x_3 = 14$
$Z = 154$
Value of objective function
Surplus/slack variables: $s_1 = 0$, $s_2 = 20$, $s_3 = 0$

(b) Constraints 1 and 3 are binding.

(c) The dual prices are 0.543, 0, and 2.11. An increase of 1 unit in the right–hand constraint 1 or 3 will increase the optimal value of the objective function by 0.543 or 2.11, respectively.

29. (a)

Variable	Range of Coefficient
x_1	4.66667 – no upper limit
x_2	no lower limit – 24
x_3	1 – 10.54545

Within these ranges, taken individually, the basis for the optimal solution will not change. The value of the objective function may change.

(continued)

29. *(continued)*

(b)

Constraint	Range of Right–Hand Side
1	17.5 – 86.45161
2	5 – 80
3	22.66667 – no upper limit

Within these ranges, taken individually, an increase of 1 unit in the right–hand side will improve the value of the objective function by the amount of the corresponding shadow price if the shadow price is positive. If the shadow price is negative, the value of the objective function will not be improved. Within these ranges, the optimal basis remains feasible and the interpretation of the dual prices remains valid.

31. (a) The added constraint is $x_3 \ge 2x_2$, or $2x_2 - x_3 \le 0$.

(b) Optimal solution: $x_1 = 0$, $x_2 = 57.14$, $x_3 = 114.29$
$Z = \$31,428.57$
Value of objective function
Since integral values are appropriate, we could produce 57 model MCW02's and 114 MCW03's.

(c) The values of the slack variables are $s_1 = 328.57$, $s_2 = 0$, $s_3 = 257.14$, $s_4 = 0$. Constraints 2 and 4 are binding.

(d) The dual prices are 0, 78.57, 0, and 35.71. An increase of 1 unit in the number of component A units available will increase the optimal profit by $78.57.

(e) The coefficients of x_1, x_2 and x_3 can vary from no lower limit to 157.14, 66.67 to no upper limit and 100 to 450, respectively. Within these individual ranges, the optimal values of the basic variables will not change.

(f) The ranges for the right–hand side of constraints 1, 2, 3, and 4 are 171.43 to no upper limit, 0 to 700, 342.86 to no upper limit, and −133.33 to 450, respectively. Within each range, taken individually, the optimal basis still contains x_1, x_2, x_3, and s_3 and the interpretation of the dual prices remains valid. The values of the basic variables may change.

33.   (a) Let x_1 = number of gallons of crude 1 used in regular gasoline,

x_2 = number of gallons of crude 1 used in premium gasoline,

x_3 = number of gallons of crude 2 used in regular gasoline,

x_4 = number of gallons of crude 2 used in premium gasoline.

Minimize
$0.20x_1 + 0.20x_2 + 0.25x_3 + 0.25x_4$ cost function
subject to
$0.15x_1 + 0.45x_3 \geq 0.30(x_1 + x_3)$ Component A
$0.5x_2 + 0.4x_4 \leq 0.45(x_2 + x_4)$ Component B
Rewrite these constraints as
$-0.15x_1 + 0.15x_3 \geq 0$
$-0.05x_2 + 0.05x_4 \geq 0$
Also, $x_1 + x_3 \geq 900,000$
$x_2 + x_4 \geq 1,200,000$ $\Big\}$ demands

(b) Optimal solution: $x_1 = 450,000, x_2 = 600,000,$
$x_3 = 450,000, x_4 = 600,000$
$Z = \$472,500$
Value of objective function

(c) The values of the surplus quantities are all zero, indicating that all constraints are binding.

(d) The dual prices are $-1/6$, $-1/2$, -0.225, and -0.225. An increase of 1 unit in the right–hand side of any of the constraints will not improve the optimal solution but will increase it by the amount of the dual price.

(e) The coefficients of $x_1, x_2, x_3,$ and x_4 can vary from -0.25 to 0.25, -0.25 to 0.25, 0.20 to no upper limit, and 0.20 to no upper limit, respectively. Within these individual ranges, the optimal values of the basic variables will not change.

(f) The ranges for the right–hand side of constraints 1, 2, 3, and 4 are $-135,000$ to $135,000$, $-60,000$ to $60,000$, 0 to no upper limit, and 0 to no upper limit, respectively. Within each range, taken individually, the optimal basis still contains $x_1, x_2, x_3,$ and x_4 and the interpretation of the dual prices remains valid. Tthe values of the basic variables may change.

CHAPTER

8

PROBABILITY

8–1 Sets

1. (a) True   (b) True

    (c) True   (d) True

    (e) True   (f) False

    (g) False   (h) False

3. $A = \{7, 9, 10\}$, $B = \{7, 10, 9\}$

    (a) $A \subset B$ (b) $B \subset A$ (c) $A = B$
       yes   yes   yes

For exercises 5 – 15, $U = \{1, 2, 3, 4, 5, 6, 7, 8, 9\}$,
                           $A = \{2, 3, 4, 5,\}$, and
                           $B = \{4, 5, 6, 7\}$

5. $A \cap B = \{4, 5\}$

7. $n(B) = 4$

9. $A \cap B' = A \cup \{1, 2, 3, 8, 9\}$
                $= \{2, 3\}$

11. $(A \cap B)' = \{1, 2, 3, 6, 7, 8, 9\}$

13. $A' \cup B' = (A \cap B)' = \{1, 2, 3, 6, 7, 8, 9\}$

15. $n(B') = 5$

17.

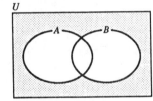

$(A \cup B)'$

19.

$(A \cap B)'$

169

21.

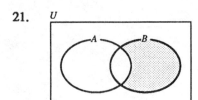

$$A' \cap B$$

23.

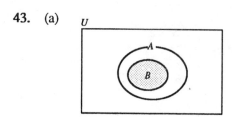

$$B' \cap A$$

For exercises 25 – 41, U = {a, b, c, d, e, f, g, h, i, j},
   A = {a, c, d, g, j}, B = {a, b, g, j},
   C = {a, b, c, g, i, j}, D = {a, d, f, g, i, j}

25. $A \cap B$ = {a, g, j}

27. $C \cap D$ = {a, g, i, j}

29. A' = {b, e, f, h, i}

31. $A' \cap B$ = {b}

33. $D' \cup A$ = {b, c, e, h} $\cup$ {a, c, d, g, j}
          = {a, b, c, d, e, g, h, j}

35. $A \cup C \cup D$ = {a, b, c, d, f, g, i, j}

37. $A \cup (B \cap D)$ = {a, c, d, g, j} $\cup$ {a, g, j}
                  = {a, c, d, g, j}

39. $A \cup (C' \cap D)$ = {a, c, d, g, j} $\cup$
                  ({d, e, f, h} $\cap$ {a, d, f, g, i, j})
                  = {a, c, d, g, j} $\cup$ { d, f}
                  = {a, c, d, f, g, j}

41. $(A \cap B) \cup (C \cap D)$ = {a, g, j} $\cup$ {a, g, i, j}
                  = {a, g, i, j}

43.   (a)

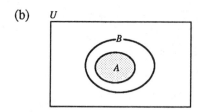

$$A \cap B$$

   (b)

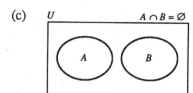

$$A \cap B$$

   (c)

$$A \cap B = \phi$$

(continued)

43. *(continued)*

(d)

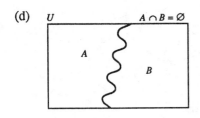

$A \cap B = \phi$

45. (a)

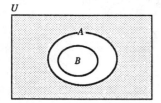

A′

(b)

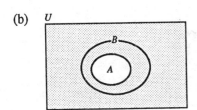

A′

(c)

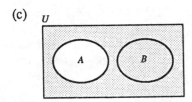

A′

(continued)

45. *(continued)*

(d)

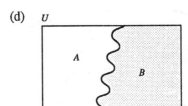

A′

47.

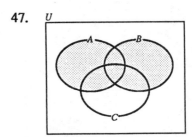

A ∪ B

49.

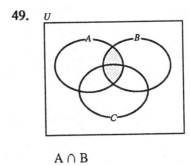

A ∩ B

51.

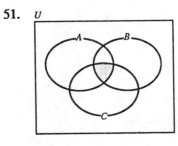

A ∩ B ∩ C

53.

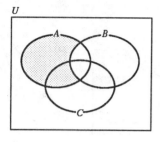

$A \cap B'$

55.

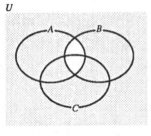

$(A \cap B)'$

57.

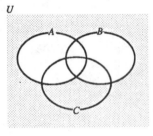

$A' \cap B'$

59.

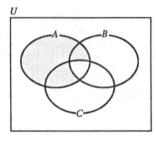

$A \cap (B \cap C)'$

61.

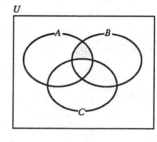

$C' \cap (A \cap B)$

63.

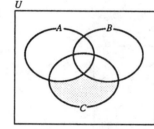

$(A \cup B)' \cap C$

65.

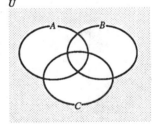

$(A \cup B \cup C)'$

67. 100 patients, of whom 40 need psychological counseling, 70 need medical care, and 25 need both. Let P = set of patients needing psychological counseling and M = set of patients needing medical care.

(continued)

67. *(continued)*

(a) n(U) = 100

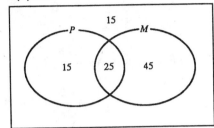

n(U) = 100

(b) n(P′) = 100 – 40 = 60

(c) n(M′) = 100 – 70 = 30

(d) n(P ∩ M)′ = 100 – 25 = 75

(e) n(P U M) = 85

(f) n(P ∩ M′) = 15

(g) n(P′ ∩ M) = 45

(h) n(P ∪ M)′ = 15

69. 200 physicians, of whom 80 are family practitioners, 130 are specialists, and 40 are both. Let F = set of family practitioners and S = set of specialists.

(a) n(U) = 200

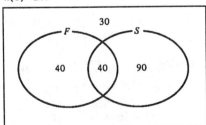

n(U) = 200

(continued)

69. *(continued)*

(b) n(F ∪ S) = 170

(c) n(F′ ∩ S′) = 30

(d) n(F ∩ S′) = 40

(e) n(F′ ∩ S) = 90

(f) n(S′) = 200 – 130 = 70

(g) $\dfrac{n(F \cap S)}{n(S)} \times 100 = \dfrac{40}{130} \times 100 = 30.77\%$

(h) $\dfrac{n(F \cap S)}{n(F)} \times 100 = \dfrac{40}{80} \times 100 = 50\%$

71. 500 families, of which 300 had savings accounts, 200 had mutual fund accounts, 115 had brokerage accounts, 150 had both savings and mutual fund accounts, 70 had both savings and brokerage accounts, 60 had both brokerage and mutual fund accounts, and 20 had all three types of accounts. Let S = set of families with savings accounts, M = set of families with mutual fund accounts, and B = set of families with brokerage accounts.

(a) n(U) = 500

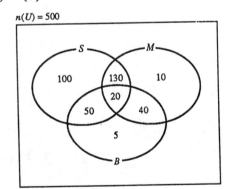

n(U) = 500

(b) n(S′ ∩ M′ ∩ B′) = 500 – 355 = 145

(c) n((S ∩ B) ∩ M′) = 70 – 20 = 50

(continued)

71. *(continued)*

(d) $n((S \cap M) \cap B') = 150 - 20 = 130$

(e) $n(S \cap B' \cap M') = 300 - 50 - 130 - 20 = 100$

(f) $n(M \cap S' \cap B') = 200 - 130 - 40 - 20 = 10$

(g) $n(S \cup M) = n(S) + n(M) - n(S \cap M)$
$= 300 + 200 - 150 = 350$

73.

	A	B	
C	90	100	190
D	60	150	210
	150	250	400

(a) $n(D) = 210$

(b) $n(A \cap C) = 90$

(c) $n(B \cap D) = 150$

(d) $n(A \cup C) = n(A) + n(C) - n(A \cap C)$
$= 150 + 190 - 90 = 250$

(e) $n(B \cup D) = n(B) + n(D) - n(B \cap D)$
$= 250 + 210 - 150 = 310$

(f) $n(B') = 400 - 250 = 150$

(g) $n(A \cap B) = n(\phi) = 0$

(h) $n(D') = 400 - 210 = 190$

(i) $n(C \cap D) = n(\phi) = 0$

(j)   % B's that are also C's $= \dfrac{n(B \cap C)}{n(B)} \times 100$

$= \dfrac{100}{250} \times 100 = 40\%$

(continued)

73. *(continued)*

(k)   % A's that are also D's $= \dfrac{n(A \cap D)}{n(A)} \times 100$

$= \dfrac{60}{150} \times 100 = 40\%$

(l)   % C's that are also A's $= \dfrac{n(C \cap A)}{n(C)} \times 100$

$= \dfrac{90}{190} \times 100 = 47.4\%$

75.

Totals	D	L	B	Totals
J	6	3	1	10
M	18	7	15	40
Totals	24	10	16	50

(a) $n(D) = 24$

(b) $n(L) = 10$

(c) $n(B) = 16$

(d) $n(D \cap J) = 6$

(e) $n(D \cap M) = 18$

(f) $n(L \cap J) = 3$

(g) $n(D \cup J) = n(D) + n(J) - n(D \cap J)$
$= 24 + 10 - 6 = 28$

(h) $n(B \cup M) = n(B) + n(M) - n(B \cap M)$
$= 16 + 40 - 15 = 41$

(i) $n(L \cup J) = n(L) + n(J) - n(L \cap J)$
$= 10 + 10 - 3 = 17$

(j) $n(B^1) = 50 - 16 = 34$

(k) $n(J^1) = 50 - 10 = 40$

(continued)

75. *(continued)*

(l) $\begin{aligned} n(D^1 \cap J) &= n(L \cap J) + n(B \cap J) \\ &= 3 + 1 = 4 \end{aligned}$

(m) % of singles who are doctors $= \dfrac{6}{10} \times 100$
$= 60\%$

(n) % doctors who are single $= \dfrac{6}{24} \times 100 = 25\%$

(o) % lawyers who are married $= \dfrac{7}{10} \times 100 = 70\%$

(p) % married who are lawyers $= \dfrac{7}{40} \times 100$
$= 17.5\%$

77. u

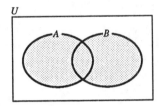

$A \cup B$ is the set that is shaded in Figure 8–21. The overlapping region includes those elements that are common to sets A and B. The simple count of $n(A) + n(B)$ counts such elements twice. To correct the count, subtract the numbers of elements in common, i.e., $n(A \cap B)$. Hence, $n(A \cup B) = n(A) + n(B) - n(A \cap B)$.

79. $\begin{aligned} n(A \cup B) &= n(A) + n(B) - n(A \cap B) \\ &= 28 + 20 - 5 = 43 \end{aligned}$

81. $n(A \cup B) = 45 + 22 - 0 = 67$

83. (a) $n(A \cup B) = 30 + 18 - 11 = 37$

(b) $\begin{aligned} n[(A \cup B)'] &= n(U) - n(A \cup B) = 70 - 37 \\ &= 33 \end{aligned}$

85. (a) $n(A' \cap C') = n[(A \cup C)'] = 500 - 410 = 90$

(b) $n(D' \cap B') = n[(D \cup B)'] = 500 - 370 = 130$

(c) $\begin{aligned} n(A' \cap D') &= n[(A \cup D)'] \\ &= 500 - (300 + 260 - 170) \\ &= 110 \end{aligned}$

(d) $n(D' \cap C') = n[(D \cup C)'] = 500 - 500 = 0$

87. (a) $n(E' \cap A') = n[(E \cup A)'] = 200 - 120 = 80$

(b) $n(B' \cap G') = n[(B \cup B)'] = 200 - 140 = 60$

(c) $\begin{aligned} n(F' \cap A') &= n[(F \cup A)'] \\ &= 200 - (60 + 80 - 50) = 110 \end{aligned}$

8–2 Counting, Permutations, and Combinations

1. $4! = 4 \cdot 3 \cdot 2 \cdot 1 = 24$

3. $1! = 1$

5. $6! = 6 \cdot 5 \cdot 4 \cdot 3 \cdot 2 \cdot 1 = 720$

7. $\dfrac{4!}{0!} = 4! = 24$

9. $\dfrac{10!}{3! \; 7!} = \dfrac{10 \cdot 9 \cdot 8 \cdot 7!}{3! \; 7!} = \dfrac{10 \cdot 9 \cdot 8}{3 \cdot 2 \cdot 1} = 120$

11. $\dfrac{10!}{4!(10 - 4)!} = \dfrac{10!}{4! \; 6!} = \dfrac{10 \cdot 9 \cdot 8 \cdot 7 \cdot 6!}{4! \; 6!}$

$= \dfrac{10 \cdot 9 \cdot 8 \cdot 7}{4 \cdot 3 \cdot 2 \cdot 1} = 210$

13. $\dfrac{18!}{0!(18-0)!} = \dfrac{18!}{(1)(18!)} = 1$

15. $\dfrac{20!}{2!(20-2)!} = \dfrac{20 \cdot 19 \cdot 18!}{2! \ 18!} = \dfrac{20 \cdot 19}{2} = 190$

17. Number of different ways $= 6 \cdot 4 = 24$

19. Number of sets $= 5^{10} = 9{,}765{,}625$

21. Number of personality types $= 6^4 = 1296$

23. (a) $26^2 \cdot 10^3 = 676{,}000$

    (b) $26 \cdot 25 \cdot 10^3 = 650{,}000$

    (c) $26^2 \cdot 10 \cdot 9 \cdot 8 = 486{,}720$ (the three digits must all be different)

25. Number of diets $= 4 \cdot 5 \cdot 6 \cdot 3 = 360$

27. $P(5,3) = \dfrac{5!}{(5-3)!} = \dfrac{5!}{2!} = 5 \cdot 4 \cdot 3 = 60$

29. $P(7,6) = \dfrac{7!}{(7-6)!} = \dfrac{7!}{1!} = 5040$

31. $P(6,4) = \dfrac{6!}{(6-4)!} = \dfrac{6!}{2!} = 6 \cdot 5 \cdot 4 \cdot 3 = 360$

33. $C(5,2) = \dfrac{5!}{2!(5-2)!} = \dfrac{5!}{2! \ 3!} = \dfrac{5 \cdot 4}{2 \cdot 1} = 10$

35. $C(9,3) = \dfrac{9!}{3!(9-3)!} = \dfrac{9!}{3! \ 6!} = \dfrac{9 \cdot 8 \cdot 7}{1 \cdot 2 \cdot 3} = 84$

37. $C(8,6) = \dfrac{8!}{6!(8-6)!} = \dfrac{8!}{6! \ 2!} = 28$

39. $C(8,8) = \dfrac{8!}{8!(8-8)!} = 1$

41. $C(4,4) = \dfrac{4!}{4!(4-4)!} = 1$

43. $\begin{bmatrix} 6 \\ 3 \end{bmatrix} = \dfrac{6!}{3!(6-3)!} = \dfrac{6!}{3! \ 3!} = \dfrac{6 \cdot 5 \cdot 4}{3 \cdot 2 \cdot 1} = 20$

45. $\begin{bmatrix} 5 \\ 5 \end{bmatrix} = \dfrac{5!}{5!(5-5)!} = 1$

47. $\begin{bmatrix} 9 \\ 2 \end{bmatrix} = \dfrac{9!}{2!(9-2)!} = \dfrac{9!}{2! \ 7!} = \dfrac{9 \cdot 8}{2 \cdot 1} = 36$

49. $C(5,1) \cdot C(4,2) = \dfrac{5!}{1!(5-1)!} \cdot \dfrac{4!}{2!(4-2)!}$

    $= \dfrac{5!}{4!} \cdot \dfrac{4!}{2! \ 2!} = \dfrac{5!}{4} = 30$

51. $\begin{bmatrix} 9 \\ 7 \end{bmatrix} \cdot \begin{bmatrix} 5 \\ 3 \end{bmatrix} = \dfrac{9!}{7!(9-7)!} \cdot \dfrac{5!}{3!(5-3)!}$

    $= \dfrac{9!}{7! \ 2!} \cdot \dfrac{5!}{3! \ 2!}$

    $= \dfrac{9 \cdot 8}{2 \cdot 1} \cdot \dfrac{5 \cdot 4}{2 \cdot 1} = 360$

53. $\begin{bmatrix} 10 \\ 6 \end{bmatrix} \cdot \begin{bmatrix} 10 \\ 4 \end{bmatrix} = \dfrac{10!}{6!(10-6)!} \cdot \dfrac{10!}{4!(10-4)!}$

    $= \dfrac{10!}{6! \ 4!} \cdot \dfrac{10!}{4! \ 6!}$

    $= \dfrac{10 \cdot 9 \cdot 8 \cdot 7}{4 \cdot 3 \cdot 2 \cdot 1} \cdot \dfrac{10 \cdot 9 \cdot 8 \cdot 7}{4 \cdot 3 \cdot 2}$

    $= 44{,}100$

55. $\begin{bmatrix} 13 \\ 6 \end{bmatrix} \cdot \begin{bmatrix} 12 \\ 10 \end{bmatrix} = \dfrac{13!}{6!(13-6)!} \cdot \dfrac{12!}{10!(12-10)!}$

$\qquad = \dfrac{13!}{6!\ 7!} \cdot \dfrac{12!}{10!\ 2!}$

$\qquad = \dfrac{13 \cdot 12 \cdot 11 \cdot 10 \cdot 9 \cdot 8}{6 \cdot 5 \cdot 4 \cdot 3 \cdot 2 \cdot 1} \cdot \dfrac{12 \cdot 11}{2 \cdot 1}$

$\qquad = 113{,}256$

57. $P(5,5) = \dfrac{5!}{(5-5)!} = 5! = 120$ ways

59. $P(5,3) = \dfrac{5!}{(5-3)!} = \dfrac{5!}{2!} = 5 \cdot 4 \cdot 3 = 60$ ways

61. $C(5,3) = \dfrac{5!}{3!(5-3)!} = \dfrac{5!}{3!\ 2!} = \dfrac{5 \cdot 4}{2 \cdot 1} = 10$ ways

63. $C(5,2) = \dfrac{5!}{2!(5-2)!} = \dfrac{5!}{2!\ 3!} = \dfrac{5 \cdot 4}{2 \cdot 1} = 10$ lines

65. $P(9,6) = \dfrac{9!}{(9-6)!} = \dfrac{9!}{3!} = 9 \cdot 8 \cdot 7 \cdot 6 \cdot 5 \cdot 4$

$\qquad = 60{,}480$ possibilities

67. $C(52,13) = \dfrac{52!}{13!\ (52-13)!} = \dfrac{52!}{13!\ 39!}$

$\qquad = 635{,}013{,}559{,}600$ hands

69. $C(7,3) = \dfrac{7!}{3!\ (7-3)!} = \dfrac{7!}{3!\ 4!} = \dfrac{7 \cdot 6 \cdot 5}{3 \cdot 2 \cdot 1}$

$\qquad = 35$ teams

71. $P(10,4) = \dfrac{10!}{(10-4)!} = \dfrac{10!}{6!} = 10 \cdot 9 \cdot 8 \cdot 7$

$\qquad = 5040$ possibilities

73. $P(52,5) = \dfrac{52!}{5!(52-5)!} = \dfrac{52!}{5!\ 47!}$

$\qquad = \dfrac{52 \cdot 51 \cdot 50 \cdot 49 \cdot 48}{5 \cdot 4 \cdot 3 \cdot 2 \cdot 1}$

$\qquad = 2{,}598{,}960$ hands

75. $C(10,3) \cdot C(9,2) = \dfrac{10!}{3!\ 7!} \cdot \dfrac{9!}{2!\ 7!} = 120 \cdot 36$

$\qquad = 4320$ ways

77. (a) The first digit has 9 possible values; the other six have ten possible values.
$9 \cdot 10^6 = 9{,}000{,}000$ numbers

(b) $9 \cdot P(9,6) = 9 \cdot \dfrac{9!}{(9-6)!} = 9 \cdot \dfrac{9!}{3!}$

$\qquad = 544{,}320$ numbers

79. (a) If repetitions are allowed, number of codes
$= 10^5 = 100{,}000$ (assuming that 00000 is acceptable)

(b) Without repititions, number of codes
$= P(10,5) = \dfrac{10!}{5!} = 30{,}240$

81. $C(4,3) \cdot C(48,10) = \dfrac{4!}{3!\ 1!} \cdot \dfrac{48!}{10!\ 38!}$

$= 4 \cdot \dfrac{48 \cdot 47 \cdot 46 \cdot 45 \cdot 44 \cdot 43 \cdot 42 \cdot 41 \cdot 40 \cdot 39}{10 \cdot 9 \cdot 8 \cdot 7 \cdot 6 \cdot 5 \cdot 4 \cdot 3 \cdot 2 \cdot 1}$

$= 26{,}162{,}863{,}584$

$C(4,4) \cdot C(48,9)$

$= 1 \cdot \dfrac{48 \cdot 47 \cdot 46 \cdot 45 \cdot 44 \cdot 43 \cdot 42 \cdot 41 \cdot 40}{9 \cdot 8 \cdot 7 \cdot 6 \cdot 5 \cdot 4 \cdot 3 \cdot 2 \cdot 1}$

$= 1{,}677{,}106{,}640$

Number of hands with at least two queens $=$
$C(4,2) \cdot C(48,11) + C(4,3) \cdot C(48,10)$
$\qquad + C(4,4) \cdot C(48,9)$
$\qquad - 163{,}411{,}172{,}432$

83. (a) $C(30,8) = \dfrac{30!}{8!\,(30-8)!} = \dfrac{30!}{3!\,22!}$

$= \dfrac{30 \cdot 29 \cdot 28 \cdot 27 \cdot 26 \cdot 25 \cdot 24 \cdot 23}{8 \cdot 7 \cdot 6 \cdot 5 \cdot 4 \cdot 3 \cdot 2 \cdot 1}$

$= 5{,}852{,}925$ samples

(b) $C(26,7) \cdot C(4,1) = \dfrac{26!}{7!\,19!} \cdot 4$

$= 4 \cdot \dfrac{26 \cdot 25 \cdot 24 \cdot 23 \cdot 22 \cdot 21 \cdot 20}{7 \cdot 6 \cdot 5 \cdot 4 \cdot 3 \cdot 2 \cdot 1}$

$= 5{,}262{,}400$ samples

(c) $C(26,6) \cdot C(4,2) = \dfrac{26!}{6!\,20!} \cdot \dfrac{4!}{2!\,2!}$

$= 6 \cdot \dfrac{26 \cdot 25 \cdot 24 \cdot 23 \cdot 22 \cdot 21}{6 \cdot 5 \cdot 4 \cdot 3 \cdot 2 \cdot 1}$

$= 1{,}381{,}380$

(d) $C(26,7) \cdot C(4,1) + C(26,6) \cdot C(4,2)$
$\quad + C(26,5) \cdot C(4,3) + C(26,4) \cdot C(4,4)$

$= 5{,}262{,}400 + 1{,}381{,}380 + \dfrac{26!}{5!\,21!} \cdot \dfrac{4!}{3!\,1!}$

$\quad + \dfrac{26!}{4!\,22!} \cdot 1$

$= 6{,}643{,}780 + 263{,}120 + 14{,}950$
$= 6{,}921{,}850$ samples

(e) $C(26,8) + C(26,7) \cdot C(4,1) + C(26,6) \cdot C(4,2)$

$= \dfrac{26!}{8!\,18!} + 5{,}262{,}400 + 1{,}381{,}380$

$= 1{,}562{,}275 + 6{,}643{,}780 = 8{,}206{,}055$ samples

85. $C(100,12) = \dfrac{100!}{12!\,88!}$

$= \dfrac{100 \cdot 99 \cdot 98 \cdot 97 \cdot 96 \cdot 95 \cdot 94 \cdot 93 \cdot 92 \cdot 91 \cdot 90 \cdot 89}{12 \cdot 11 \cdot 10 \cdot 9 \cdot 8 \cdot 7 \cdot 6 \cdot 5 \cdot 4 \cdot 3 \cdot 2 \cdot 1}$

$\approx 1.050421 \times 10^{15}$ committees

8–3 Probability; Addition and Complement Rules; Odds

1.   (a)   S = {mo, mn, me, my, om, on, oe, oy, nm, no, ne, ny, em, eo, en, ey, ym, yo, yn, ye}

(b)   Probability of getting "o" on first selection
$= \dfrac{n(E)}{n(S)} = \dfrac{4}{20} = \dfrac{1}{5}$, or 20%

(c)   Probability of getting "n" on the second selection $= \dfrac{n(E)}{n(S)} = \dfrac{4}{20} = \dfrac{1}{5}$, or 20%

(d)   Probability of getting same letter twice
$= \dfrac{n(E)}{n(S)} = \dfrac{0}{20} = 0$

3.   (a)   S = {pn, pd, pq, np, nd, nq, dp, dn, dq, qp, qn, qd}

(b) $P(E) = \dfrac{n(E)}{n(S)} = \dfrac{2}{12} = \dfrac{1}{6}$, or 16.7% where the event is that the sum of the two coins is 35 cents.

(c)   Probability of getting the same coin twice
$= \dfrac{n(E)}{n(S)} = \dfrac{0}{12} = 0$

5.   (a)   {hhhh, hhht, hhth, hhtt, hthh, htht, htth, httt, thhh, thht, thth, thtt, tthh, thth, ttth, tttt}

(b) $P(3 \text{ heads}) = \dfrac{n(E)}{n(S)} = \dfrac{4}{16} = \dfrac{1}{4}$, or 25%

(c) $(\leq 3 \text{ heads}) = \dfrac{n(E)}{n(S)} = \dfrac{15}{16}$, or 93.75%

(d) $P(\geq 3 \text{ heads}) = \dfrac{n(E)}{n(S)} = \dfrac{5}{16}$, or 31.25%

(e) $P(4 \text{ heads or 4 tails}) = \dfrac{n(E)}{n(S)} = \dfrac{2}{16} = \dfrac{1}{8}$, or 12.5%

For Exercises 7 – 11, there are 100 applicants, of whom 65 are white, 15 black, 10 Hispanic, 5 Asian, and 5 American Indian. One applicant is selected.

7. $P(\text{non–white}) = \dfrac{100 - 65}{100} = \dfrac{35}{100}$, or 35%

9. $P(\text{non–black}) = \dfrac{100 - 15}{100} = \dfrac{85}{100}$, or 85%

11. $P(\text{Asian or American Indian}) = P(\text{Asian}) +$
     $P(\text{American Indian}) = \dfrac{5}{100} + \dfrac{5}{100} = \dfrac{10}{100}$, or 10%

For Exercises 13 – 19, two fair dice are rolled. Use the sample space S given in Example 8–31.

13. $P(\text{3 on second die}) = \dfrac{n(E)}{n(S)} = \dfrac{6}{36} = \dfrac{1}{6}$, or 16.7%

15. $P(\text{even number on first die and odd number on second die}) = \dfrac{n(E)}{n(S)} = \dfrac{9}{36} = \dfrac{1}{4}$, or 25%

17. $P(\text{a sum of 13}) = 0$

19. $P(\text{a sum not equal to 7}) = 1 - P(\text{a sum of 7})$
     $= 1 - 0.167 = 0.833$, or 83.3%

For Exercises 21 – 29, a card is selected from an ordinary deck of 52 cards.

21. $P(\text{a queen}) = \dfrac{n(E)}{n(S)} = \dfrac{4}{52} = \dfrac{1}{13}$, or 7.69%

23. $P(\text{not a red card}) = \dfrac{n(E)}{n(S)} = \dfrac{26}{52} = \dfrac{1}{2}$, or 50%

25. $P(\text{not a jack}) = 1 - \dfrac{1}{13} = \dfrac{12}{13}$, or 92.31%

27. $P(\text{the ace of spades}) = \dfrac{n(E)}{n(S)} = \dfrac{1}{52}$, or 1.92%

29. $P(\text{either a jack or a queen}) = P(\text{a jack}) +$
     $P(\text{a queen}) = \dfrac{4}{52} + \dfrac{4}{52} = \dfrac{2}{13}$, or 15.38%

31. $P(E) = \dfrac{5}{5 + 3} = \dfrac{5}{8}$, or 62.5%

33. $P(E) = \dfrac{3}{3 + 8} = \dfrac{3}{11}$, or 27.3%

35. $P(E) = \dfrac{2}{2 + 5} = \dfrac{2}{7}$, or 28.6%

37. Odds in favor of $E = \dfrac{P(E)}{P(E')} = \dfrac{0.80}{1 - 0.80} = \dfrac{0.80}{0.20}$
     $= 4$ to 1

39. Odds in favor of $E = \dfrac{P(E)}{P(E')} = \dfrac{5/9}{1 - 5/9} = \dfrac{5}{4} = 5$ to 4

41. Odds in favor of $E = \dfrac{P(E)}{P(E')} = \dfrac{6/7}{1 - 6/7} = \dfrac{6}{1} = 6$ to 1

43. $P(\text{Democratic victory}) = \dfrac{2}{2 + 5} = \dfrac{2}{7}$, or 28.6%

45. Odds in favor of snow $= \dfrac{0.70}{1 - 0.70} = \dfrac{7}{3} = 7$ to 3

47. Odds in favor of growth in consumer spending
     $= \dfrac{0.10}{1 - 0.10} = \dfrac{1}{9} = 1$ to 9

49. The probability distribution is in error, since $e_2 = -0.2$, a negative probability, is not possible.

51. The probability distribution is in error, since the sum of the probabilities > 1.0.

53. The probability is in error, since the sum of the probabilities is greater than 1.0.

55. (a) $P(\{e_1, e_3\}) = .10 + .25 = .35$

(b) $P(\{e_1, e_4, e_5\}) = .10 + .30 + .20 = .60$

57. (a) Divide each value by the total in the sample, 200.

Opinion	Probability
strongly in favor	.20
somewhat in favor	.50
neutral	.05
somewhat opposed	.10
strongly opposed	.15

(b) $P(\text{strongly in favor}) = .20$

(c) $P(\text{opposed}) = .10 + .15 = .25$

(d) $P(\text{in favor}) = .20 + .50 = .70$

59. (a) $S = \{(1,1), (1,2), (1,3), (1,4), (1,5), (1,6), (2,1), (2,2), (2,3), (2,4), (2,5), (2,6), (3,1), (3,2), (3,3), (3,4), (3,5), (3,6), (4,1), (4,2), (4,3), (4,4), (4,5), (4,6), (5,1), (5,2), (5,3), (5,4), (5,5), (5,6), (6,1), (6,2), (6,3), (6,4), (6,5), (6,6)\}$

(b) $S_1 = \{(1,1), (1,2), (1,3), (1,4), (2,1), (2,2), (2,3), (3,1), (3,2), (4,1)\}$

$P(A) = 10/36$, or 27.8%

(c) $S_2 = \{(1,1), (2,2), (3,3), (4,4), (5,5), (6,6)\}$
$P(B) = 6/36 = 1/6$, or 16.7%

(continued)

59. (continued)

(d) The events are not mutually exclusive, since (1,1) and (2,2) occur in both S_1 and S_2.

(e) $P(A \text{ or } B) = P(A) + P(B) - P(A \cap B)$
$= 10/36 + 6/36 - 2/36 = 14/36$, or 38.9%

(f) $S_3 = \{(1,5), (2,4), (3,3), (4,2), (5,1)\}$
$P(C) = 6/36$, or 16.7%

(g) $S_4 = \{(3,6), (4,5), (5,4), (6,3)\}$
$P(D) = 4/36$, or 11.1%

(h) Events C and D are mutually exclusive, since sets S_3 and S_4 are disjoint.

(i) $P(C \text{ or } D) = P(C) + P(D) = 6/36 + 4/36$
$= 10/36$, or 27.8%

61. (a) $P(\text{red}) = \dfrac{20}{100} = \dfrac{1}{5}$, or 20%

(b) $P(\text{blue}) = \dfrac{10}{100} = \dfrac{1}{10}$, or 10%

(c) $P(\text{red or blue}) = P(\text{red}) + P(\text{blue}) = \dfrac{30}{100} = \dfrac{3}{10}$, or 30%
Yes, these events are mutually exclusive.

(d) $P(\text{green or orange}) = P(\text{green}) + P(\text{orange})$
$= \dfrac{30}{100} + \dfrac{40}{100} = \dfrac{70}{100} = \dfrac{7}{10}$, or 70%
These events are mutually exclusive.

63. $P(\text{rain or snow})$
$= P(\text{rain}) + P(\text{snow}) - P(\text{rain and snow})$
$= .30 + .50 - .10$
$= .70$

65. Let W = set of readers of the Wall Street Journal, B = set of readers of Barron's.

S   n(s) = 60

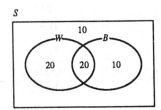

(a) $P(\text{reader} \in W) = \dfrac{n(W)}{n(S)} = \dfrac{40}{60} = .667,\ \text{or}\ 66.7\%$

(b) $P(\text{reader} \in B) = \dfrac{n(B)}{n(S)} = \dfrac{30}{60} = .50,\ \text{or}\ 50\%$

(c) $P(\text{reader} \in (W \cap B)) = \dfrac{n(W \cap B)}{n(S)} = \dfrac{20}{60}$

   $= .333,\ \text{or}\ 33.3\%$

(d) $P(\text{reader} \in (W \cup B))$

   $= \dfrac{n(W) + n(B) - n(W \cap B)}{n(S)} = \dfrac{40 + 30 - 20}{60}$

   $= \dfrac{50}{60} = .833,\ \text{or}\ 83.3\%$

(e) $P(\text{reader} \in [(W \cup B) \cap (W \cap B)']) = \dfrac{50}{60} - \dfrac{20}{60}$

   $= \dfrac{30}{60} = .50,\ \text{or}\ 50\%$

(f) $P(\text{reader} \in (W' \cap B')) = P[(W \cup B)']$
   $= 1 - .833 = .167,\ \text{or}\ 16.7\%$

67. Let D = event that a doctor is selected, L = event that a lawyer is selected, B = event that a business executive is selected, 2R = event that an occupant of a 2–room suite is selected, 3R = event that an occupant of a 3–room suite is selected, and 4R = event that an occupant of a 4–room suite is selected.

(a) $P(D) = \dfrac{20}{100} = \dfrac{1}{5},\ \text{or}\ 20\%$

(b) $P(B) = \dfrac{35}{100} = \dfrac{7}{20},\ \text{or}\ 35\%$

(c) $P(2R) = \dfrac{36}{100},\ \text{or}\ 36\%$

(d) $P(4R) = \dfrac{24}{100},\ \text{or}\ 24\%$

(e) $P(D \cap 4R) = \dfrac{15}{100},\ \text{or}\ 15\%$

(f) $P(L \cap 2R) = \dfrac{30}{100},\ \text{or}\ 30\%$

(g) $P(B \cap 3R) = \dfrac{25}{100},\ \text{or}\ 25\%$

(h) $P(D \cap 2R) = 0$

(i) $P(D \cup 4R) = P(D) + P(4R) - P(D \cap 4R)$

   $= \dfrac{20}{100} + \dfrac{24}{100} - \dfrac{15}{100} = \dfrac{29}{100},\ \text{or}\ 29\%$

(j) $P(L \cup 2R) = P(L) + P(2R) - P(L \cap 2R)$
   $= \dfrac{45}{100} + \dfrac{36}{100} - \dfrac{30}{100} = \dfrac{51}{100},\ \text{or}\ 51\%$

(k) $P(B \cup 3R) = P(B) + P(3R) - P(B \cap 3R)$
   $= \dfrac{35}{100} + \dfrac{40}{100} - \dfrac{25}{100} = \dfrac{50}{100},\ \text{or}\ 50\%$

(continued)

67. *(continued)*

(l) $P(D \cup L) = P(D) + P(L) = \dfrac{20}{100} + \dfrac{45}{100}$

$= \dfrac{65}{100}$, or 65%

(m) $P(2R \cup 4R) = P(2R) + P(4R) = \dfrac{36}{100} + \dfrac{24}{100}$

$= \dfrac{60}{100}$, or 60%

(n) $P[(D \cup 3R) \cap (D \cap 3R)']$

$= \dfrac{(20 + 40 - 5)}{100} - \dfrac{5}{100} = \dfrac{50}{100}$, or 50%

(o) $P[(L \cup 4R) \cap (L \cap 4R)']$

$= \dfrac{(45 + 24 - 5)}{100} - \dfrac{5}{100}$

$= \dfrac{59}{100}$, or 59%

(p) $P(L' \cap 3R') = P[(L \cup 3R)']$

$= 1 - \dfrac{45 + 40 - 10}{100} = 1 - .75 = .25$, or 25%

(q) $P(B' \cap 2R') = P[(B \cup 2R)']$

$= 1 - \dfrac{35 + 36 - 6}{100} = 1 - .65 = .35$, or 35%

69. Let U = set of families surveyed, V = set of families that like vanilla, S = set of families that like strawberry, and C = set of families that like coffee flavor. Let *f* = a family selected.

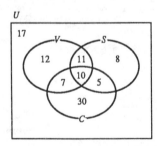

(continued)

69. *(continued)*

(a) $P(f \in V) = \dfrac{n(V)}{n(U)} = \dfrac{40}{100} = 0.4$, or 40%

(b) $P(f \in (V \cap S)) = \dfrac{n(V \cap S)}{n(U)} = \dfrac{21}{100} = .21,$

or 21%

(c) $P(f \in (V \cup S)) = \dfrac{n(V) + n(S) - n(V \cap S)}{n(U)}$

$= \dfrac{40 + 34 - 21}{100} = \dfrac{53}{100} = .53$, or 53%

(d) $P(f \in (V \cup S \cup C))$

$= \dfrac{n(V) + n(S) + n(C) - n(V \cap S) - n(V \cap C)}{n(U)}$

$\dfrac{- n(S \cap C) + n(S \cap V \cap C)}{n(U)}$

$= \dfrac{40 + 34 + 52 - 21 - 17 - 15 + 10}{100}$

$= \dfrac{83}{100} = .83$, or 83%

Note: A shorter solution is
$P = 1 - P[f \in (V \cup S \cup C)'] = 1 - .17 = .83.$

71. (a) $P(\text{all aces}) = \dfrac{C(4,4)}{C(52,4)} = \dfrac{1 \cdot 4!(52 - 4)!}{52!}$

$= \dfrac{4!\ 48!}{52!}$

$= \dfrac{4 \cdot 3 \cdot 2 \cdot 1}{52 \cdot 51 \cdot 50 \cdot 49}$

$= \dfrac{1}{270,725} \approx 3.694 \times 10^{-6}$

(b) $P(\text{all red cards}) = \dfrac{C(26,4)}{C(52,4)}$

$= \dfrac{26!}{4!(26 - 4)!} \cdot \dfrac{4!(52 - 4)!}{52!} = \dfrac{26!\ 48!}{22!\ 52!}$

$= \dfrac{26 \cdot 25 \cdot 24 \cdot 23}{52 \cdot 51 \cdot 50 \cdot 49}$

$= \dfrac{46}{833} \approx .05522$, or 5.522%

73. (a) $P(2 \text{ queens}) = \dfrac{C(4,2) \cdot C(48,11)}{C(52,13)}$

$= \dfrac{4!}{2!\ 2!} \cdot \dfrac{48!}{11!\ 37!} \cdot \dfrac{13!\ 39!}{52!}$

$= 6 \cdot \dfrac{13 \cdot 12 \cdot 39 \cdot 38}{52 \cdot 51 \cdot 50 \cdot 49}$

$= \dfrac{4446}{20,825} \approx .2135,\ \text{or } 21.35\%$

(b) $P(4 \text{ queens}) = \dfrac{C(4,4) \cdot C(48,9)}{C(52,13)}$

$= 1 \cdot \dfrac{48!}{9!\ 39!} \cdot \dfrac{13!\ 39!}{52!} = \dfrac{13 \cdot 12 \cdot 11 \cdot 10}{52 \cdot 51 \cdot 50 \cdot 49}$

$= \dfrac{11}{4165} \approx .002641,\ \text{or } .2641\%$

(c) $P(\geq 2 \text{ queens})$

$= P(2 \text{ queens}) + P(3 \text{ queens}) + P(4 \text{ queens})$

$P(3 \text{ queens}) = \dfrac{C(4,3) \cdot C(48,10)}{C(52,13)}$

$= \dfrac{4!\ 48!\ 13!\ 39!}{3!\ 10!\ 38!\ 52!}$

$= \dfrac{4 \cdot 13 \cdot 12 \cdot 11 \cdot 39}{52 \cdot 51 \cdot 50 \cdot 49}$

$= \dfrac{858}{20,825}$

$P(\geq 2 \text{ queens}) = \dfrac{4446}{20,825} + \dfrac{858}{20,825} + \dfrac{11}{4165}$

$= \dfrac{5359}{20,825} \approx .2573,\ \text{or } 25.73\%$

75. (a) $P(\text{start with } 1) = \dfrac{1 \cdot 10^4}{10^5} = .1,\ \text{or } 10\%,$ repetitions allowed

(b) $P(\text{start with } 1) = \dfrac{1 \cdot P(9,4)}{P(10,5)} = \dfrac{9!\ 5!}{5!\ 10!} = \dfrac{1}{10}$

$= .10,\ \text{or } 10\%,\ \text{no repetitions}$

77. (a) $P(1 \text{ red}) = \dfrac{C(6,1) \cdot C(4,4)}{C(10,5)} = \dfrac{6!\ 1!\ 5!\ 5!}{5!\ 10!} \cdot 1$

$= \dfrac{6 \cdot 5 \cdot 4 \cdot 3 \cdot 2 \cdot 1}{10 \cdot 9 \cdot 8 \cdot 7 \cdot 6}$

$= \dfrac{1}{42} \approx .02381,\ \text{or } 2.381\%$

(b) $P(2 \text{ red}) = \dfrac{C(6,2) \cdot C(4,3)}{C(10,5)}$

$= \dfrac{6!\ 4!\ 5!\ 5!}{2!\ 4!\ 3!\ 1!\ 10!}$

$= \dfrac{6 \cdot 5}{2 \cdot 1} \cdot 4 \cdot \dfrac{5 \cdot 4 \cdot 3 \cdot 2 \cdot 1}{10 \cdot 9 \cdot 8 \cdot 7 \cdot 6}$

$= \dfrac{5}{21} \approx .2381,\ \text{or } 23.81\%$

(c) $P(\leq 2 \text{ red}) = P(1 \text{ red}) + P(2 \text{ red})$

$= \dfrac{1}{42} + \dfrac{10}{42} = \dfrac{11}{42} \approx .2619,\ \text{or } 26.19\%$

(d) $P(> 2 \text{ red}) = P(3 \text{ red}) + P(4 \text{ red}) + P(5 \text{ red})$
$= 1 - P(1 \text{ red}) - P(2 \text{ red})$
$= 1 - .2619 = .7381,\ \text{or } 73.81\%$

79. (a) $n(S) = C(30,8) = \dfrac{30!}{8!\ (30-8)!}$

$= \dfrac{30 \cdot 29 \cdot 28 \cdot 67 \cdot 26 \cdot 25 \cdot 24 \cdot 23}{8 \cdot 7 \cdot 6 \cdot 5 \cdot 4 \cdot 3 \cdot 2 \cdot 1}$

$= 5,852,925$

(b) $P(\text{none in error}) = \dfrac{C(26,8)}{C(30,8)} = \dfrac{26!\ 8!\ 22!}{8!\ 18!\ 30!}$

$= \dfrac{22 \cdot 21 \cdot 20 \cdot 19}{30 \cdot 29 \cdot 28 \cdot 27}$

$= \dfrac{209}{783} \approx .2669,\ \text{or } 26.69\%$

(continued)

79. *(continued)*

(c) $P(1 \text{ in error}) = \dfrac{C(4,1) \cdot C(26,6)}{C(30,8)}$

$= 4 \cdot \dfrac{26!}{7! \ 19!} \cdot \dfrac{8! \ 22!}{30!}$

$= \dfrac{4 \cdot 8 \cdot 22 \cdot 21 \cdot 20}{30 \cdot 29 \cdot 28 \cdot 27}$

$= \dfrac{352}{783} \approx .4496, \text{ or } 44.96\%$

(d) $P(2 \text{ in error}) = \dfrac{C(4,2) \cdot C(26,6)}{C(30,8)}$

$= \dfrac{4!}{2! \ 2!} \cdot \dfrac{26!}{6! \ 20!} \cdot \dfrac{8! \ 22!}{30!}$

$= \dfrac{6 \cdot 8 \cdot 7 \cdot 22 \cdot 21}{30 \cdot 29 \cdot 28 \cdot 27}$

$= \dfrac{308}{783} \approx .2360, \text{ or } 23.60\%$

(e) $P(\le 2 \text{ in error})$
$= P(\text{none in error}) + P(1 \text{ in error})$
$\quad + P(2 \text{ in error})$

$= \dfrac{209}{783} + \dfrac{352}{783} + \dfrac{308}{1305}$

$= \dfrac{5,574,855}{5,852,925} \approx .9525, \text{ or } 95.25\%$

(f) $P(\ge 3 \text{ in error}) = 1 - P(\le 2 \text{ in error})$
$= 1 - .9525$
$= .0475, \text{ or } 4.75\%$

81.   If 5 people are selected,

(a) $n(S) = 365^5$

(b) $365 \cdot 364 \cdot 363 \cdot 362 \cdot 361$ possible ways in which all 5 selected birthdays are different.

(c) $P(E^1) = \dfrac{n(E')}{n(S)} = \dfrac{365 \cdot 364 \cdot 363 \cdot 362 \cdot 361}{365 \cdot 365 \cdot 365 \cdot 365 \cdot 365}$
$\approx .97286$
$P(E) = 1 - P(E') = .02714$
Probability that at least 2 selected people have the same birthday $= .02714$

8–4 Conditional Probability and Independence

For exercises 1 – 3, let
$\quad$ C = the set of buyers of compact cars,
$\quad$ F = the set of buyers of full–size cars
$\quad$ S = the set of buyers who paid cash
$\quad$ R = the set of buyers who used credit.

1. $P(F \mid R) = \dfrac{n(F \cap R)}{n(R)} = \dfrac{20}{120} = .167, \text{ or } 16.7\%$

3. $P(F \cap R) = \dfrac{n(F \cap R)}{n(S)} = \dfrac{20}{200} = .1, \text{ or } 10\%$

5. $P(M \mid W) = \dfrac{P(M \cap W)}{P(W)} = \dfrac{.15}{.40} = .375, \text{ or } 37.5\%$

7. $P(F \mid B) = \dfrac{P(F \cap B)}{P(B)} = \dfrac{.10}{.60} = .167, \text{ or } 16.7\%$

9. $P(B \text{ and } F) = P(B \cap F) = .10, \text{ or } 10\%$

11. $P(M \mid B) = \dfrac{P(M \cap B)}{P(B)} = \dfrac{.50}{.60} = .833, \text{ or } 83.3\%$

13. $P(M \text{ and } B) = P(M \cap B) = .50, \text{ or } 50\%$

15. $P(M \text{ and } W) = P(M \cap W) = .15$

Use the following figure for problems 17 to 25.

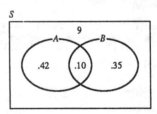

17. $P(A) = .42 + .10 = .52$

19. $P(A \cup B) = P(A) + P(B) - P(A \cap B)$
$= .52 + .45 - .10$
$= .87$

21. $P(B \mid A) = \dfrac{P(B \cap A)}{P(A)} = \dfrac{.10}{.52} = .192$

23. $P(A') = 1 - .52 = .48$

25. $P[(A \cap B)'] = 1 - .10 = .90$

27. Let W = set of customers buying wa?hers, D = set of customers buying dryers.

(a) $P(D \mid W) = \dfrac{P(D \cap W)}{P(W)} = \dfrac{.20}{.40} = .50,$
or 50%

(b) $P(W \mid D) = \dfrac{P(W \cap D)}{P(D)} = \dfrac{.20}{.30} = .667,$
or 66.7%

29. (a) $P(R_1) = \dfrac{40}{100} = .40$

(b) $P(R_2 \mid R_1) = \dfrac{33}{99} \approx .3939$

(c) $P(G_2 \mid R_1) = \dfrac{60}{99} \approx .6061$

31. Let D = selection of defective plug, N = selection of non–defective plug.

(a) $P(D_1) = \dfrac{20}{200} = .10$

(b) $P(D_2 \mid D_1) = \dfrac{19}{199} \approx .0955$

(c) $P(D_3 \mid D_2 \mid D_1) = \dfrac{18}{198} \approx .0909$

(d) $P(D_3 \mid N_2 \mid N_1) = \dfrac{20}{198} \approx .1010$

33. (a) $P(R_1) = \dfrac{26}{52} = .50,$ or 50%

(b) $P(R_2 \mid R_1) = \dfrac{25}{51} \approx .490,$ or 49.0%

(c) $P(R_3 \mid R_2 \mid R_1) = \dfrac{24}{50} \approx .48,$ or 48%

(d) $P(R_3 \mid B_2 \mid B_1) = \dfrac{26}{50} \approx .52,$ or 52%

35.

	Male	Female	
<20, T	.20	.15	.35
≥20, A	.60	.05	.65
	.80	.20	1.00

(a) $P(F \cap A) = .05,$ or 5%

(b) $P(T \mid F) = \dfrac{.15}{.20} = .75,$ or 75%

37. (a) $P(R_2 \mid R_1) = \left[\dfrac{40}{200}\right]^2 = .04,$ or 4.0%

(b) $P(W_2 \mid W_1) = \left[\dfrac{50}{200}\right]^2 = .0625,$ or 6.25%

(c) $P(G_2 \mid R_1) = \dfrac{110}{200} \cdot \dfrac{40}{200} = .11,$ or 11.0%

(d) $P(R_2 \mid G_1) = \dfrac{40}{200} \cdot \dfrac{110}{200} = .11,$ or 11.0%

(e) $P(R \cap G) = .11 + .11 = .22,$ or 22.0%

39. $P(M) = .10$, $P(> \$10) = .20$, $P(> \$10 \mid M) = .80$

(a) $P(M \cap > \$10) = P(M)P(> \$10 \mid M)$
$$= (.10)(.80) = .08$$

	M	F	
≤ $10	.02		
> $10	.08	.12	.20
	.10		

(b) $P(> \$10 \cup M)$
$$= P(> \$10) + P(M) - P(> \$10 \cap M)$$
$$= .20 + .10 - .08 = .22$$

(c) $P(M \mid > \$10) = \dfrac{.08}{.20} = .40$

41.

	M	F	
≤ 40	.30	.15	.45
> 40	.10	.45	.55
	.40	.60	1.00

(a) $P(M \cap \leq 40)$ $= P(M) - P(M \cap > 40)$
$$= [1 - P(F)] - P(M \cap > 40)$$
$$= 1 - .60 - .10 = .30$$

(b) $P(F \cap > 40)$ $= P(> 40) - P(M \cap > 40)$
$$= .55 - .10 = .45$$

(c) $P(F \cap \leq 40)$ $= P(F) - P(F \cap > 40)$
$$= .60 - .45 = .15$$

(d) $P(> 40) = .55$

43.   Given that events A and B are independent, $P(A) = .3$, and $P(B) = .6$:

(a) $P(A \text{ and } B) = P(A \cap B) = (.3)(.6) = .18$

(continued)

43.   *(continued)*

(b) $P(A \text{ or } B) = P(A \cup B)$
$$= P(A) + P(B) - P(A \cap B)$$
$$= .3 + .6 - .18 = .72$$

(c) $P(A \mid B) = \dfrac{P(A \cap B)}{P(B)} = \dfrac{.18}{.60} = .30$

(d) $P(B \mid A) = \dfrac{P(B \cap A)}{P(A)} = \dfrac{.18}{.30} = .60$

45.   (a) $P(B_1 \cap B_2 \cap B_3 \cap B_4)$
$$= P(B_1)P(B_2 \mid B_1)P(B_3 \mid B_1 \cap B_2)P(B_4 \mid B_1 \cap B_2 \cap B_3)$$
$$= \frac{80}{100} \cdot \frac{79}{99} \cdot \frac{78}{98} \cdot \frac{77}{97} \approx .4033$$

(b) $P(R_1 \cap R_2 \cap R_3 \cap R_4)$
$$= P(R_1)P(R_2 \mid R_1)P(R_3 \mid R_1 \cap R_2)P(B_4 \mid R_1 \cap R_2 \cap R_3)$$
$$= \frac{20}{100} \cdot \frac{19}{99} \cdot \frac{18}{98} \cdot \frac{80}{97} \approx .005815$$

(c) $P(R_1 \cap B_2 \cap B_3 \cap B_4)$
$$= P(R_1)P(B_2 \mid R_1)P(B_3 \mid R_1 \cap B_2)P(B_4 \mid R_1 \cap B_2 \cap B_3)$$
$$= \frac{20}{100} \cdot \frac{80}{99} \cdot \frac{79}{98} \cdot \frac{78}{97} \approx .1048$$

(d) $P(B_1 \cap R_2 \cap R_3 \cap B_4)$
$$= P(B_1)P(R_2 \mid B_1)P(R_3 \mid B_1 \cap R_2)P(B_4 \mid B_1 \cap R_2 \cap R_3)$$
$$= \frac{80}{100} \cdot \frac{20}{99} \cdot \frac{19}{98} \cdot \frac{79}{97} \approx .02552$$

47.   Let E represent the selection of an erroneous invoice and C the selection of a correct invoice.

(a) $P(E_1 \cap E_2 \cap E_3 \cap E_4)$
$$= P(E_1)P(E_2 \mid E_1)P(E_3 \mid E_1 \cap E_2)P(E_4 \mid E_1 \cap E_2 \cap E_3)$$
$$= \frac{10}{80} \cdot \frac{9}{79} \cdot \frac{8}{78} \cdot \frac{7}{77} \approx .000133$$

(continued)

47. *(continued)*

(b) $P(E_1 \cap E_2 \cap E_3 \cap C_4)$
$= P(E_1)P(E_2|E_1)P(E_3|E_1 \cap E_2)P(C_4|E_1 \cap E_2 \cap E_3)$
$= \dfrac{10}{80} \cdot \dfrac{9}{79} \cdot \dfrac{8}{78} \cdot \dfrac{70}{77} \approx .00133$

(c) $P(E_1 \cap E_2 \cap C_3 \cap E_4)$
$= P(E_1)P(E_2|E_1)P(C_3|E_1 \cap E_2)P(E_4|E_1 \cap E_2 \cap C_3)$
$= \dfrac{10}{80} \cdot \dfrac{9}{79} \cdot \dfrac{70}{78} \cdot \dfrac{8}{77} \approx .00133$

(d) $P(C_1 \cap C_2 \cap C_3 \cap C_4)$
$= P(C_1)P(C_2|C_1)P(C_3|C_1 \cap C_2)P(C_4|C_1 \cap C_2 \cap C_3)$
$= \dfrac{70}{80} \cdot \dfrac{69}{79} \cdot \dfrac{68}{78} \cdot \dfrac{67}{77} \approx .5797$

49. Let D represent the selection of a defective plug and N represent the selection of a non–defective plug.

(a) $P(D_1 \cap D_2 \cap D_3 \cap D_4)$
$= P(D_1)P(D_2|D_1)P(D_3|D_1 \cap D_2)P(D_4|D_1 \cap D_2 \cap D_3)$
$= \dfrac{9}{60} \cdot \dfrac{8}{59} \cdot \dfrac{7}{58} \cdot \dfrac{6}{57} \approx .000258$

(b) $P(N_1 \cap N_2 \cap D_3 \cap D_4)$
$= P(N_1)P(N_2|N_1)P(D_3|N_1 \cap N_2)P(D_4|N_1 \cap N_2 \cap D_3)$
$= \dfrac{51}{60} \cdot \dfrac{50}{59} \cdot \dfrac{9}{58} \cdot \dfrac{8}{57} \approx .0157$

(c) $P(D_1 \cap N_2 \cap N_3 \cap N_4)$
$= P(D_1)P(N_2|D_1)P(N_3|D_1 \cap N_2)P(N_4|D_1 \cap N_2 \cap N_3)$
$= \dfrac{9}{60} \cdot \dfrac{51}{59} \cdot \dfrac{50}{58} \cdot \dfrac{49}{57} \approx .0961$

(d) $P(N_1 \cap N_2 \cap N_3 \cap N_4)$
$= P(N_1)P(N_2|N_1)P(N_3|N_1 \cap N_2)P(N_4|N_1 \cap N_2 \cap N_3)$
$= \dfrac{51}{60} \cdot \dfrac{50}{59} \cdot \dfrac{49}{58} \cdot \dfrac{48}{57} \approx .512$

8–5 Bayes' Formula

In Exercises 1 – 7, let B_1 = even of selecting box 1, B_2 = event of selecting box 2, R = event of drawing a red marble, G = event of drawing a green marble, Y = event of drawing a yellow marble.

1. $P(R) = P(B_1 \cap R) + P(B_2 \cap R)$
$= P(B_1)P(R|B_1) + P(B_2)P(R|B_2)$
$= (.50)\left[\dfrac{5}{8}\right] + (.50)\left[\dfrac{7}{10}\right]$
$= .50\left[\dfrac{5}{8} + \dfrac{7}{10}\right] = \dfrac{53}{80} \approx .66255$

3. $P(Y) = P(B_1 \cap Y) + P(B_2 \cap Y)$
$= P(B_1)P(Y|B_1) + P(B_2)P(Y|B_2)$
$= (.50)(0) + (.50)\left[\dfrac{1}{10}\right]$
$= \dfrac{1}{20} = .05$

5. $P(B_1|G) = \dfrac{P(B_1 \cap G)}{P(G)} = \dfrac{(0.50)(3/8)}{23/80}$ (using results
$= \dfrac{15}{23} \approx .6522$ from Exercise 2)

7. $P(B_2|G) = 1 - P(B_1|G) = 1 - \dfrac{15}{23} = \dfrac{8}{23} \approx .3478$

In Exercises 9 – 11,
let B_1 = event that moped came from plant 1,
B_2 = event that moped came from plant 2,
B_3 = event that moped came from plant 3,
D = event that moped is defective.

$P(B_1) = .40,\ P(B_2) = .35,\ P(B_3) = .25,$
$P(D|B_1) = .02,\ P(D|B_2) = .01,\ P(D|B_3) = .03$
$P(D) = P(B_1 \cap D) + P(B_2 \cap D) + P(B_3 \cap D)$
$= P(B_1)P(D|B_1) + P(B_2)P(D|B_2)$
$\quad P(B_3)P(D|B_3)$
$= (.40)(.02) + (.35)(.01) + (.25)(.03)$
$= .008 + .0035 + .0075 = .0190$

9. $P(B_1 | D) = \dfrac{P(B_1 \cap D)}{P(D)} = \dfrac{.008}{.0190} \approx .4211$

11. $P(B_3 | D) = \dfrac{P(B_3 \cap D)}{P(D)} = \dfrac{.0075}{.0190} \approx .3947$

For Exercise 13, let S_i = event that consumer is in submarket i, i = 1,2,3,4.   Also, let F = event that consumer prefers foreign–made cars.

$P(S_1) = .30$, $P(S_2) = .25$, $P(S_3) = .21$,
$P(S_4) = .24$, $P(F | S_1) = .40$, $P(F | S_2) = .20$,
$P(F | S_3) = .35$, $P(F | S_4) = .50$

13. $P(F) = P(S_1 \cap F) + P(S_2 \cap F) + P(S_3 \cap F)$
$\quad + P(S_4 \cap F)$
$\quad = P(S_1)P(F | S_1) + P(S_2)P(F | S_2) + P(S_3)P(F | S_3)$
$\quad\quad + P(S_4)P(F | S_4)$
$\quad = (.30)(.40) + (.25)(.20) + (.21)(.35)$
$\quad\quad + (.24)(.50)$
$\quad = .12 + .05 + .0735 + .12 = .3635$

15. $P(<21 | A) = \dfrac{P(<21 \cap A)}{P(A)} = \dfrac{.0080}{.0370} \approx .216$

17. Let G = event that a customer is a good payer,
$\quad\quad$ H = event that a customer is a poor payer,
$\quad\quad$ R = event that a customer returns an item.

$P(G) = .90$, $P(H) = .10$, $P(R | G) = .02$,
$P(R | H) = .70$
$P(R) = P(G \cap R) + P(H \cap R)$
$\quad = P(G)P(R | G) + P(H)P(R | H)$
$\quad = (.90)(.02) + (.10)(.70) = .18 + .07$
$\quad = .25$
$P(H | R) = \dfrac{P(H \cap R)}{P(R)} = \dfrac{.07}{.088} = .7\cdot$

For Exercise 19, let
$\quad\quad$ B_1 = event that a tire of brand B_1 is sold,
$\quad\quad$ B_2 = event that a tire of brand B_2 is sold,
$\quad\quad$ B_3 = event that a tire of brand B_3 is sold,
$\quad\quad$ D = event that a tire is defective.

$P(B_1) = .50$, $P(B_2) = .30$, $P(B_3) = .20$,
$P(D | B_1) = .10$, $P(D | B_2) = .05$, $P(D | B_3) = .08$

19. $P(D) = P(B_1 \cap D) + P(B_2 \cap D) + P(B_3 \cap D)$
$\quad = P(B_1)P(D | B_1) + P(B_2)P(D | B_2)$
$\quad\quad + P(B_3)P(D | B_3)$
$\quad = (.50)(.10) + (.30)(.05) + (.20)(.08)$
$\quad = .05 + .015 + .016 = .081$

For Exercises 21 and 23, the data are represented in the tree diagram.

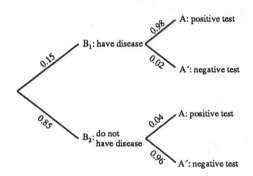

Let B_1 = event that a person has the disease,
$\quad\quad$ B_2 = event that a person does not have the disease,
$\quad\quad$ A = event that a person tests positive,
$\quad\quad$ A$'$ = event that a person tests negative.

$P(B_1) = .15$, $P(B_2) = .85$, $P(A | B_1) = .98$,
$P(A | B_2) = .04$

$P(A) = P(B_1 \cap A) + P(B_2 \cap A)$

$\quad = P(B_1)P(A | B_1) + P(B_2)P(A | B_2)$
$\quad = (.15)(.98) + (.85)(.04)$
$\quad = .147 + .034$
$\quad = .181$

21. $P(B_1 \mid A) = \dfrac{P(B_1 \cap A)}{P(A)} = \dfrac{.147}{.181} = .812$

23. $P(B_1 \mid A') = \dfrac{P(B_1 \cap A')}{P(A')} = \dfrac{P(B_1)P(A' \mid B_1)}{P(A')}$

$= \dfrac{(.15)(.02)}{(1 - .181)} \approx .00366$

Extra Dividends System Reliability

1. The components are connected in serial order.

$P(S) = P(A)P(B)P(C)P(D)$

$= (.98)(.96)(.94)(.99) \approx .876$

3. $P(S) = P(A)[1 - P(B')P(C')][1 - P(D')P(E')]$

$= (.96)[1 - (.05)(.02)][1 - (.06)(.07)]$

$= (.96)(.999)(.9958) \approx .955$

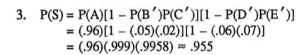

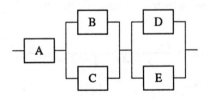

5. For the system shown,

$P(S) = P[(A \cap B) \cup C]$

$1 - [1 - P(A)P(B)]P(C')$

$= 1 - [1 - P(A)P(B)](1 - P(C))$

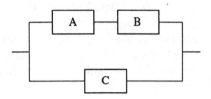

Review Exercises

1. (a) $n(U) = 50$

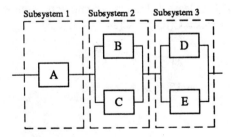

(b) $n(S') = n(U) - n(S) = 50 - 20 = 30$

(c) $n(S \cup R) = n(S) + n(R) - n(S \cap R)$

$= 20 + 35 - 9 = 46$

(d) $n(S' \cap R') = n[(S \cup R)'] = 50 - 46 = 4$

(e) $n(S \mid R) = 9$; $P(S \mid R) = \dfrac{n(S \cap R)}{n(R)}$

$\% = \dfrac{9}{35} \times 100 \approx 25.7\%$

(f) $n(R \mid S) = 9$; $P(R \mid S) = \dfrac{n(R \cap S)}{n(S)}$

$\% = \dfrac{9}{20} \times 100 \approx 45.0\%$

(g) $n(S \cap R') = 20 - 9 = 11$

3. a) Column Percent Table

	Good	Average	Poor	Totals
Male	66.7	40.0	47.6	56.0
Female	33.3	60.0	52.4	44.0
Total	100.0	100.0	100.0	100.0

Column 1 lists the distribution of good sentiments among male and female employees. Columns 2 and 3 similarly list the distributions of average and poor sentiments.

(b) Row Percent Table

	Good	Average	Poor	Totals
Male	57.14	7.14	35.72	100.0
Female	36.36	13.64	50.00	100.0
Total	48.0	10.0	42.0	100.0

Row 1 lists the distribution of good, average, and poor sentiments among male employees. Row 2 lists the distribution among female employees.

(c) $P(F \cap P) = .22$, or 22%

(d) $P(M \cup A) = \dfrac{n(M \cup A)}{n(S)}$

$= \dfrac{n(M) + n(A) - n(M \cap A)}{n(S)}$

$= \dfrac{560 + 100 - 40}{1000}$

$= .62$, or 62%

(e) $P(G^1) = 1 - P(G) = 1 - \dfrac{480}{1000} = .52$, or 52%

(f) $P(G \cup A) = P(G) + P(A) - P(G \cap A)$
$= .48 + .10 - 0$
$= .58$, or 58%

5. $1! = 1$

7. $4! = 4 \cdot 3 \cdot 2 \cdot 1 = 24$

9. $P(7,2) = \dfrac{7!}{(7-2)!} = \dfrac{7!}{5!} = 7 \cdot 6 = 42$

11. $C(8,5) = \dfrac{8!}{5!(8-5)!} = \dfrac{8!}{5! \ 3!} = \dfrac{8 \cdot 7 \cdot 6}{3 \cdot 2 \cdot 1} = 56$

13. (a) $P(7,4) = \dfrac{7!}{(7-4)!} = \dfrac{7!}{3!} = 7 \cdot 6 \cdot 5 \cdot 4$

$= 840$ ways

(b) $C(7,4) = \dfrac{7!}{4! \ 3!} = \dfrac{7 \cdot 6 \cdot 5}{3 \cdot 2 \cdot 1} = 35$ ways

15. (a) $C(30,5) \ \dfrac{30!}{5!(30-5)!} = \dfrac{30!}{5! \ 25!}$

$= \dfrac{30 \cdot 29 \cdot 28 \cdot 27 \cdot 26}{5 \cdot 4 \cdot 3 \cdot 2 \cdot 1}$

$= 142,506$ samples

(b) With 2 defective computers, number of samples $= C(6,2) \cdot C(24,3)$

$= \dfrac{6!}{2! \ 4!} \cdot \dfrac{24!}{3! \ 21!} = \dfrac{6 \cdot 5}{2 \cdot 1} \cdot \dfrac{24 \cdot 23 \cdot 22}{3 \cdot 2 \cdot 1}$

$= 30,360$

(c) With no defective computers, number of samples $= C(24,5)$

$= \dfrac{24!}{3! \ 19!} = \dfrac{24 \cdot 23 \cdot 22 \cdot 21 \cdot 20}{5 \cdot 4 \cdot 3 \cdot 2 \cdot 1}$

$= 42,504$

(d) With at least 4 defective computers, number of samples $= C(6,4) \cdot C(24,1) + C(6,5)$

$= \dfrac{6!}{4! \ 2!} \cdot 24 + \dfrac{6!}{5! \ 1!}$

$= 15 \cdot 24 + 6 = 366$

17. $n(U) = 200$

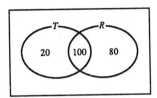

Let T = set of employees getting a promotion
R = set of employees getting a raise

(a) $P(R) = \dfrac{n(R)}{n(U)} = \dfrac{180}{200} = .90$

(b) $P(T) = \dfrac{n(T)}{n(U)} = \dfrac{120}{200} = .60$

(c) $P(R \cap T) = \dfrac{100}{200} = .50$

(d) $P(R \cup T) = P(R) + P(T) - P(R \cap T)$
$$= .90 + .60 - .50$$
$$= 1.0$$

(e) $P[(R \cup T) \cap (R \cap T)'] = \dfrac{20 + 80}{200} = .50$

(f) $P(R' \cap T') = P[(R \cup T)'] = 1 - 1 = 0$

19. (a) $P(\text{no defectives}) = \dfrac{C(32,6)}{C(40,6)} = \dfrac{32!}{6! \ 26!} \cdot \dfrac{6! \ 34!}{40!}$

$$= \dfrac{32 \cdot 31 \cdot 30 \cdot 29 \cdot 28 \cdot 27}{40 \cdot 39 \cdot 38 \cdot 37 \cdot 36 \cdot 35} = \dfrac{10,788}{45,695}$$

$$\approx .2361$$

(continued)

19. *(continued)*

(b) $P(\text{3 defectives}) = \dfrac{C(8,3) \ \cdot \ C(32,3)}{C(40,6)}$

$$= \dfrac{8!}{3! \ 5!} \cdot \dfrac{32!}{3! \ 29!} \cdot \dfrac{6! \ 34!}{40!}$$

$$= \dfrac{8 \cdot 7 \cdot 6}{3 \cdot 2 \cdot 1} \cdot \dfrac{32 \cdot 31 \cdot 30}{3 \cdot 2 \cdot 1}$$

$$\cdot \dfrac{6 \cdot 5 \cdot 4 \cdot 3 \cdot 2 \cdot 1}{40 \cdot 39 \cdot 38 \cdot 37 \cdot 36 \cdot 35}$$

$$= \dfrac{1984}{27,417} \approx .07236$$

(c) $P(\le 3 \text{ defectives}) = P(\text{no defectives})$
+ P(1 defective) + P(2 defectives)
+ P(3 defectives)

$$= .2361 + \dfrac{C(8,1) \ \cdot \ C(32,5)}{C(40,6)}$$

$$+ \dfrac{C(8,2) \ \cdot \ C(32,4)}{C(40,6)} + .07236$$

$$= .2361 + \dfrac{8 \cdot 32! \ 6! \ 34!}{5! \ 27! \ 40!} + \dfrac{8! \ 32! \ 6! \ 34!}{2! \ 6! \ 4! \ 28! \ 40!}$$

$$+ .07236$$
$$= .2361 + .4197 + .2623 + .0724 = .9905$$

(d) $P(\ge 5 \text{ defectives})$
= P(5 defectives) + P(6 defectives)

$$= \dfrac{C(8,5)C(32,1)}{C(40,6)} + \dfrac{C(8,6)}{C(40,6)}$$

$$= \dfrac{8!}{5! \ 3!} \cdot 32 \cdot \dfrac{6! \ 34!}{40!} + \dfrac{8! \ 6! \ 34!}{6! \ 2! \ 40!}$$

$$\approx .0004669 + .0000073 = .000474$$

21. $P(E) = \dfrac{1}{1 + 5} \approx .1667$

23. $P(E) = \dfrac{3}{3 + 7} = .30$

25. Odds in favor $= \dfrac{.80}{1-.80} = \dfrac{.80}{.20} = 4$ to 1

27. Odds in favor $= \dfrac{.30}{1-.30} = \dfrac{.30}{.70} = 3$ to 7

29. $P(E) = \dfrac{5}{5+2} = \dfrac{5}{7} \approx .7143$

31. Odds in favor of rain $= \dfrac{.60}{1-.60} = \dfrac{.60}{.40} = 3$ to 2

33.

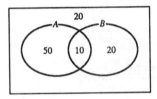

Percentages

Let $A =$ set of shoppers who buy brand A detergent, $B =$ set of shoppers who buy brand B detergent.

(a) $P(B\,|\,A) = \dfrac{P(B \cap A)}{P(A)} = \dfrac{.10}{.60} \approx .1667$

(b) $P(A\,|\,B) = \dfrac{P(A \cap B)}{P(B)} = \dfrac{.10}{.30} \approx .3333$

(c) $P(A \cup B) = P(A) + P(B) - P(A \cap B)$
$= .60 + .30 - .10$
$= .80,$ or 80%

35. Let $D =$ event of selecting a defective tire,
$N =$ event of selecting a non–defectve tire.

(a) $P(D_1) = \dfrac{40}{500} = .08$

(b) $P(D_2\,|\,N_1) = \dfrac{40}{499} \approx .0802$

(c) $P(D_1 \cap D_2 \cap D_3)$
$= P(D_1)P(D_2\,|\,D_1)P(D_3\,|\,D_1 \cap D_2)$

$= \dfrac{40}{500} \cdot \dfrac{39}{499} \cdot \dfrac{38}{498} \approx .000477$

37. Let $A =$ event that a refrigerator of brand A is sold,
$B =$ event that a refrigerator of brand B is sold,
$C =$ event that a refrigerator of brand C is sold,
$D =$ event that a defective refrigerator is sold.

$P(A) = .40,\ P(B) = .50,\ P(C) = .10,\ P(D\,|\,A) = .06,$
$P(D\,|\,B) = .10,\ P(D\,|\,C) = .07.$

(a) $P(D) = P(A \cap D) + P(B \cap D) + P(C \cap D)$
$= P(A)P(D\,|\,A) + P(B)P(D\,|\,B)$
$+ P(C)P(D\,|\,C)$
$= (.40)(.06) + (.50)(.10)$
$+ (.10)(.07)$
$= .024 + .050 + .007$
$= .081$

(b) $P(A\,|\,D) = \dfrac{P(A \cap D)}{P(D)} = \dfrac{.024}{.081} \approx .296$

CHAPTER

9

DIFFERENTIATION

9–1 Limits

1. $\lim_{x \to 5-} f(x) = 6$; $\lim_{x \to 5+} f(x) = 6$. $\therefore \lim_{x \to 5} f(x) = 6$.

3. $\lim_{x \to 5-} f(x) = 8$; $\lim_{x \to 5+} f(x) = 8$. $\therefore \lim_{x \to 5} f(x) = 8$.

5. $\lim_{x \to 7-} h(x) = 2$; $\lim_{x \to 7+} h(x) = 2$. $\therefore \lim_{x \to 7} h(x) = 2$.

7. $\lim_{x \to 7-} h(x) = 9$; $\lim_{x \to 7+} h(x) = 9$. $\therefore \lim_{x \to 7} h(x) = 9$.

9. $\lim_{x \to \infty} f(x) = 0$

11. $\lim_{x \to -\infty} f(x)$ does not exist.

13. $\lim_{x \to -\infty} g(x) = 0$

15. $\lim_{x \to -\infty} g(x)$ does not exist.

17. $\lim_{x \to 2} (4x + 7) = 4 \lim_{x \to 2} x + \lim_{x \to 2} 7 = 4 \cdot 2 + 7 = 15$

19. $\lim_{x \to 5} \dfrac{x}{x - 5}$ does not exist.

21. $\lim_{x \to 0} \dfrac{x^2 - 5x}{x} = \lim_{x \to 0} \dfrac{x(x - 5)}{x} = \lim_{x \to 0} (x - 5) = -5$

23. $\lim_{x \to 3} \dfrac{x^2 - x - 6}{x - 3} = \lim_{x \to 3} \dfrac{(x + 2)(x - 3)}{(x - 3)}$
$= \lim_{x \to 3} (x + 2) = 5$

25. $\lim_{x \to \infty} \dfrac{9}{x^3} = 0$

27. $\lim_{x \to \infty} \dfrac{4}{3x - 7} = 0$

29. $\lim\limits_{x \to 1} (8x - 5) = 8 \lim\limits_{x \to 1} x - \lim\limits_{x \to 1} 5 = 8 \cdot 1 - 5 = 3$

31. $\lim\limits_{x \to -3} \dfrac{x}{x + 3}$ does not exist.

33. $\lim\limits_{x \to 2} \dfrac{x^2 + 3x - 10}{x - 2} = \lim\limits_{x \to 2} \dfrac{(x + 5)(x - 2)}{(x - 2)} = \lim\limits_{x \to 2} (x + 5) = 7$

35. $\lim\limits_{x \to 0} \dfrac{x^3 - 4x^2 + 5x}{x} = \lim\limits_{x \to 0} \dfrac{x(x^2 - 4x + 5)}{x} = \lim\limits_{x \to 0} (x^2 - 4x + 5) = 5$

37. $\lim\limits_{x \to -\infty} \dfrac{1}{x^2} = 0$

39. $\lim\limits_{x \to \infty} \dfrac{3x^2 - 2x}{5x^3 + x} = \lim\limits_{x \to \infty} \dfrac{3/x - 2/x^2}{5 + 1/x^2} = \dfrac{0}{5} = 0$

41.

x	-2	-1.5	-1	-0.5	-0.0001	0.0001	0.5	1	1.5	2
$f(x) = \dfrac{\lvert x \rvert}{x}$	-1	-1	-1	-1	-1	1	1	1	1	1

$\lim\limits_{x \to 0-} f(x) = -1$; $\lim\limits_{x \to 0+} f(x) = 1$. $\therefore$ $\lim\limits_{x \to 0} f(x)$ does not exist.

43.

x	3	3.5	3.75	3.95	4.05	4.25	4.5
$f(x) = \dfrac{\sqrt{x} - 2}{x - 4}$	≈ 0.268	≈ 0.258	≈ 0.254	≈ 0.250	≈ 0.249	≈ 0.246	≈ 0.243

$\lim\limits_{x \to 4-} f(x) = 0.250$; $\lim\limits_{x \to 4+} f(x) = 0.250$. $\therefore$ $\lim\limits_{x \to 4} f(x) = 0.250$

Note: The limit can also be found algebraically:

$$\lim\limits_{x \to 4} \dfrac{\sqrt{x} - 2}{x - 4} = \lim\limits_{x \to 4} \dfrac{(\sqrt{x} - 2)}{(\sqrt{x} - 2)(\sqrt{x} + 2)} = \lim\limits_{x \to x} \dfrac{1}{(\sqrt{x} + 2)} = \dfrac{1}{4}$$

45.

x	$1{,}000$	$10{,}000$	$100{,}000$	$1{,}000{,}000$	$10{,}000{,}000$
$f(x) = (1 + 1/x)^x$	2.717	2.7181	2.7183	2.71828	2.718281828

$\lim\limits_{x \to \infty} f(x) \to e$

47.

x	100	1,000	10,000	100,000	1,000,000
$f(x) = 1/x^2$	10^{-4}	10^{-6}	10^{-8}	10^{-10}	10^{-12}

$$\lim_{x \to \infty} f(x) \to 0$$

49. $T(x) = \begin{cases} 0.20\,x & \text{for } 0 \le x \le 40{,}000 \\ 0.30x - 3800 & \text{for } x > 40{,}000 \end{cases}$

(a)

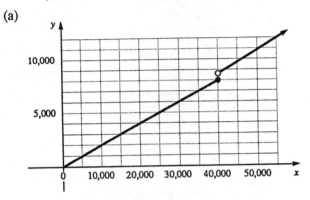

(b) $\lim_{x \to 40{,}000-} T(x) = 0.20(40{,}000) = 8{,}000$

$\lim_{x \to 40{,}000+} T(x) = 0.30(40{,}000) - 3800 = 8{,}200$

$\therefore \lim_{x \to 40{,}000} T(x)$ does not exist.

51.

x	5.9	5.99	5.999	6.001	6.01	6.10
$f(x) = \dfrac{x^2 - 36}{x - 6}$	11.9	11.99	11.999	12.001	12.01	12.10

$$\lim_{x \to 6} f(x) = 12$$

53.

x	1	0.1	0.01	0.001	0.0001	0.00001
$f(x) = x^x$	1	0.794	0.955	0.993	0.99908	0.99988

$$\lim_{x \to 0+} f(x) = 1$$

55.

x	2.9	2.99	2.999	3.001	3.01	3.1
$f(x) = \dfrac{x^3 + 2x^2 - 15x}{x^3 - x^2 - 6x}$	1.612	1.601	1.6001	1.5999	1.599	1.588

$$\lim_{x \to 3} f(x) = 1.6$$

9–2 Continuity

1-5. The functions whose graphs are given in Figures 1 and 3 are continuous at $x = a$. Those given in Figure 5 are not continuous at $x = a$.

7-11. The functions whose graphs are given in Figures 7 and 9 are continuous at $x = a$. Those given in Figure 11 are not continuous at $x = a$.

13.

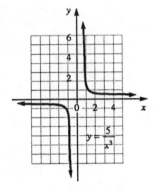

The functions is discontinuous at $x = 0$.

15. $f(x) = \begin{cases} x^3 & \text{if } x \neq 2 \\ 10 & \text{if } x = 2 \end{cases}$

The function is discontinuous at $x = 2$.

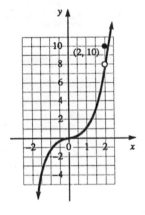

17. $f(x) = \dfrac{(x^2 - 25)}{(x - 5)}$

The function is discontinuous at $x = 5$.

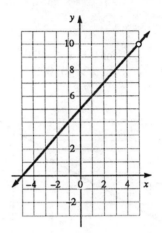

19. (a) $P_0 = \$100,000$
$P_1 = 100,000(1 + 0.03) = \$103,000$
$P_2 = 100,000(1 + 0.03)^2 = \$106,090$
$P_3 = 100,000(1 + 0.03)^3 = \$109,272.70$
$P_4 = 100,000(1 + 0.03)^4 = \$112,550.88$

(b)

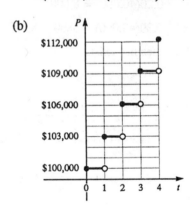

(c) The graph is discontinuous at $t = 1, 2, 3, 4$ (quarters).

21. (a) $\text{Cost}(x) = \begin{cases} 40,000 + 30x & \text{if } 0 \leq x \leq 50,000 \\ 50,000 + 30x & \text{if } 50,000 < x \end{cases}$

x = number of units produced

(b)

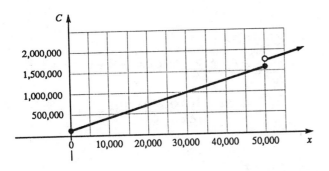

(c) The cost function is discontinuous at $x = 50,000$.

23. (a) The tax function is discontinuous at $x = 29,300$.

(b) The functions is continuous on the intervals $0 \leq x < 29,300$ and $x > 29,300$.

25. $f(x) = \begin{cases} -(x - 120)^2 + 600 & \text{for } 100 \leq x \leq 120 \\ -(x - 140)^2 + 1050 & \text{for } 120 < x \leq 140 \end{cases}$

(a)

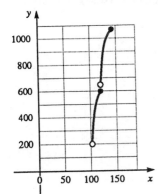

(b) The function is discontinuous at $x = 120$.

9–3 Average Rate of Change

1. (a) Average rate of change

$$= \frac{\Delta y}{\Delta t} = \frac{y(6) - y(2)}{6 - 2} = \frac{18\frac{1}{2} - 16\frac{1}{2}}{4} = 0.5$$

(b) $\dfrac{\Delta y}{\Delta t} = \dfrac{y(9) - y(1)}{9 - 1} = \dfrac{20 - 15\frac{1}{4}}{8} = 0.59375$

3. Average rate of change

$$= \frac{2494 - 2000}{28} = \frac{494}{28} \approx 17.64 \text{ points/week}$$

5. (a) $\dfrac{\Delta y}{\Delta x} = \dfrac{9.00 - 8.25}{10 - 2} = \dfrac{0.75}{8} = 0.09375\%/\text{year}$ to maturity

(b) $\dfrac{\Delta y}{\Delta x} = \dfrac{9.125 - 6.75}{30 - 1/4} = \dfrac{2.375}{29.75} \approx 0.07983\%/\text{year}$ to maturity

7. Average rate of change of inflation

$$= \frac{\Delta(\text{inflation rate})}{\Delta \text{ years}}$$

(a) $\dfrac{21 - 2.5}{8 - 3} = \dfrac{18.5}{5} = 3.7\%/\text{year}$

(b) $\dfrac{21 - 8}{8 - 6} = \dfrac{13}{2} = 6.5\%/\text{year}$

(c) $\dfrac{13 - 2.5}{5 - 3} = \dfrac{10.5}{2} = 5.25\%/\text{year}$

9. (a) $\dfrac{\Delta P}{\Delta Y} = \dfrac{100 - 18}{6 - 2} = \dfrac{82}{4} = 20.5$ thousands/year

(b) $\dfrac{\Delta P}{\Delta Y} = \dfrac{120 - 28}{7 - 3} = \dfrac{92}{4} = 23$ thousands/year

(c) $\dfrac{\Delta P}{\Delta Y} = \dfrac{134 - 70}{10 - 5} = \dfrac{64}{5} = 12.8$ thousands/year

11. $f(x) = 3x$;

$\dfrac{\Delta y}{\Delta x} = \dfrac{f(x + \Delta x) - f(x)}{\Delta x} = \dfrac{3(x + \Delta x) - 3x}{\Delta x} = 3$

13. $f(x) = -2x + 5$;

$\dfrac{\Delta y}{\Delta x} = \dfrac{-2(x + \Delta x) + 5 - (-2x + 5)}{\Delta x} = -2$

15. $f(x) = 4x^2$;

$\dfrac{\Delta y}{\Delta x} = \dfrac{4(x + \Delta x)^2 - 4x^2}{\Delta x} = \dfrac{8x\,\Delta x + 4(\Delta x)^2}{\Delta x}$

$= 8x + 4\Delta x$

17. $f(x) = x^2 - 5x + 8$;

$\dfrac{\Delta y}{\Delta x} = \dfrac{(x + \Delta x)^2 - 5(x + \Delta x) + 8 - (x^2 - 5x + 8)}{\Delta x}$

$= \dfrac{2x\,\Delta x + (\Delta x)^2 - 5\Delta x}{\Delta x} = 2x - 5 + \Delta x$

19. $f(x) = -5x^2 + 9$;

$\dfrac{\Delta y}{\Delta x} = \dfrac{-5(x + \Delta x)^2 + 9 - (-5x^2 + 9)}{\Delta x}$

$= \dfrac{-10x\,\Delta x - 5(\Delta x)^2}{\Delta x} = -10x - 5\Delta x$

21. $f(x) = x^2 - 2x + 1$;

$\dfrac{\Delta y}{\Delta x} = \dfrac{(x + \Delta x)^2 - 2(x + \Delta x) + 1 - (x^2 - 2x + 1)}{\Delta x}$

$= \dfrac{2x\,\Delta x + (\Delta x)^2 - 2\Delta x}{\Delta x} = 2x - 2 + \Delta x$

23. $f(x) = 2x^2 - x + 4$;

$\dfrac{\Delta y}{\Delta x} = \dfrac{2(x + \Delta x)^2 - (x + \Delta x) + 4 - (2x^2 - x + 4)}{\Delta x}$

$= \dfrac{4x\,\Delta x + 2(\Delta x)^2 - \Delta x}{\Delta x} = 4x - 1 + 2\Delta x$

25. $f(x) = x^3$;

$\dfrac{\Delta y}{\Delta x} = \dfrac{(x + \Delta x)^3 - x^3}{\Delta x}$

$= \dfrac{3x^2\,\Delta x + 3x(\Delta x)^2 + (\Delta x)^3}{\Delta x}$

$= 3x^2 + 3x\,\Delta x + (\Delta x)^2$

27. $f(x) = x^4$;

$\dfrac{\Delta y}{\Delta x} = \dfrac{(x + \Delta x)^4 - x^4}{\Delta x}$

$= \dfrac{4x^3\,\Delta x + 6x^2(\Delta x)^2 + 4x(\Delta x)^3 + (\Delta x)^4}{\Delta x}$

$= 4x^3 + 6x^2\,\Delta x + 4x(\Delta x)^2 + (\Delta x)^3$

29. $f(x) = x^4 + x^2 + 4x$;

$$\frac{\Delta y}{\Delta x} = \frac{(x + \Delta x)^4 + (x + \Delta x)^2 + 4(x + \Delta x) - (x^4 + x^2 + 4x)}{\Delta x} = \frac{4x^3\,\Delta x + 6x^2(\Delta x)^2 + 4x(\Delta x)^3 + (\Delta x)^4 - 6x\,\Delta x - 3(\Delta x)^2}{\Delta x}$$

$$= 4x^3\Delta x\ 2x + 4 + (6x^2 + 1)\Delta x + 4x(\Delta x)^2 + (\Delta x)^3$$

31. $f(x) = x^2 - 4x + 5$ from $x = 2$ to $x = 6$;

$$\frac{\Delta y}{\Delta x} = \frac{(x + \Delta x)^2 - 4(x + \Delta x) + 5 - (x^2 - 4x + 5)}{\Delta x} = \frac{2x\,\Delta x + (\Delta x)^2 - 4\Delta x}{\Delta x} = 2x - 4 + \Delta x$$

From $x = 2$ to $x = 6$, $\frac{\Delta y}{\Delta x} = 2(2) - 4 + (6 - 2) = 4$. Graphically, $\frac{\Delta y}{\Delta x}$ is the slope of secant line from $(2,1)$ to $(6,17)$.

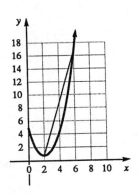

33. $f(x) = 4x + 7$ from $x = 2$ to $x = 3$;

$$\frac{\Delta y}{\Delta x} = \frac{4(x + \Delta x) + 7 - (4x + 7)}{\Delta x} = \frac{4\Delta x}{\Delta x} = 4$$

From $x = 2$ to $x = 3$, $\frac{\Delta y}{\Delta x} = 4$ graphically $\frac{\Delta y}{\Delta x}$ is the slope of the secant line from $(2,15)$ to $(3,19)$.

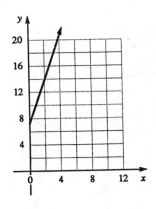

35. $f(x) = -3x^2 - 2x + 1$ from $x = 2$ to $x = 5$;

$$\frac{\Delta y}{\Delta x} = \frac{-3(x + \Delta x)^2 - 2(x + \Delta x) + 1 - (-3x^2 - 2x + 1)}{\Delta x}$$

$$= \frac{-6x\,\Delta x - 3(\Delta x)^2 - 2\Delta x}{\Delta x} = -6x - 2 - 3\Delta x$$

From $x = 2$ to $x = 5$, $\dfrac{\Delta y}{\Delta x} = -12 - 2 - 3(5 - 2) = -23$

Graphically, $\dfrac{\Delta y}{\Delta x}$ is the slope of the secant line from

$(2,-15)$ to $(5,-84)$.

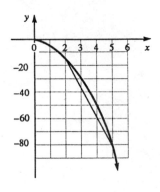

37. $R = f(x) = x^2 - 6x + 9$ $(x \geq 3)$;

$$\frac{\Delta R}{\Delta X} = \frac{R(6) - R(4)}{6 - 4} = \frac{9 - 1}{2} = 4$$

Graphically, $\dfrac{\Delta R}{\Delta X}$ is the slope of the secant line from

$(4,1)$ to $(6,9)$.

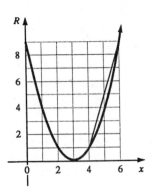

39. $C(x) = 0.5x^2 + 10,000$ $(0 \leq x \leq 1000)$

$$\frac{\Delta C}{\Delta X} = \frac{C(220) - C(200)}{220 - 200} = \frac{34,200 - 30,000}{20} = 210$$

The slope of the secant line from 200 to 220 is 210.

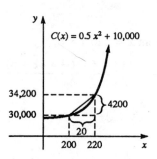

41. $P(x) = -0.1x^2 + 4x - 30$ $(10 \leq x \leq 30)$

$$\frac{\Delta P}{\Delta X} = \frac{P(15) - P(12)}{15 - 12} = \frac{7.5 - 3.6}{3} = 1.3 \text{ million}$$

dollars/1,000 units
This is the slope of the secant line from 12 to 15.

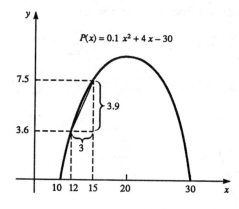

43. $f(x) = \sqrt{x}$

x	Δx	$\dfrac{f(x + \Delta x) - f(x)}{\Delta x}$
4	0.75	$\dfrac{\sqrt{4.75} - 2}{0.75} = 0.239266$
4	0.50	$\dfrac{\sqrt{4.50} - 2}{0.5} = 0.242641$
4	0.25	$\dfrac{\sqrt{4.25} - 2}{0.25} = 0.246211$

45. $f(x) = \dfrac{1}{x^2}$

x	Δx	$\dfrac{f(x + \Delta x) - f(x)}{\Delta x}$
1	0.60	$\dfrac{\dfrac{1}{1.60^2} - 1}{0.60} = -1.015625$
1	0.40	$\dfrac{\dfrac{1}{1.40^2} - 1}{0.40} = -1.224490$
1	0.20	$\dfrac{\dfrac{1}{1.20^2} - 1}{0.20} = -1.527777$

9–4 The Derivative

1. $f(x) = x^2$;

$$f'(x) = \lim_{\Delta x \to 0} \frac{f(x + \Delta x) - f(x)}{\Delta x}$$

$$= \lim_{\Delta x \to 0} \frac{(x + \Delta x)^2 - x^2}{\Delta x}$$

$$= \lim_{\Delta x \to 0} \frac{2x \, \Delta x + (\Delta x)^2}{\Delta x}$$

$$= \lim_{\Delta x \to 0} 2x + \Delta x = 2x$$

3. $f(x) = 6x^3$;

$$f'(x) = 6 \lim_{\Delta x \to 0} \frac{(x + \Delta x)^3 - x^3}{\Delta x}$$

$$= 6 \lim_{\Delta x \to 0} \frac{3x^2 \, \Delta x + 3x(\Delta x)^2 + (\Delta x)^3}{\Delta x}$$

$$= 6 \lim_{\Delta x \to 0} (3x^2 + 3x \, \Delta x + (\Delta x)^2)$$

$$= 18x^2$$

5. $f(x) = x^4$;

$$f'(x) = \lim_{\Delta x \to 0} \frac{(x + \Delta x)^4 - x^4}{\Delta x}$$

$$= \lim_{\Delta x \to 0} \frac{4x^3 \, \Delta x + 6x^2(\Delta x)^2 + 4x(\Delta x)^3 + (\Delta x)^4}{\Delta x}$$

$$= \lim_{\Delta x \to 0} (4x^3 + 6x^2 \, \Delta x + 4x(\Delta x)^2 + (\Delta x)^3)$$

$$= 4x^3$$

7. $f(x) = 3x$;

$$f'(x) = \lim_{\Delta x \to 0} \frac{3(x + \Delta x) - 3x}{\Delta x} = 3 \lim_{\Delta x \to 0} \frac{\Delta x}{\Delta x}$$

$$= 3 \cdot 1 = 3$$

9. $f(x) = x^2 - 5x + 7$;
Since the limit of a sum is the sum of the limits, $f'(x) = 2x - 5$, using the results from Exercises 1 and 8 and the fact that the derivative of a constant is zero.

Proof: If $f(x) = c$, $f'(x) = \lim_{\Delta x \to 0} \dfrac{f(x + \Delta x) - f(x)}{\Delta x}$

$$= \lim_{\Delta x \to 0} \frac{c - c}{\Delta x} = 0.$$

11. $f(x) = -3x^2 + 4x$;
$f'(x) = -3(2x) + 4(1) = -6x + 4$

13. $f(x) = 5x^2 + 6$;
$f'(x) = 5(2x) + 0 = 10x$

15. $f(x) = x^3 - x^2 + 5x$;
$f'(x) = 3x^2 - 2x + 5(1) = 3x^2 - 2x + 5$

17. $f(x) = x^2 - 4x + 1$;
$f'(x) = 2x - 4$; $f'(1) = 2(1) - 4 = -2$,
$f'(2) = 0$, $f'(-2) = -8$

19. $f(x) = x^2 + 6x$;
$f'(x) = 2x + 6$; $f'(1) = 8$,
$f'(2) = 10$, $f'(-2) = 2$

21. $f(x) = -3x^2 + 5$;
$f'(x) = -6x$; $f'(1) = -6$,
$f'(2) = -12$, $f'(-2) = 12$

23. $f(x) = x^2 - 5x + 4$;
$f'(x) = 2x - 5$; $f'(1) = -3$,
$f'(2) = -1$, $f'(-2) = -9$

25. $f(x) = x^3 - 5x^2 + 7$;
$f'(x) = 3x^2 - 10x$; $f'(1) = -7$,
$f'(2) = -8$, $f'(-2) = 32$

27. $f(x) = 2x^3 - 3x^2 + 9$;
$f'(x) = 6x^2 - 6x$; $f'(1) = 0$,
$f'(2) = 12$, $f'(-2) = 36$

29. $y = f(x) = x^2 - 10x$;
$y' = 2x - 10$; $y'(2) = -6$, $y'(3) = -4$

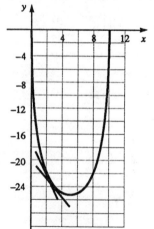

31. $y = f(x) = 5x + 7$;
$y' = 5$; $y'(4) = 5$, $y'(5) = 5$

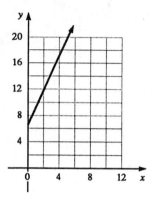

33. $y = f(x) = 3x^2 - 5$; $y' = 6x$

35. $y = f(x) = 5x^2 - 3x + 1$;
$y' = 5(2x) - 3(1) = 10x - 3$

37. $y = f(x) = x^2 - 3x + 5$; $y' = 2x - 3$

39. $y = f(x) = x^2 - 4x + 5$;
$f'(x) = 2x - 4$; $f'(2) = 4 - 4 = 0$

Tangent line: $\dfrac{y - y_0}{x - x_0} = m$; $\dfrac{y - y(2)}{x - 2} = 0$

$$y = y(2) = 1$$

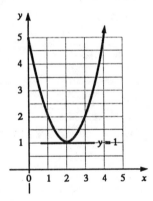

41. $y = f(x) = 4x + 7$;
$f'(x) = 4; f'(2) = 4$

Tangent line: $\dfrac{y - y(2)}{x - 2} = 4; y - 15 = 4x - 8$

$$y = 4x + 7$$

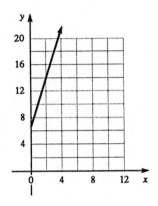

43. $y = f(x) = x^2 - 16x; f'(x) = 2x - 16; f'(1) = -14$

Tangent line: $\dfrac{y - y(1)}{x - 1} = -14;$

$$y - (-15) = -14x + 14$$
$$y = -14x - 1$$

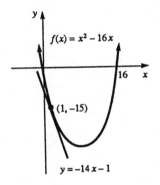

45. $y = f(x) = -3x^2 + 60x \quad (0 \leq x \leq 20)$

(a) $f'(x) = -6x + 60$

(b) $f'(3) = -18 + 60 = 42$

(c) Tangent line: $\dfrac{y - y(3)}{x - 3} = 42;$

$$y - 153 = 42x - 126$$
$$y = 42x + 27$$

47. $f(x) = \sqrt{x}$

x	Δx	$\dfrac{f(x + \Delta x) - f(x)}{\Delta x}$
4	0.1	$\dfrac{\sqrt{4.1} - 2}{0.1} = 0.2484567$
4	0.01	$\dfrac{\sqrt{4.01} - 2}{0.01} = 0.2498439$
4	0.001	$\dfrac{\sqrt{4.001} - 2}{0.001} = 0.2499843$
4	0.0001	$\dfrac{\sqrt{4.0001} - 2}{0.0001} = 0.2499980$
4	0.00001	$\dfrac{\sqrt{4.00001} - 2}{0.00001} = 0.2500000$

x	Δx	$\dfrac{f(x + \Delta x) - f(x)}{\Delta x}$
4	−0.1	$\dfrac{\sqrt{3.9} - 2}{-0.1} = 0.2515823$
4	−0.01	$\dfrac{\sqrt{3.99} - 2}{-0.01} = 0.2501564$
4	−0.001	$\dfrac{\sqrt{3.999} - 2}{-0.001} = 0.2500156$
4	−0.0001	$\dfrac{\sqrt{3.9999} - 2}{-0.0001} = 0.2500012$
4	−0.00001	$\dfrac{\sqrt{3.99999} - 2}{-0.00001} = 0.2500000$

$f'(4) = 0.25$

The difference quotient appears to be approaching 0.25.

49. $f(x) = \dfrac{1}{x^2}$

x	Δx	$\dfrac{f(x + \Delta x) - f(x)}{\Delta x}$
1	0.1	$\dfrac{\dfrac{1}{1.1^2} - 1}{0.1} = -1.7355372$
1	0.01	$\dfrac{\dfrac{1}{1.01^2} - 1}{0.01} = -1.9703951$
1	0.001	$\dfrac{\dfrac{1}{1.001^2} - 1}{0.001} = -1.9970040$
1	0.0001	$\dfrac{\dfrac{1}{1.0001^2} - 1}{0.0001} = -1.9997001$
1	0.00001	$\dfrac{\dfrac{1}{1.00001^2} - 1}{0.00001} = -1.9999710$

x	Δx	$\dfrac{f(x + \Delta x) - f(x)}{\Delta x}$
1	−0.1	$\dfrac{\dfrac{1}{1.9^2} - 1}{-0.1} = -2.3456790$
1	−0.01	$\dfrac{\dfrac{1}{1.99^2} - 1}{-0.01} = -2.0304051$
1	−0.001	$\dfrac{\dfrac{1}{1.999^2} - 1}{-0.001} = -2.0030040$
1	−0.0001	$\dfrac{\dfrac{1}{1.9999^2} - 1}{-0.0001} = -2.0003000$
1	−0.00001	$\dfrac{\dfrac{1}{1.99999^2} - 1}{-0.00001} = -2.0000300$

$f'(1) = -2.0$

The difference quotient appears to be approaching −2.

51. $f(x) = 7x - 2$, $x = -1$;

$$f'(-1) \approx \frac{7(-1.00001) - 2 - [7(-1) - 2]}{-0.00001} = 7.0$$

53. $f(x) = 3x^2 - 2x + 5$, $x = 1$

$$f'(1) \approx \frac{[3(1.00001)^2 - 2(1.00001) + 5] - [3(1)^2 - 2(1) + 5]}{0.00001} = 4.00003$$

$$\therefore f'(1) \approx 4.0$$

55. $f(x) = \dfrac{4x - 7}{x + 3}$, $x = 2$

$$f'(2) \approx \left[\frac{4(2.00001) - 7}{2.0001 + 3} - \frac{4(2) - 7}{2 + 3} \right] / 0.00001 = 0.76$$

57. $f(x) = \sqrt{2x + 10}$, $x = 3$

$$f'(3) \approx \frac{\sqrt{2(3.00001) + 10} - \sqrt{2 \cdot 3 + 10}}{0.00001} = 0.25$$

59. $f(x) = \dfrac{5}{\sqrt{x} + 8}$, $x = 9$

$$f'(9) \approx \left[\frac{5}{\sqrt{9.00001} + 8} - \frac{5}{\sqrt{9} + 8} \right] / 0.00001$$

$$\approx -0.006887$$

9–5 Differentiability and Continuity

1. The functions whose graphs are shown in (A) and (D) are differentiable at $x = a$.

3. $f(x) = \dfrac{1}{(x-5)^2}; f'(x) = \dfrac{-2}{(x-5)^3}$

(a)

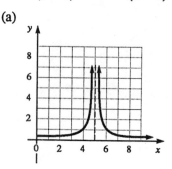

(b) f is not continuous at $x = 5$.

(c) f is not differentiable at $x = 5$.

5. $f(x) = x^{\frac{1}{3}}; f'(x) = \dfrac{1}{3x^{2/3}}$

f is not differentiable at $x = 0$.

7. $f(x) = \begin{cases} x^2 + 5 & \text{if } x \neq 4 \\ 20 & \text{if } x = 4 \end{cases}$

(a)

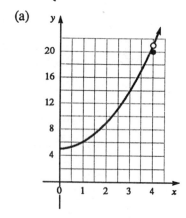

(b) f is discontinuous at $x = 4$.

(c) f is not differentiable at $x = 4$.

9. $f(x) = \dfrac{x^2 - 81}{x + 9}$

(a)

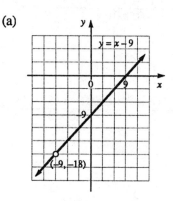

(b) f is discontinuous at $x = -9$.

(c) f is not differentiable at $x = -9$.

11. $f(x) = \sqrt{x - 5}; f'(x) = \dfrac{1}{2\sqrt{x - 5}}$

f is not differentiable at $x = 5$, within its domain. (f is not defined for $x \leq 5$.)

13. $f(x) = \dfrac{1}{x - 2}; f'(x) = \dfrac{-1}{(x - 2)^2}$

f is not defined at $x = 2$, and therefore it is not differentiable there.

15. $f(x) = \dfrac{x + 5}{x - 1}; f'(x) = \dfrac{-6}{(x - 1)^2}$

f is not defined at $x = 1$, and therefore it is not differentiable there.

17. $f(x) = \dfrac{1}{\sqrt{x - 3}}; f'(x) = \dfrac{-1}{2\sqrt{(x - 3)^3}}$

f is not differentiable at $x \leq 3$, within its domain.

19. $f(x) = (7 - x)^{4/5}; f'(x) = \dfrac{-4}{5(7 - x)^{1/5}}$

f is not differentiable at $x = 7$.

21. $f(x) = (4x - 1)^{3/5}; f'(x) = \dfrac{12}{5(4x - 1)^{2/5}}$

f is not differentiable at $x = 1/4$.

23. $f(x) = (x - 6)^{5/2}; f'(x) = \dfrac{5}{2}(x - 6)^{3/2}$

f is not defined for $x < 6$, and therefore it is not differentiable for $x < 6$.

9–6 Rules for Finding Derivatives

1. $y = x^3; \dfrac{dy}{dx} = 3x^2$ by the power rule.

3. $y = x^{20}; \dfrac{dy}{dx} = 20x^{19}$

5. $y = 4x; \dfrac{dy}{dx} = 4(1) = 4$ by the constant multiplier rule.

7. $y = 1/x^6; \dfrac{dy}{dx} = \dfrac{-6}{x^7}$

9. $y = x^5; \dfrac{dy}{dx} = 5x^4$

11. $y = 1/x^3; \dfrac{dy}{dx} = \dfrac{-3}{x^4}$

13. $y = x^{-1/4}; \dfrac{dy}{dx} = -\dfrac{1}{4}x^{-5/4}$

15. $y = 6; \dfrac{dy}{dx} = 0$

17. $f(x) = -x^8; f'(x) = -8x^7$

19. $f(x) = -3/x^2 = -3x^{-2}; f'(x) = 6x^{-3}$

21. $f(x) = 5/\sqrt{x^3} = 5/x^{3/2}; f'(x) = \dfrac{-15}{2x^{5/2}}$

23. $f(x) = 1/6; f'(x) = 0$

25. $f(x) = 5/\sqrt{x}; f'(x) = \dfrac{-5}{2x^{3/2}}; f'(4) = \dfrac{-5}{2 \cdot 8} = \dfrac{-5}{16}$

27. $f(x) = 8\sqrt{x^3} = 8x^{3/2}; f'(x) = 12x^{1/2}; f'(4) = 24$

29. $f(x) = -2/x^3; f'(x) = \dfrac{6}{x^4}; f'(4) = \dfrac{6}{256} = \dfrac{3}{128}$

31. $y = x^2 - 8x; \dfrac{dy}{dx} = 2x - 8$

33. $y = 8x^2 + 10; \dfrac{dy}{dx} = 16x$

35. $y = x^3 + 2x^2 + 9; \dfrac{dy}{dx} = 3x^2 + 4x$

37. $y = x^6 - x^5 + 4/x^2 + 9; y' = 6x^5 - 5x^4 - \dfrac{8}{x^3}$

39. $y = -5x^3 - 6x^2 + 8x - 4; y' = -15x^2 - 12x + 8$

41. $y = x^3(x^2 - 6x + 8) = x^5 - 6x^4 + 8x^3;$
$y' = 5x^4 - 24x^3 + 24x^2$

43. $\dfrac{d}{dx}(3x^2 + 7) = 6x$

45. $\dfrac{d}{dx}(-2x^2 + 1) = -4x$

47. $\dfrac{d}{dx}(-3x^3 + 4) = -9x^2$

49. $\dfrac{d}{dx}(x^3 - 4x^2 + 5) = 3x^2 - 8x$

51. $\dfrac{d}{dx}\left[\dfrac{1}{\sqrt{x}} + 4\sqrt{x}\right] = -\dfrac{1}{2x^{3/2}} + \dfrac{2}{\sqrt{x}}$

53. $\dfrac{d}{dt}(-3t^2 + 8t + 7) = -6t + 8$

55. $\dfrac{d}{dw}(w^3 - 5w + 8) = 3w^2 - 5$

57. $y = -2x^3 + 4x;\ \dfrac{dy}{dx} = -6x^2 + 4;\ \dfrac{dy}{dx}\Big|_{x=1} = -6 + 4 = -2$

59. $y = \dfrac{-6}{\sqrt{x}} + 7;\ \dfrac{dy}{dx} = \dfrac{3}{x^{3/2}};\ \dfrac{dy}{dx}\Big|_{x=1} = 3$

61. $y = -3x^3 + 7x;$

$\dfrac{dy}{dx} = -9x^2 + 7;\ \dfrac{dy}{dx}\Big|_{x=4}\ -144 + 7 = -137$

63. $y = \sqrt{x} - \dfrac{4}{\sqrt{x}};\ \dfrac{dy}{dx} = \dfrac{1}{2\sqrt{x}} + \dfrac{2}{x^{3/2}};\ \dfrac{dy}{dx}\Big|_{x=4} = \dfrac{1}{4} + \dfrac{2}{8} = \dfrac{1}{2}$

65. $y = -3x^2 + 2x + 1;\ y' = -6x + 2$

67. $y = x^3 + 6x^2 + 8x + 7;\ y' = 3x^2 + 12x + 8$

69. $y = x^5 - 6x^3 + 6x + 5;\ y' = 5x^4 - 18x^2 + 6$

71. $D_x(5x^{10}) = 50x^9$

73. $D_x(-3/x^2) = \dfrac{6}{x^3}$

75. $D_x(4x^3 - 6x^2 + 4x + 8) = 12x^2 - 12x + 4$

77. $f(x) = 2x^3 + 5;\ f'(x) = 6x^2;\ f'(1) = 6$

Tangent line: $\dfrac{y - f(1)}{x - 1} = 6;\ \dfrac{y - 7}{x - 1} = 6;$

$$y - 7 = 6x - 6$$
$$y = 6x + 1$$

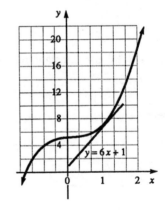

79. $P(x) = 0.03x^2 + 5 \ (x \geq 0)$

(a) $P'(x) = 0.06x$;

$P'(2) = 0.12$ million dollars/year

(b) $P'(3) = 0.18$ million dollars/year

(c) If $P'(x) = 0.18$ for $x \geq 3$,

$\dfrac{P(x) - P(3)}{x - 3} = 0.18$; $P(x) - 5.27 = 0.18x - 0.54$

$P(x) = 0.18x + 4.73$
$P(7) = 0.18(7) + 4.73 = 5.99$ million dollars

81. $y = f(x) = 6x^2 \ (x \geq 0)$

(a) $f'(x) = 12x$

(b) $f'(2) = 24; f'(3) = 36;$
The rate of change is not constant.

(c) If $f'(x) = 36$ for $x \geq 3$,

$\dfrac{y - f(3)}{x - 3} = 36$; $y - 54 = 36x - 108$;

$y = 36x - 54$

$y(5) = 180 - 54 = 126$ million dollars

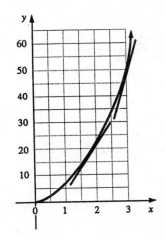

83. $R(x) = -2x^2 + 60x \ (0 \leq x \leq 30)$

(a) $R'(x) = -4x + 60$

(b) $R'(10) = -4(10) + 60 = 20$
This is approximately the additional revenue obtained by selling one more unit beyond 10 units. It is the slope of the line tangent to the graph of $R(x)$ at $x = 10$.

(c)

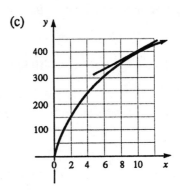

85. $S(x) = -16x^2 + 64x \ (0 \leq x \leq 4)$

(a) $S'(x) = -32x + 64$
$S'(1) = -32 + 64 = 32$ feet/second

(b) $S'(2) = -32(2) + 64 = 0$ feet/second

(c) $S'(3) = -32(3) + 64 = -32$ feet/second
(The ball is moving downward.)

(d)

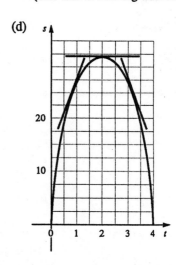

87. $P(x) = 4x - \dfrac{x^3}{1,000,000}$ $(0 \le x \le 2000)$

(a) Marginal profit $= P'(x) = 4 - \dfrac{3x^2}{1,000,000}$

(b) $P'(100) = 4 - \dfrac{3(10,000)}{1,000,000} = 4 - \dfrac{3}{100} = 3.97$

This is the approximate additional profit gained from selling one more unit when at the level of $x = 100$.

(c) $P'(1900) = 4 - \dfrac{3(3,610,000)}{1,000,000} = 4 - 10.83$

$\qquad = -6.83$

The approximate change in profit in moving from a sales level of 1900 units to one additional unit is negative.

89. $y = 90\sqrt{x}$ $(x \ge 0)$

(a) $y' = \dfrac{45}{\sqrt{x}}$ units/hour

(b) $y'(4) = \dfrac{45}{\sqrt{2}} = 22.5$ units/hour

(c) $y'(9) = \dfrac{45}{\sqrt{9}} = 15$ units/hour

91. $P(t) = 4000t^2 + 200,000$

(a) $P'(t) = 8000t$ people/year

(b) $P'(3) = 8000(3) = 24,000$ people/year

(c) $P'(6) = 8000(6) = 48,000$ people/year

9–7 The Product and Quotient Rules

1. $y = (x^2 + 4x + 5)(x^3 - 2x^2 + 7);$
$y = (x^2 + 4x + 5)(3x^2 - 4x) + (2x + 4)(x^3 - 2x^2 + 7)$
$\quad = (3x^4 + 8x^3 - x^2 - 20x) + (2x^4 - 8x^2 + 14x + 28)$
$\quad = 5x^4 + 8x^3 - 9x^2 - 6x + 28$

3. $f(x) = (x^3 - 2x + 6)(x^2 - 3x + 2)$
$f'(x) = (x^3 - 2x + 6)(2x - 3)$
$\qquad + (x^2 - 3x + 2)(3x^2 - 2)$

5. $y = (x^3 - x^2 - 2)(x^2 - x - 1)$
$y' = (x^3 - x^2 - 2)(2x - 1) + (x^2 - x - 1)(3x^2 - 2x)$

7. $y = (x^7 - 6x^2)(4x^2 - 3x + 4)$

$\dfrac{dy}{dx} = (x^7 - 6x^2)(8x - 3) + (4x^2 - 3x + 4)(7x^6 - 12x)$

9. $y = (5x^2 - 3x + 1)(3x^4 - 7)$

$\dfrac{dy}{dx} = (5x^2 - 3x + 1)(12x^2) + (3x^4 - 7)(10x - 3)$

11. $y = \dfrac{x^2 - 4x + 5}{x^3 - 3x + 9}$

$y' = \dfrac{(x^3 - 3x + 9)(2x - 4) - (x^2 - 4x + 5)(3x^2 - 3)}{(x^3 - 3x + 9)^2}$

13. $f(x) = \dfrac{x^3}{x + 3}$

$f'(x) = \dfrac{(x + 5)(3x^2) - x^3}{(x + 5)^2} = \dfrac{2x^3 + 15x^2}{(x + 5)^2}$

15. $g(x) = \dfrac{x^3 + 7}{x^4}$

$g'(x) = \dfrac{x^4(3x^2) - (x^3 + 7)(4x^3)}{x^8} = \dfrac{3x^3 - 4(x^3 + 7)}{x^5}$

$\qquad = \dfrac{-x^3 - 28}{x^5}$

17. $y = 8 - \dfrac{6}{x-2} + \dfrac{3x}{5x+1} = 8 - 6(x-2)^{-1} + 3x(5x+1)^{-1}$

$y' = 0 + 6(x-2) - 2 + 3(5x+1)^{-1} - 3x(5x+1)^{-2}(5) = \dfrac{6}{(x-2)^2} + \dfrac{3(5x+1) - 15x}{(5x+1)^2} = \dfrac{6}{(x-2)^2} + \dfrac{3}{(5x+1)^2}$

19. $y = \dfrac{x^3+9}{x^2-9}$

$y' = \dfrac{(x^2-9)(3x^2) - (x^3+9)(2x)}{(x^2-9)^2} = \dfrac{3x^4 - 27x^2 - 2x^4 - 18x}{(x^2-9)^2} = \dfrac{x^4 - 27x^2 - 18x}{(x^2-9)^2}$

21. $y = \dfrac{1}{x_2 - 3x} = (x^2 - 3x)^{-1}$

$\dfrac{dy}{dx} = -(x^2-3x)^{-2}(2x-3) = \dfrac{-2x+3}{(x^2-3x)^2}$

23. $y = \dfrac{4x+1}{3x^2-2x}$

$\dfrac{dy}{dx} = \dfrac{(3x^2-2x)(4) - (4x+1)(6x-2)}{(3x^2-2x)^2} = \dfrac{-2(6x^2+3x-1)}{(3x^2-2x)^2}$

25. $y = \dfrac{5x^3-2x+1}{\sqrt[5]{x}} = \dfrac{5x^3}{\sqrt[5]{x}} - \dfrac{2x}{\sqrt[5]{x}} + \dfrac{1}{\sqrt[5]{x}} = 5x^{\frac{14}{5}} - 2x^{\frac{4}{5}} + x^{-\frac{1}{5}}$

$\dfrac{dy}{dx} = 14x^{\frac{9}{5}} - \dfrac{8}{5}x^{-\frac{1}{5}} - \dfrac{1}{5}x^{-\frac{6}{5}} = \dfrac{14x^3 - \frac{8}{5}x - \frac{1}{5}}{x^{\frac{6}{5}}}$

27. $y = \dfrac{x^4 - 8x^3 + 5x + 1}{5x^3 - 7x^2 + 3}$

$\dfrac{dy}{dx} = \dfrac{(5x^3 - 7x^2 + 3)(4x^3 - 24x^2 + 5) - (x^4 - 8x^3 + 5x + 1)(15x^2 - 14x)}{(5x^3 - 7x^2 + 3)^2}$

29. $y = \dfrac{\sqrt{x}+x}{x - 1/\sqrt{x}}$

$\dfrac{dy}{dx} = \dfrac{(x - 1/\sqrt{x})\left[\dfrac{1}{2\sqrt{x}} + 1\right] - (\sqrt{x}+x)\left[1 + \dfrac{1}{2x^{3/2}}\right]}{(x - 1/\sqrt{x})^2}$

30. $y = \dfrac{x + 1/\sqrt{x}}{\sqrt{x} - 1}$

$$\frac{dy}{dx} = \frac{(\sqrt{x} - 1)\left[1 - \dfrac{1}{2x^{3/2}}\right] - \left[x + \dfrac{1}{\sqrt{x}}\right]\left[\dfrac{1}{2\sqrt{x}}\right]}{(\sqrt{x} - 1)^2}$$

31. $y = \dfrac{x^2 - 3x + 1}{x^3 - 2x}$

$$\frac{dy}{dx} = \frac{(x^3 - 2x)(2x - 3) - (x^2 - 3x + 1)(3x^2 - 2)}{(x^3 - 2x)^2}$$

$$\left.\frac{dy}{dx}\right|_{x=1} = \frac{(-1)(-1) - (-1)(1)}{(-1)^2} = \frac{1 + 1}{1} = 2$$

33. $y = \dfrac{\sqrt{x} + 6x}{x - 2/\sqrt{x}}$

$$\frac{dy}{dx} = \frac{(x - 2/\sqrt{x})(1/(2\sqrt{x}) + 6) - (\sqrt{x} + 6x)(1 + 1/x^{3/2})}{(x - 2/\sqrt{x})^2}$$

$$\left.\frac{dy}{dx}\right|_{x=1} = \frac{(-1)(13/2) - (7)(2)}{(-1)^2} = \frac{-13/2 - 14}{1} = -\frac{41}{2}$$

35. $y = (x^3 - 4x^2 + 5)(x^2 - 8x + 7)$
$y' = (x^3 - 4x^2 + 5)(2x - 8) + (3x^2 - 8x)(x^2 - 8x + 7)$

37. $y = \dfrac{(x^3 - 4x + 7)(x^2 - 2x)}{x^3 - 6x + 1}$

$$y' = \frac{[x^3 - 6x + 1][(x^3 - 4x + 7)(2x - 2) + (3x^2 - 4)(x^2 - 2x)] - (x^3 - 4x + 7)(x^2 - 2x)(3x^2 - 6)}{(x^3 - 6x + 1)^2}$$

39. $y = \dfrac{(x^3 - 6x^2 + 5)(x^2 - 4x)}{(x^5 - 8x^2)(x^3 + 1)}$

$$y' = \dfrac{\begin{array}{l}(x^5 - 8x^2)(x^3 + 1)[(x^3 - 6x^2 + 5)(2x - 4) + (3x^2 - 12x)(x^2 - 4x)] \\ \quad -(x^3 - 6x^2 + 5)(x^2 - 4x)[(x^5 - 8x^2)(3x^2) + (5x^4 - 16x)(x^3 + 1)]\end{array}}{[(x^5 - 9x^2)(x^3 + 1)]^2}$$

41. $g(t) = (t^5 - 4)\left[\dfrac{t^3 + 6}{t^4 + 7}\right]$

$$g'(t) = (t^5 - 4)\left[\dfrac{(t^4 + 7)(3t^2) - (t^3 + 6)(4t^3)}{(t^4 + 7)^2}\right] + (5t^4)\left[\dfrac{t^3 + 6}{t^4 + 7}\right]$$

43. $y = (x^2 - 4x + 1)\left[\dfrac{x^3 - 5x}{x^2 + 6}\right]$

$$y' = (x^2 - 4x + 1)\left[\dfrac{(x^2 + 6)(3x^2 - 5) - (x^3 - 5x)(2x)}{(x^2 + 6)^2}\right] + (2x - 4)\left[\dfrac{x^3 - 5x}{x^2 + 6}\right]$$

45. $y = x^3(x^6 - 8x^2) = x^9 - 8x^5$

(a) $y' = 9x^8 - 40x^4 = x^4(9x^4 - 40)$

(b) $y' = x^3(6x^5 - 16x) + 3x^2(x^6 - 8x^2) = 6x^8 - 16x^4 + 3x^8 - 24x^4 = 9x^8 - 40x^4$

47. $f(x) = (x^2 + 2)(x^3 - 2x + 5)$
$f'(x) = (x^2 + 2)(3x^2 - 2) + (2x)(x^3 - 2x + 5)$
$f'(1) = (3)(1) + (2)(4) = 11$

49. $f(x) = (x^2 - 9)(x + 5)$
$f'(x) = (x^2 - 9)(1) + (2x)(x + 5)$
$f'(1) = -8 + (2)(6) = 4$

Tangent line: $\dfrac{y - f(1)}{x - 1} = 4;\ \dfrac{y - (-48)}{x - 1} = 4$

$y + 48 = 4x - 4;\ y = 4x - 52$

51. $f(x) = \dfrac{1}{x^2}$

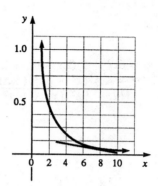

$f'(x) = \dfrac{-2}{x^3}; f'(7) = \dfrac{-2}{343}$

Tangent line: $\dfrac{y - f(7)}{x - 7} = -\dfrac{2}{343}; \dfrac{y - 1/49}{x - 7} = -\dfrac{2}{343}$

$y - \dfrac{1}{49} = -\dfrac{2}{343}x + \dfrac{2}{49}; y = -\dfrac{2}{343}x + \dfrac{3}{49}$

53. $f(x) = \dfrac{(x^2 - 36)(x + 4)}{x^2 - 81}$

$f'(x) = \dfrac{(x^2 - 81)[(x^2 - 36) + (2x)(x + 4)] - (x^2 - 36)(x + 4)(2x)}{(x^2 - 81)^2}$

$f'(1) = \dfrac{(-80)[-35 + 10] - (-35)(5)(2)}{80^2} = \dfrac{2000 + 350}{6400} = \dfrac{47}{128}$

Tangent line: $\dfrac{y - f(1)}{x - 1} = \dfrac{47}{128}; \dfrac{y - 35/16}{x - 1} = \dfrac{47}{128}$

$y - \dfrac{35}{16} = \dfrac{47}{128}x - \dfrac{47}{128}$

$y = \dfrac{47}{128}x + \dfrac{233}{128}$, or $128y - 47x = 233$

55. (a) $R(x) = p(t) \times (t) = (10 + 2t)(20 + 0.2t^2)$

   (b) $\dfrac{dR}{dt} = (10 + 2t)(0.4t) + 2(20 + 0.2t^2)$

   $= 40 + 4t + 1.2t^2$

   (c) $\dfrac{dR}{dt}\Big|_{t=2} = (14)(0.8) + (2)(20.8) = 52.8$

   At $t = 2$, the marginal revenue is \$52.80/unit.

57. (a) $\bar{C}(x) = \dfrac{C(x)}{x} = \dfrac{10x + 500}{x} = 10 + \dfrac{500}{x}$

(b) $\bar{C}'(x) = \dfrac{-500}{x^2}$

(c) $\bar{C}'(4) = \dfrac{-500}{16} = -31.25;\quad \bar{C}'(5) = \dfrac{-500}{25} = -20;$

$\bar{C}'(6) = \dfrac{-500}{36} \approx -13.89$

In each case $\bar{C}'(x_0)$ represents the rate of change of the average cost at $x = x_0$. At $x = 5$, the average cost per unit in going to $x = 6$ decreases by approximately \$20.

59. (a) $\bar{C}(x) = \dfrac{C(x)}{x} = \dfrac{-x^2 + 30x}{x} = -x + 30$

(b) $\bar{C}'(x) = -1$

(c) $\bar{C}'(4) = \bar{C}'(5) = \bar{C}'(6) = -1$

The average cost per unit decreases by \$1 when the production is increased by one unit.

61. $y = \dfrac{30}{110 - x} \quad (0 \le x \le 100)$

(a) $\dfrac{dy}{dx} = \dfrac{30}{(110 - x)^2}$

(b) $\dfrac{dy}{dx}\bigg|_{x=60} = \dfrac{30}{50^2} = 0.012$

$\dfrac{dy}{dx}\bigg|_{x=80} = \dfrac{30}{(30)^2} \approx 0.0333$

$\dfrac{dy}{dx}\bigg|_{x=90} = \dfrac{30}{20^2} = 0.075$

In each case $\dfrac{dy}{dx}\bigg|_{x=x_0}$ represents the rate of change in the cost of y as x is increased. The cost rises more rapidly as x_0 approaches 110.

9–8 The Chain Rule

1. $z = t^5 - 4t^3 + 7t - 8$

   $\dfrac{dz}{dt} = 5t^4 - 12t^2 + 7$

3. $y = u^5$

   $\dfrac{dy}{du} = 5u^4$

5. $y = 3u^6 - 8u^5 + 4u - 8$

   $\dfrac{dy}{du} = 18u^5 - 40u^4 + 4$

7. $f(u) = -7u^3 - 8u^2 + 6$
   $f'(u) = -21u^2 - 16u$

9. $y = (x^3 + 5)^{11}$
   $y' = 11(x^3 + 5)^{10}(3x^2) = 33x^2(x^3 + 5)^{10}$

11. $y = (x^3 - 6x)^{-4}$

    $y' = -4(x^3 - 6x)^{-5}(3x^2 - 6) = \dfrac{-4(3x^2 - 6)}{(x^3 - 6x)^5}$

13. $f(x) = (x^4 - 9)^{15}$
    $f'(x) = 15(x^4 - 9)^{14}(4x^3) = 60x^3(x^4 - 9)^{14}$

15. $y = \dfrac{1}{\sqrt[3]{(x^6 - 8x)^2}} = (x^6 - 8x)^{-2/3}$

    $y' = -\dfrac{2}{3}(x^6 - 8x^2)^{-5/3}(6x^5 - 8) = \dfrac{-2(6x^5 - 8)}{3(x^6 - 8x)^{5/3}}$

17. $y = (x^3 - 4x^2 + 5)^{20}$
    $y' = 20(x^3 - 4x^2 + 5)^{19}(3x^2 - 8x) = 20(3x^2 - 8x)(x^3 - 4x^2 + 5)^{19}$

19. $y = (x^3 - 4x)^{10}(x^2 + 5)$
$y' = 10(x^3 - 4x)^9(3x^2 - 4)(x^2 + 5) + (x^3 - 4x)^{10}(2x)$

21. $y = \left[\dfrac{5x - 3}{8 - x^4}\right]^7$

$y' = 7\left[\dfrac{5x - 3}{8 - x^4}\right]^6\left[\dfrac{(8 - x^4)(5) - (5x - 3)(-4x^3)}{(8 - x^4)}\right] = 7\left[\dfrac{5x - 3}{8 - x^4}\right]^6\left[\dfrac{15x^4 - 12x^3 + 40}{(8 - x^4)^2}\right]$

23. $y = \dfrac{(5x - 3)^7}{(8 - x^4)^7} = \left[\dfrac{5x - 3}{8 - x^4}\right]^7$

$y' = 7\left[\dfrac{5x - 3}{8 - x^4}\right]^6\left[\dfrac{(8 - x^4)(5) - (5x - 3)(-4x^3)}{(8 - x^4)^2}\right]$

25. $y = [(5x - 3)(x^2 - 1)]^{10}$
$y' = 10[(5x - 3)(x^2 - 1)]^9[(5x - 3)(2x) + 5(x^2 - 1)] = 10[(5x - 3)(x^2 - 1)]^9[15x^2 - 6x - 5]$

27. $y = (5x - 3)^{10}(x^2 - 1)^{10} = [(5x - 3)(x^2 - 1)]^{10}$
$y' = 10[(5x - 3)(x^2 - 1)]^9[(5x - 3)(2x) + 5(x^2 - 1)]$

29. $y = (x^3 - 4x^2)^5(x^2 - 1)^3$
$y' = (x^3 - 4x^2)^5(3)(x^2 - 1)^2(2x) + 5(x^3 - 4x^2)^4(3x^2 - 8x)(x^2 - 1)^3$

31. $y = (x^3 + 1)^5 + (x^2 - 2)^7$
$y' = 5(x^3 + 1)^4(3x^2) + 7(x^2 - 2)^6(2x) = 15x^2(x^3 + 1)^4 + 14x(x^2 - 2)^6$

33. $y = (\sqrt{x} + 9)^5$

$y' = 5(\sqrt{x} + 9)^4\left[\dfrac{1}{2\sqrt{x}}\right] = \dfrac{5}{2\sqrt{x}}(\sqrt{x} + 9)^4$

35. $y = x\sqrt{x^3 + 1}$

$y' = x \cdot \dfrac{3x^2}{2\sqrt{x^3 + 1}} + \sqrt{x^3 + 1} = \dfrac{3x^3 + 2(x^3 + 1)}{2\sqrt{x^3 + 1}} = \dfrac{5x^3 + 2}{2\sqrt{x^3 + 1}}$

37. $y = \dfrac{x^3}{\sqrt{x+1}}$

$$y' = \dfrac{\sqrt{x+1}(3x^2) - \dfrac{x^3}{2\sqrt{x+1}}}{x+1} = \dfrac{2(x+1)(3x^2) - x^3}{2(x+1)^{3/2}} = \dfrac{5x^3 + 6x^2}{2(x+1)^{3/2}}$$

39. $y = \dfrac{1}{x^3 + 4}$

$$\dfrac{dy}{dx} = \dfrac{-3x^2}{(x^3+4)^2}; \ \dfrac{dy}{dx}\Big|_{x=1} = \dfrac{-3}{(1+4)^2} = -\dfrac{3}{25}$$

41. $y = \dfrac{x}{\sqrt{x^3 + 9}}$

$$\dfrac{dy}{dx}\Big| = \dfrac{\sqrt{x^3+9} - x \cdot \dfrac{3x^2}{2\sqrt{x^3+9}}}{x^3+9}$$

$$= \dfrac{2(x^3+9) - 3x^3}{2(x^3+9)^{3/2}} = \dfrac{18 - x^3}{2(x^3+9)^{3/2}}$$

$$\dfrac{dy}{dx}\Big|_{x=1} = \dfrac{18-1}{2(1+9)^{\frac{3}{2}}} = \dfrac{17}{20\sqrt{10}} \text{ or } \dfrac{17\sqrt{10}}{200}$$

43. $f(x) = (x-7)^5$
$f'(x) = 5(x-7)^4; \ f'(3) = 5(-4)^4 = 5(25b) = 1280$

Tangent line: $\dfrac{y - f(3)}{x-3} = 1280; \ \dfrac{y - (-1024)}{x-3} = 1280$

$$y + 1024 = 1280x - 3840$$
$$y = 1280x - 4864$$

45. $C(x) = x^3 + 9 \ (x \geq 0)$
$\quad\ x = 5t + 4 \ (t \geq 0)$

(a) $\dfrac{dC}{dt} = \dfrac{dC}{dx}\dfrac{dx}{dt} = 3x^2 \cdot 5 = 15(5t+4)^2$

(b) $\dfrac{dC}{dt}\Big|_{t=6} = 15(30+4)^2 = 17{,}340$, the instantaneous rate of change with respect to t at $t = 6$.

47. $C(x) = 0.01x^3 + 8 \ (x \geq 0)$

$x = 50t + 90; \ x = 200 \Rightarrow t = 110/50 = 2.2$

$$\frac{dC}{dt} = \frac{dC}{dx}\frac{dx}{dt} = (0.03x^2)(50) = 1.5x^2 = 1.5(50t + 90)^2$$

$$\left.\frac{dC}{dt}\right|_{t=2.2} = 1.5(200)^2$$

$$= 60,000 \text{ millions of dollars/month}$$

49. $S = 10,000\left[1 + \dfrac{r}{400}\right]^{28}$

(a) $\dfrac{dS}{dr} = 280,000\left[1 + \dfrac{r}{400}\right]^{27}\left[\dfrac{1}{400}\right] = 700\left[1 + \dfrac{r}{400}\right]^{27}$

(b) $\left.\dfrac{dS}{dr}\right|_{r=8} = 700\left[1 + \dfrac{8}{400}\right]^{27} = 1194.82$

(c) $\left.\dfrac{dS}{dr}\right|_{r=10} = 700\left[1 + \dfrac{10}{400}\right]^{27} = 1363.46$

$\dfrac{dS}{dr}$ represents the instantaneous rate of change of S with respect to r. At $r = 10$, for example, the rate of change is $\$1363.46$/unit change in r.

51. $S = \left[\dfrac{600,000}{r}\right]\left[\left[1 + \dfrac{r}{100}\right]^8 - 1\right]$

(a) $\dfrac{dS}{dr} = \left[\dfrac{600,000}{r}\right](8)\left[1 + \dfrac{r}{100}\right]^7\left[\dfrac{1}{100}\right] - \left[\dfrac{600,000}{r^2}\right]\left[\left[1 + \dfrac{r}{100}\right]^8 - 1\right]$

$= \left[\dfrac{48,000}{r}\right]\left[1 + \dfrac{r}{100}\right]^7 - \left[\dfrac{600,000}{r^2}\right]\left[\left[1 + \dfrac{r}{100}\right]^8 - 1\right]$

(b) $\left.\dfrac{dS}{dr}\right|_{r=8} = \left[\dfrac{48,000}{8}\right](1.08)^7 - \left[\dfrac{600,000}{64}\right](1.08^8 - 1) = 10,282.95 - 7977.47 = 2305.48$

An increase of 1% in the interest rate will increase the value of S by approximately $\$2305.48$.

(c) $\left.\dfrac{dS}{dr}\right|_{r=10} = \left[\dfrac{48,000}{10}\right](1.10)^7 - \left[\dfrac{600,000}{100}\right](1.10^8 - 1) = 9353.84 - 6861.53 = 2492.31$

An increase of 1% in the interest rate will increase the value of S by approximately $\$2492.31$.

Review Exercises

1. $f(x) = \dfrac{1}{x-2}$; $\displaystyle\lim_{x \to 2} f(x)$ does not exist.

3. $f(x) = 4x + 8$; $\displaystyle\lim_{x \to 2} f(x) = 8 + 8 = 16$

5. $f(x) = \dfrac{\sqrt{x} - \sqrt{2}}{x - 2}$

$$\lim_{x \to 2} f(x) = \lim_{x \to 2} \frac{(\sqrt{x} - \sqrt{2})}{(\sqrt{x} - \sqrt{2})(\sqrt{x} + \sqrt{2})} = \lim_{x \to 2} \frac{1}{(\sqrt{x} + \sqrt{2})} = \frac{1}{2\sqrt{2}}$$

7. $f(x) = \dfrac{1}{x-2}$ is discontinuous at $x = 2$.

9. $f(x) = 4x + 8$ is continuous everywhere.

11. $f(x) = \dfrac{\sqrt{x} - \sqrt{2}}{x - 2}$ is discontinuous at $x = 2$.

13. $P = P_0\left[1 + \dfrac{r}{2}\right]^n$

(a) $P_0 = \$1{,}000$
$P_1 = 1000(1 + 0.04) = \$1040$
$P_2 = 1000(1 + 0.04)^2 = \$1081.60$
$P_3 = 1000(1 + 0.04)^3 = \$1124.86$
$P_4 = 1000(1 + 0.04)^4 = \$1169.86$

(continue

13. *(continued)*

(b)

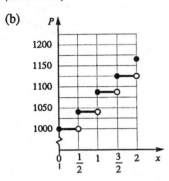

(c) The graph is discontinuous at $t = \frac{1}{2}, 1, \frac{3}{4}$, and 2.

15. $f(x) = x^2 - 3x + 5$

$$\frac{\Delta f(x)}{\Delta x} = \frac{(x + \Delta x)^2 - 3(x + \Delta x) + 5 - (x^2 - 3x + 5)}{\Delta x} = \frac{2x\,\Delta x + (\Delta x)^2 - 3\Delta x}{\Delta x}$$

$$= 2x + \Delta x - 3 = 2x - 3 + \Delta x$$

17. The average rate of change of $y = f(x)$ from $x = 1$ to $x = 3$ is

$$\frac{\Delta y}{\Delta x} = 2(1) - 3 + (3 - 1) = 1.$$

19. $R(x) = -x^2 + 8000x \ (0 \le x \le 8000)$

$$\frac{\Delta R}{\Delta x} = \frac{-(x + \Delta x)^2 + 8000(x + \Delta x) - (-x^2 + 8000x)}{\Delta x} = \frac{-2x\,\Delta x - (\Delta x)^2 + 8000\,\Delta x}{\Delta x} = -2x + 8000 - \Delta x$$

(a) The average rate of change of R with respect to x over the interval $1000 \le x \le 3000$ is
$-2(1000) + 8000 - (3000 - 1000) = 4000.$

(b) Over the interval $1000 \le x \le 2000$ the average rate of change is $-2(1000) + 8000 - (2000 - 1000) = 5000$

(c) The value calculated is the slope of the secant line from (1000, 7,000,000) to (3000, 15,000,000).

(d) The value calculated is the slope of the secant line from (1000, 7,000,000) to (2000, 12,000,000).

(e) and (f)

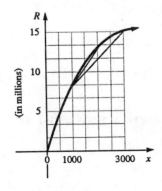

21. $f(x) = -5x^2 + 2x + 7$

    (a) $f'(x) = -10x + 2$

    (b) $f'(1) = -10 + 2 = -8$

    (c) $f'(-2) = -10(-2) + 2 = 22$

23. $f(x) = x^8 - 8x$

    (a) $f'(x) = 8x^7 - 8$

    (b) $f'(1) = 8 - 8 = 0$

    (c) $f'(-2) = 8(-128) - 8 = -1032$

25. $P(t) = 5000t^2 + 300,000$

    (a) $P'(t) = 10,000t$ people year

    (b) $P'(4) = 40,000$

    (c) $P'(6) = 60,000$

    (d) $P'(9) = 90,000$

27. $y = (x^2 + 4x + 7)(x^3 + 3x^2 + 8)$

    $\dfrac{dy}{dx} = (x^2 + 4x + 7)(3x^2 + 6x)$

    $\qquad + (2x + 4)(x^3 + 3x^2 + 8)$

29. $y = (x^4 - 6x + 1)^6;\ \dfrac{dy}{dx} = 6(x^4 - 6x + 1)^5(4x^3 - 6)$

31. $y = 1/(x^2 + 8x + 2)^9;\ \dfrac{dy}{dx} = \dfrac{-9(2x + 8)}{(x^2 + 8x + 2)^{10}}$

33. $y = (x^3 + x^2 + 4)^5(x^2 + 4x + 5);\ \dfrac{dy}{dx} = (x^3 + x^2 + 4)^5(2x + 4) + 5(x^3 + x^2 + 4)^4(3x^2 + 2x)(x^2 + 4x + 5)$

35. $y = \dfrac{x^4 + 5x^2 + 7}{x^3 - 6x + 8}; \quad \dfrac{dy}{dx} = \dfrac{(x^3 - 6x + 8)(4x^3 + 10x) - (x^4 + 5x^2 + 7)(3x^2 - 6)}{(x^3 - 6x + 8)^2}$

37. $y = \dfrac{(x^4 + 4x + 7)^9}{x^3 - 4x + 2}; \quad \dfrac{dy}{dx} = \dfrac{9(x^3 - 4x + 2)(x^4 + 4x + 7)^8(4x^3 + 4) - (x^4 + 4x + 7)^9(3x^2 - 4)}{(x^3 - 4x + 2)^2}$

39. $y = (x^8 + 9)\left[\dfrac{x^2 - 9}{x^4 + 7}\right]^6; \quad \dfrac{dy}{dx} = 6(x^8 + 9)\left[\dfrac{x^2 - 9}{x^4 + 7}\right]^5\left[\dfrac{(x^4 + 7)(2x) - (x^2 - 9)(4x^3)}{(x^4 + 7)^2}\right] + 8x^7\left[\dfrac{x^2 - 9}{x^4 + 7}\right]^6$

41. $y = [(x + 6)/(x - 5)]^8; \quad \dfrac{dy}{dx} = 8\left[\dfrac{x + 6}{x - 5}\right]^7\left[\dfrac{(x - 5) - (x + 6)}{(x - 5)^2}\right] = 8\left[\dfrac{x + 6}{x - 5}\right]^7\left[\dfrac{-11}{(x - 5)^2}\right] = \dfrac{-88(x + 6)^7}{(x - 5)^7(x - 5)^2} = \dfrac{-88(x + 6)^7}{(x - 5)^9}$

43. $C(x) = 8x + 240$

(a) Cost per unit $= \dfrac{C(x)}{x} = 8 + \dfrac{240}{x} = \bar{C}(x)$

(b) $\bar{C}'(x) = \dfrac{-240}{x^2}$

(c) $\bar{C}'(5) = \dfrac{-240}{25} = -9.6$

$\bar{C}'(20) = \dfrac{-240}{400} = -0.6$

$\bar{C}'(50) = \dfrac{-240}{2500} = -0.096$

A value of $\bar{C}'(x_0)$ represents the instantaneous rate of change of the average cost per unit at a production level of x_0.

45. $S = \left[\dfrac{900,000}{r}\right]\left[\left[1 + \dfrac{r}{100}\right]^6 - 1\right]$

(a) $\dfrac{dS}{dr} = \dfrac{-900,000}{r^2}\left[\left[1 + \dfrac{r}{100}\right]^6 - 1\right]$

$\qquad + \dfrac{6(900,000)}{r}\left[1 + \dfrac{r}{100}\right]^5\left[\dfrac{1}{100}\right]$

(b) $\dfrac{dS}{dr}\bigg|_{r=8} = \dfrac{-900,000}{64}(0.586874)$

$\qquad + \dfrac{54,000}{8}(1.469328)$

$\qquad = 1665.04$

At $r = 8$, a change of 1 unit in r will result in a change of approximately \$1665.04 in S.

(c) $\dfrac{dS}{dr}\bigg|_{r=10} = \dfrac{-900,000}{100}(1.10^6 - 1) + \dfrac{54,000}{10}(1.10)^5$

$\qquad = 1752.71$

At $r = 10$, a change of 1 unit in r will result in an approximate change of \$1752.71 in S.

47. $f(x) = \sqrt{x - 5}$

(a) f is not defined for $x < 5$ because $f(x)$ does not exist for $x < 5$.

(b) f is not differentiable for $x \leq 5$.

49. $f(x) = x^{1/5}$ is not differentiable at $x = 0$.

CHAPTER
10
CURVE SKETCHING AND OPTIMIZATION

10–1 Higher–Order Derivatives

1. $f(x) = x^3 - 4x^2 + 7x - 9$
    $f'(x) = 3x^2 - 8x + 7$
    $f''(x) = 6x - 8$

3. $f(x) = x^5 - 8x^3 + 2x^2 + 4x + 1$
    $f'(x) = 5x^4 - 24x^2 + 4x + 4$
    $f''(x) = 20x^3 - 48x + 4$

5. $f(x) = (x^3 + 1)(x^2 - x + 4)$
    $f'(x) = (x^3 + 1)(2x - 1) + (3x^2)(x^2 - x + 4)$
    $f''(x) = 2(x^3 + 1) + 3x^2(2x - 1) + (2x - 1)(3x^2) + 6x(x^2 - x + 4)$
    $\quad = 2x^2 + 2 + 6x^3 - 3x^2 + 6x^3 - 3x^2 + 6x^3 - 6x^2 + 24x$
    $\quad = 20x^3 - 12x^2 + 24x + 2$

7. $f(x) = (x + 1)^4(x - 2) + 6$
    $f'(x) = (x + 1)^4 + 4(x + 1)^3(x - 2)$
    $f''(x) = 4(x + 1)^3 + 4(x + 1)^3 + 12(x + 1)^2(x - 2)$
    $\quad = 8(x + 1)^3 + 12(x + 1)^2(x - 2)$

9. $y = x^3 - 2x^2 + 8x - 3$

$$\frac{dy}{dx} = 3x^2 - 4x + 8$$

$$\frac{d^2y}{dx^2} = 6x - 4$$

11. $y = \dfrac{x^4 + 6}{x^3 - 7}$

$$\frac{dy}{dx} = \frac{(x^3 - 7)(4x^3) - (x^4 + 6)(3x^2)}{(x^3 - 7)^2}$$

$$= \frac{x^6 - 28x^3 - 18x^2}{(x^3 - 7)^2}$$

$$\frac{d^2y}{dx^2} = \frac{(x^3 - 7)^2(6x^5 - 84x^2 - 36x) - 6(x^6 - 28x^3 - 18x^2)(x^3 - 7)(x^2)}{(x^3 - 7)^4}$$

$$= \frac{(x^3 - 7)(6x^5 - 84x^2 - 36x) - (6x^2)(x^6 - 28x^3 - 18x^2)}{(x^3 - 7)^3}$$

13. $y = (x - 4)^3(x - 7) + 8$

$$\frac{dy}{dx} = (x - 4)^3 + 3(x - 4)^2(x - 7)$$

$$\frac{d^2y}{dx^2} = 3(x - 4)^2 + 3(x - 4)^2 + 6(x - 4)(x - 7)$$

$$= 6(x - 4)^2 + 6(x - 4)(x - 7)$$
$$= 6(x - 4)(2x - 11)$$

15. $g(x) = \dfrac{x^2 - 1}{x - 5}$

$$g'(x) = \frac{2x(x - 5) - (x^2 - 1)}{(x - 5)^2} = \frac{x^2 - 10x + 1}{(x - 5)^2}$$

$$g''(x) = \frac{(x - 5)^2(2x - 10) - 2(x^2 - 10x + 1)(x - 5)}{(x - 5)^4}$$

$$= \frac{2(x - 5)^2 - 2(x^2 - 10x + 1)}{(x - 5)^3}$$

$$g''(2) = \frac{2(9) - 2(-15)}{-27} = \frac{-48}{27} \approx -1.778$$

17. $y = x^4 - 5x^3 + 6x^2 + 3x + 1$

$\dfrac{dy}{dx} = 4x^3 - 15x^2 + 12x + 3$

$\dfrac{dy}{dx}\bigg|_{x=1} = 4 - 15 + 12 + 3 = 4$

$\dfrac{d^2y}{dx^2} = 12x^2 - 30x + 12$

$\dfrac{d^2y}{dx^2}\bigg|_{x=1} = 12 - 30 + 12 = -6$

19. $f(x) = x^4 - 6x^3 + 8x^2 - 4x - 5$
$f'(x) = 4x^3 - 18x^2 + 16x - 4$
$f'(1) = 4 - 18 + 16 - 4 = -2$
$f''(x) = 12x^2 - 36x + 16$
$f''(3) = 12(9) - 36(3) + 16 = 16$

21. $f(x) = x^6 - 8x^5 + 6x^4 - 2x^3 + x + 5$
$f'(x) = 6x^5 - 40x^4 + 24x^3 - 6x^2 + 1$
$f''(x) = 30x^4 - 160x^3 + 72x^2 - 12x$
$f'''(x) = 120x^3 - 480x^2 + 144x - 12$
$f^{(4)}(x) = 360x^2 - 960x + 144$
$f^{(4)}(3) = 360(9) + 960(3) + 144 = 504$

23. $f(x) = x^4 - 6x^3 + 4x^2 - 8x + 1$
$f'(x) = 4x^3 - 18x^2 + 8x - 8$
$f''(x) = 12x^2 - 36x + 8$
$f'''(x) = 24x - 36$
$f^{(4)}(x) = 24$
$f^{(5)}(x) = 0$
$f^{(6)}(x) = 0$

25. $S = -16t^2 + 192t \ (0 \le t \le 12)$

(a) $\dfrac{dS}{dt} = -32t + 192$

$\dfrac{dS}{dt}\bigg|_{t=2} = -64 + 192 = 128$ ft/sec

$\dfrac{dS}{dt}\bigg|_{t=3} = -96 + 192 = 96$ ft/sec

$\dfrac{dS}{dt}\bigg|_{t=8} = -256 + 192 = -64$ ft/sec

(downward)

(b) $\dfrac{d^2S}{dt^2} = -32$ ft/sec² at $t = 2, 3,$ and 8

10–2 Critical Values and the First Derivative

1. (a) f is increasing over the intervals $x_1 < x < x_2$ and $x_3 < x < x_4$.

    (b) f is decreasing over the intervals $x_2 < x < x_3$ and $x_4 < x < x_5$.

    (c) Relative maxima of f are $f(x_2)$ and $f(x_4)$. The only relative minimum of f is $f(x_3)$.

    (d) The absolute maximum of f is $f(x_4)$; the absolute minimum is $f(x_1)$.

3. (a) f is increasing over the interval $-2 < x < 2$.

    (b) f is decreasing over the intervals $-\infty < x < -2$ and $2 < x < \infty$.

    (c) A relative minimum occurs at $x = -2$; a relative maximum occurs at $x = 2$.

5. (a) f is increasing over the interval $-\infty < x < -4$ and $-4 < x < 5$.

(b) f is decreasing over the interval $5 < x < \infty$.

(c) A relative maximum of f occurs at $x = 5$; there are no relative minima.

7. (a) f is increasing over the intervals $-3 < x < -1$, $-1 < x < 2$, and $4 < x < \infty$.

(b) f is decreasing over the intervals $-\infty < x < -3$ and $2 < x < 4$.

(c) Relative minima of f occur at $x = -3$ and $x = 4$; a relative maximum occurs at $x = 2$.

9. (a) f is increasing over the intervals $-\infty < x < -1$ and $5 < x < \infty$.

(b) f is decreasing over the interval $-1 < x < 5$

(c) A relative minimum of f occurs at $x = 5$; a relative maximum occurs at $x = -1$.

11. (a) Profit is increasing over the intervals $0 < x < 3000$ and $7000 < x < 10,000$.

(b) Profit is decreasing over the interval $3000 < x < 7000$.

(c) 45,000 is a relative maximum; 35,000 is a relative minimum.

(d) 10,000 units should be produced and sold to maximize profit.

(e) The maximum profit is $83,000. This is not a relative maximum. It is an absolute maximum.

13. (a) At $x = 3000$ the average cost per unit is minimal.

(b) The minimum value of $\bar{C}$ is 2. This is both a relative and an absolute minimum.

15. $y = f(x) = 3x + 5$
$y' = 3$

(a) $y' > 0$ for all values of x. $\therefore$ f is increasing over the interval $-\infty < x < \infty$.

(b) f is decreasing nowhere.

(c) There are no relative maxima or minima of f.

17. $y = f(x) = -x^2 + 8x$
$y' = -2x + 8 = 2(-x + 4)$

(a) $y' > 0 \Rightarrow -x + 4 > 0 \Rightarrow x < 4$
f is increasing over the interval $-\infty < x < 4$.

(b) $y' < 0 \Rightarrow -x + 4 < 0 \Rightarrow x > 4$
f is decreasing over the interval $4 < x < \infty$.

(c) $f(4) = 16$ is a relative maximum.

19. $f(x) = 3x^4 - 4x^3 + 10$
$f'(x) = 12x^3 - 12x^2 = 12x^2(x - 1)$

(a) $f'(x) > 0 \Rightarrow x - 1 > 0 \Rightarrow x > 1$
f is increasing over the interval $1 < x < \infty$.

(b) $f' < 0 \Rightarrow x - 1 < 0 \Rightarrow x < 1$
f is decreasing over the interval $-\infty < x < 0$ and $0 < x < 1$.

(c) $f(1) = 9$ is a relative minimum.

21. $y = f(x) = \sqrt{x^2 + 4}$

$f'(x) = \dfrac{x}{\sqrt{x^2 + 4}}$

(a) $f'(x) > 0 \Rightarrow x > 0$
f is increasing over the interval $0 < x < \infty$.

(b) $f'(x) < 0 \Rightarrow x < 0$
f is decreasing over the interval $-\infty < x < 0$.

(c) $f(0) = 2$ is a relative minimum.

23. $y = f(x) = (x - 5)^3$
$f'(x) = 3(x - 5)^2 \geq 0$

(a) Since $f'(x) \geq 0$ for all values of x, f is increasing over the intervals $-\infty < x < 5$ and $5 < x < \infty$.

(b) f is decreasing nowhere.

(c) There are no relative maxima or minima.

25. $y = f(x) = \dfrac{x + 3}{x + 1}$, not defined at $x = -1$.

$f'(x) = \dfrac{x + 1 - x - 3}{(x + 1)^2} = \dfrac{-2}{(x + 1)^2} < 0$

(a) f is increasing nowhere.

(b) f is decreasing over the intervals $-\infty < x < -1$ and $-1 < x < \infty$.

(c) There are no relative maxima or minima.

27. $y = f(x) = x^2 + 8x + 9$
$f'(x) = -2x + 8 = 2(-x + 4)$

(a) $f'(x) > 0 \Rightarrow -x + 4 > 0 \Rightarrow x < 4$
f is increasing over the interval $-\infty < x < 4$

(b) $f'(x) < 0 \Rightarrow x > 4$
f is decreasing over the interval $4 < x < \infty$

(c) $f(4) = 25$ is a relative maximum.

(d)

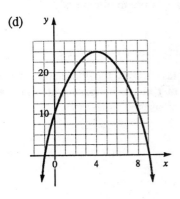

29. $y = f(x) = x^3 - 6x^2$
$f'(x) = 3x^2 - 12x = 3x(x - 4)$

sign of $f'(x)$

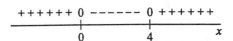

(a) f is increasing over the intervals $-\infty < x < 0$ and $4 < x < \infty$.

(b) f is decreasing over the interval $0 < x < 4$.

(c) $f(0) = 0$ is a relative maximum; $f(4) = -32$ is a relative minimum.

(d)

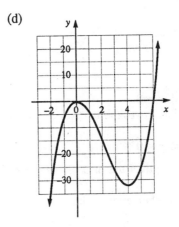

31. $y = f(x) = x^4 - 8x^2$
$f'(x) = 4x^3 - 16x = 4x(x^2 - 4)$

sign of $f'(x)$

$$\underline{\quad - - - - - - \ 0 + + + + + + \ 0 \ - - - - - - \ 0 \ + + + + + +\quad}$$
$$\qquad\qquad -2 \qquad\qquad 0 \qquad\qquad 2 \qquad\qquad x$$

(a) f is increasing over the intervals $-2 < x < 0$ and $2 < x < \infty$.

(b) f is decreasing over the intervals $-\infty < x < -2$ and $0 < x < 2$.

(continued)

31. *(continued)*

(c) $f(-2) = -16$ and $f(2) = -16$ are relative minima; $f(0) = 0$ is a relative maximum.

(d)

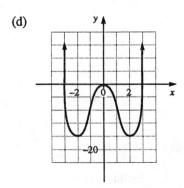

33. $y = f(x) = x^4 + 5$
$f'(x) = 4x^3$

(a) $f'(x) > 0 \Rightarrow x > 0$
f' is increasing over the interval $0 < x < \infty$.

(b) $f'(x) < 0 \Rightarrow x < 0$
f is decreasing over the interval $-\infty < x < 0$.

(c) $f(0) = 5$ is a relative minimum.

(d)

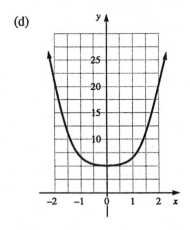

35. $f(x) = x^3 + 6x^2 - 15x + 1$ $(-10 \le x \le 11)$
$f'(x) = 3x^2 + 12x - 15 = 3(x-1)(x+5)$

sign of $f'(x)$

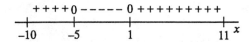

$f(-5) = 101$ is a relative maximum; $f(1) = -7$ is a relative minimum
$f(-10) = -249$ is the absolute minimum;
$f(11) = 1893$ is the absolute maximum.

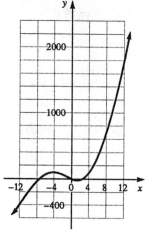

37. $f(x) = ax^2 + bx + c$
$f'(x) = 2ax + b$; $f'(x) = 0 \Rightarrow x = -b/2a$
For $x < -b/2a$, $f'(x) < 0$; for $x > -b/2a$, $f'(x) > 0$.
$\therefore f(-b/2a)$ is a relative minimum of f.

39. (a) f is increasing over the interval $x_1 < x < x_4$.

(b) $f'(x_3) < f'(x_2)$. Over the internal $x_1 < x < x_4$, the values of $f(x)$ are increasing at a increasing rate as x increases.

13. $f(x) = 5x^4 - 20x^3 + 7$
$f'(x) = 20x^3 - 60x^2 = 20x^2(x - 3)$
$\quad f'(x) > 0$ for $x > 3 \Rightarrow f$ increasing for $x > 3$
$\quad f'(x) < 0$ for $x < 3 \Rightarrow f$ decreasing for $x < 3$
$\quad f(3) = -128$ is a relative minimum
$f''(x) = 60x^2 - 120x = 60x(x - 2)$

sign of $f''(x)$

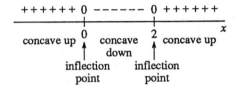

$f(x)$

No asymptotes.
y–intercept: (0,7)

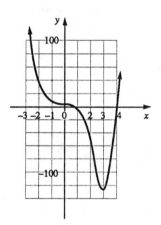

15. $y = f(x) = x^3 - 3x + 2$
$f'(x) = 3x^2 - 3 = 3(x^2 - 1) = 3(x - 1)(x + 1)$

sign of $f'(x)$

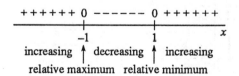

$f(x)$

(continued)

15. *(continued)*

$f(-1) = 4; f(1) = 0$

$f''(x) = 6x$
$f''(x) > 0$ if $x > 0 \Rightarrow f(x)$ concave up
$f''(x) < 0$ if $x < 0 \Rightarrow f(x)$ concave down
inflection point at $x = 0$
No asymptotes
y–intercept: (0,2)

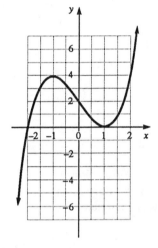

17. $f(x) = (x - 5)^3$
$f'(x) = 3(x - 5)^2$
$f'(x) = 0$ at $x = 5; f'(x) > 0$ elsewhere; no relative
extrema
f increasing over the interval $-\infty < x < \infty$
$f''(x) = 6(x - 5)$
$f''(x) > 0$ if $x > 5 \Rightarrow f$ concave up
$f''(x) < 0$ if $x < 5 \Rightarrow f$ concave down
inflection point at $x = 5$

(continued)

17. *(continued)*

No asymptotes
y–intercept: $(0, -125)$

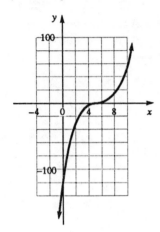

19. $y = f(x) = (x + 3)^{10} + 2$
$f'(x) = 10(x + 3)^9$
  $f'(x) > 0$ if $x > -3 \Rightarrow f$ increasing
  $f'(x) < 0$ if $x < -3 \Rightarrow f$ decreasing
  $f(-3) = 2$, a relative minimum

$f''(x) = 90(x + 3)^8$
  $f''(x) = 0$ if $x = -3$; $f''(x) > 0$ elsewhere $\Rightarrow f$
  concave up
  no inflection points
No asymptotes
y–intercept: $(0, 59{,}051)$

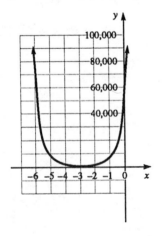

21. $y = f(x) = (x + 2)(x - 5)^4 + 6$
$f'(x) = 4(x + 2)(x - 5)^3 + (x - 5)^4$
   $= (x - 5)^3(5x + 3)$

sign of $f'(x)$

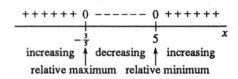

$f(x)$

$f\left[-\dfrac{3}{5}\right] = 1382.8; \; f(5) = 6$

$f''(x) = 5(x - 5)^3 + 3(x - 5)^2(5x + 3)$

sign of $f''(x)$

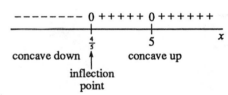

$f(x)$

inflection point at $x = \dfrac{4}{5}$

No asymptotes
y–intercept: $(0, 1256)$

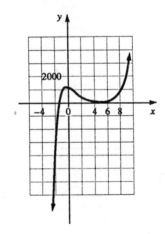

23. $f(x) = (x - 4)^3(x + 2)^2$

$f'(x) = 2(x - 4)^3(x + 2) + 3(x - 4)^2(x + 2)^2$

$\quad = (x - 4)^2(x + 2)(5x - 2)$

sign of $f'(x)$

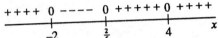

relative   relative
maximum minimum

$f(x)$

$$f(-2) = 0; f\left[\frac{2}{5}\right] \approx -268.74$$

$f''(x) = 5(x - 4)^2(x + 2) + (x - 4)^2(5x - 2)$

$\quad\quad + 2(x - 4)(x + 2)(5x - 2)$

$\quad = (x - 4)[5(x - 4)(x + 2) + (x - 4)(5x - 2)$

$\quad\quad + 2(x + 2)(5x - 2)]$

$\quad = (x - 4)(20x^2 - 16x - 40)$

$\quad = 4(x - 4)(5x^2 - 4x - 10)$

$$20(x - 4)\left[x - \frac{2}{5} - \frac{3\sqrt{6}}{5}\right]\left[x - \frac{2}{5} + \frac{3\sqrt{6}}{5}\right]$$

sign of $f''(x)$

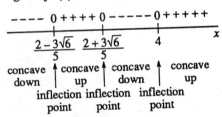

inflection inflection inflection
point   point   point

$f(x)$

inflection points at $x = \dfrac{2 - 3\sqrt{6}}{5}$, $x = \dfrac{2 + 3\sqrt{6}}{5}$,

and $x = 4$

No asymptotes

y–intercept: $(0, -256)$

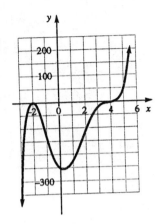

25. $y = f(x) = x^3$

$f'(x) = 3x^2$

$\quad f'(x) = 0$ at $x = 0$;

$\quad$ elsewhere $f'(x) > 0 \Rightarrow f$ increasing.

$f''(x) = 6x$

$\quad f''(x) > 0$ if $x > 0 \Rightarrow f$ concave up

$\quad f''(x) < 0$ if $x < 0 \Rightarrow f$ concave down

$\quad$ inflection point at $x = 0$

No asymptotes

y–intercept: $(0, 0)$

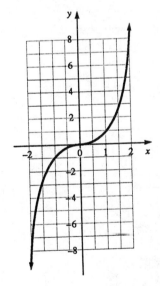

27. $f(x) = x^{4/3}$

$f'(x) = \frac{4}{3}x^{1/3}$

$f'(x) > 0$ for $x > 0 \Rightarrow f$ increasing
$f'(x) < 0$ for $x < 0 \Rightarrow f$ decreasing
$f(0) = 0$ is a relative minimum

$f''(x) = \frac{4}{9}\frac{1}{x^{2/3}}$

$f''(x) > 0$ for $x \neq 0 \Rightarrow f$ concave up
$f''(x)$ does not exist at $x = 0$
no inflection points

No asymptotes
y–intercept: (0, 0)

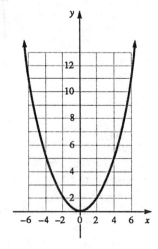

29. $y = f(x) = \frac{-6}{x^2 + 2}$

$f'(x) = \frac{12x}{(x^2 + 2)^2}$

$f'(x) > 0$ for $x > 0 \Rightarrow f$ increasing
$f'(x) < 0$ for $x < 0 \Rightarrow f$ decreasing
$f(0) = -3$ is a relative minimum

$f''(x) = \frac{12}{(x^2 + 2)^2} - \frac{48x^2}{(x^2 + 2)^3} = \frac{-36x^2 + 24}{(x^2 + 2)^3}$

$\qquad = \frac{12(-3x^2 + 2)}{(x^2 + 2)^3}$

sign of $f''(x)$

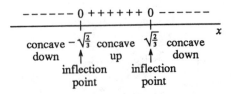

inflection points at $x = -\sqrt{\dfrac{2}{3}}$ and $x = \sqrt{\dfrac{2}{3}}$

(continued)

29. *(continued)*

Horizontal asymptote: $y = 0$
y–intercept: (0, –3)

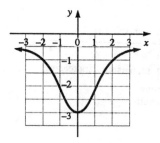

31. $f(x) = \frac{x}{x^2 + 1}$

$f'(x) = \frac{x^2 + 1 - 2x^2}{(x^2 + 1)^2} = \frac{-x^2 + 1}{(x^2 + 1)^2}$

$f'(x) < 0$ for $-\infty < x < -1$ and $1 < x < \infty \Rightarrow f$ decr
$f'(x) > 0$ for $-1 < x < \infty \Rightarrow f$ increasing

$f(-1) = -\frac{1}{2}$ is a relative minimum; $f(1) = \frac{1}{2}$ is a rel

$f''(x) = \frac{-2x(x^2 + 1)^2 - 4x(-x^2 + 1)(x^2 + 1)}{(x^2 + 1)^4}$

$\qquad = \frac{-2x(x^2 + 1) - 4x(-x^2 + 1)}{(x^2 + 1)^3} = \frac{2x(x^2 - 3)}{(x^2 + 1)^3}$

sign of $f''(x)$

```
---- 0 ++++ 0 ----- 0 +++++
――――――――――――――――――――――――――― x
     −√3      0      √3
concave   concave   concave   concave
down        up       down       up
      inflection inflection inflection
        point     point      point
```

$f(x)$

inflection points at $x = -\sqrt{3}$, $x = 0$, and $x = \sqrt{3}$

(continued)

31. (continued)

Horizontal asymptote: $y = 0$
y–intercept: $(0, 0)$

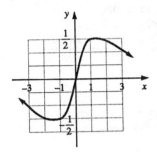

33. $f(x) = \dfrac{800}{x} + 2x$

$f'(x) = -\dfrac{800}{x^2} + 2$

$f'(x) > 0 \Rightarrow \dfrac{800}{x^2} < 2 \Rightarrow x^2 > 400 \Rightarrow x > 20$

  or $x < -20$

sign of $f'(x)$

undefined

++++ 0 − − − − − ↓ − − − − − 0 ++++

−20 0 20 $\qquad x$

increasing decreasing decreasing increasing

$f(x)$

  $f(-20) = -80$ is a local maximum
  $f(20) = 80$ is a local minimum

$f''(x) = \dfrac{1600}{x^3}$

$f''(x) > 0$ if $x > 0$; f concave up
$f''(x) < 0$ if $x < 0$; f concave down

33. *(continued)*

At $x = 0$, f is undefined.
Vertical asymptote: $x = 0$

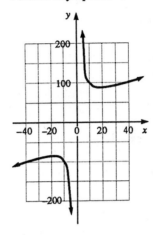

35. $f(x) = \dfrac{(x - 3)^2}{x + 1}$

$f'(x) = \dfrac{2(x + 1)(x - 3) - (x - 3)^2}{(x + 1)^2} = \dfrac{(x - 3)(x + 5)}{(x + 1)^2}$

sign of $f'(x)$

undefined

++++ 0 − − − − − ↓ − − − − − 0 ++++

−5 −1 3 $\qquad x$

increasing decreasing decreasing increasing

$f(x)$

  $f(-5) = -16$ is a relative maximum
  $f(3) = 0$ is a relative minimum

$f''(x) = \dfrac{(x + 1)^2(2x + 2) - 2(x - 3)(x + 5)(x + 1)}{(x + 1)^4}$

$ = \dfrac{2(x + 1)^2 - 2(x - 3)(x + 5)}{(x + 1)^3} = \dfrac{32}{(x + 1)^3}$

(continued)

(continued)

35. (continued)

sign of $f''(x)$

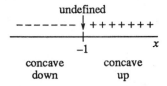

$f(x)$

Vertical asymptote at $x = -1$

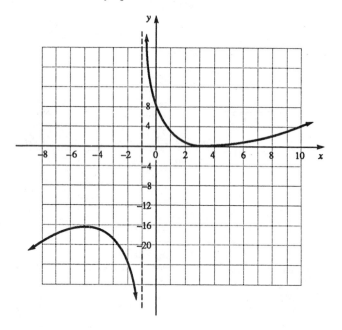

37. $f(x) = \dfrac{x^2(x - 2)}{(x - 1)^3}$

$f'(x) = \dfrac{(x - 1)^3[x^2 + 2x(x - 2)] - 3x^2(x - 2)(x - 1)^2}{(x - 1)^6}$

$= \dfrac{x(4 - x)}{(x - 1)^4}$

sign of $f'(x)$

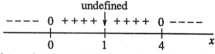

(continued)

---

37. *(continued)*

$f(x)$

$f(0) = 0$ is a relative minimum
$f(4) \approx 1.185$ is a relative maximum

$f''(x) = \dfrac{(x - 1)^4(4 - 2x) - 4x(4 - x)(x - 1)^3}{(x - 1)^8}$

$= \dfrac{2(x^2 - 5x - 2)}{(x - 1)^5}$

$= \dfrac{2\left[x - \dfrac{5}{2} - \dfrac{\sqrt{33}}{2}\right]\left[x - \dfrac{5}{2} + \dfrac{\sqrt{33}}{2}\right]}{(x - 1)^5}$

sign of $f''(x)$

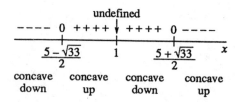

$f(x)$

inflection points at $x = \dfrac{5 - \sqrt{33}}{2} \approx -0.37$ and

$x = \dfrac{5 + \sqrt{33}}{2} \approx 5.37$

Vertical asymptote at $x = 1$;
horizontal asymptote $y = 1$

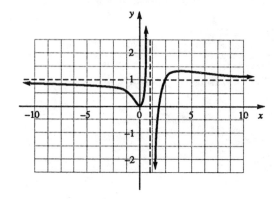

39. $f(x) = \dfrac{20}{x} + x + 9$

$f'(x) = \dfrac{-20}{x^2} + 1 = \dfrac{x^2 - 20}{x^2}$

sign of $f'(x)$

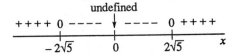

increasing decreasing decreasing increasing

$f(x)$

$f(-2\sqrt{5}) = 9 - 4\sqrt{5} \approx 0.06$ is a relative maximum

$f(2\sqrt{5}) = 9 + 4\sqrt{5} \approx 17.9$ is a relative minimum

$f''(x) = \dfrac{40}{x^3}$

$f''(x) > 0$ for $x > 0 \Rightarrow f$ concave up

$f''(x) < 0$ for $x < 0 \Rightarrow f$ concave down

Vertical asymptote at $x = 0$

Oblique asymptote $y = x + 9$

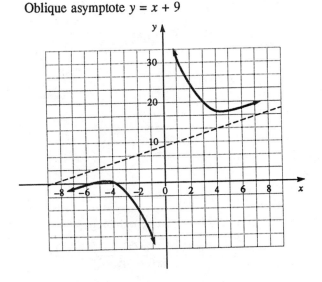

41. $f(x) = x^3 - 16x^2 (-5 \le x \le 6)$

$f'(x) = 3x^2 - 32x = x(3x - 32)$

sign of $f'(x)$

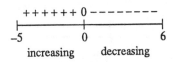

increasing   decreasing

$f(x)$

$f(0) = 0$ is a relative maximum

$f''(x) = 6x - 32 = 2(3x - 16)$

sign of $f''(x)$

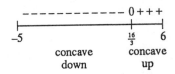

concave   concave
down   up

$f(x)$

inflection point at $x = \dfrac{16}{3}$

No asymptotes

y–intercept: $(0, 0)$

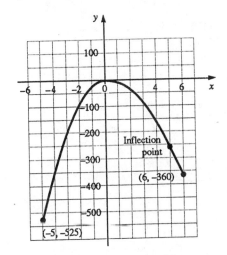

43. $f(x) = x^3 - 5x^2 + 7x - 3 \quad (0 \le x \le 4)$

$f'(x) = 3x^2 - 10x + 7 = 3(x - 1)\left[x - \dfrac{7}{3}\right]$

sign of $f'(x)$

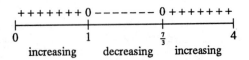

$$+ + + + + + + 0 - - - - - - - 0 + + + + + + +$$

0   1 $\frac{7}{3}$ 4

    increasing   decreasing   increasing

$f(x)$

    $f(1) = 0$ is a relative maximum

    $f\left[\dfrac{7}{3}\right] = -\dfrac{32}{27} \approx -1.18519$ is a relative minimum

$f''(x) = 6x - 10 = 2(3x - 5)$

sign of $f''(x)$

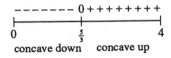

$$- - - - - - - 0 + + + + + + + + +$$

0             $\frac{5}{3}$ 4

concave down   concave up

$f(x)$

    inflection point at $x = \dfrac{5}{3}$

No asymptotes
y–intercept: $(0, -3)$

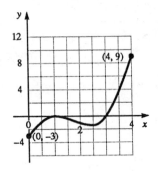

45. $f(x) = x^{1/3}$

    $f'(x) = \dfrac{1}{3}x^{-2/3}$

    $f''(x) = -\dfrac{4}{9}x^{-5/3}$

(a) $f(x)$ is concave for $-\infty < x < 0$.

(b) $f(x)$ is concave down for $0 < x < \infty$.

(c) $f(x)$ changes concavity at $x = 0$. $f''(x)$ is not defi

(d)

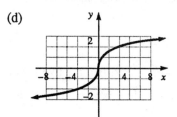

47. $f(x) = \dfrac{x^2 - x + 4}{x^2 + 4} = 1 - \dfrac{x}{x^2 + 4} (x \ge 0)$

(a) $f'(x) = \dfrac{-(x^2 + 4) + 2x^2}{(x^2 + 4)^2} = \dfrac{x^2 - 4}{(x^2 + 4)^2}$

    sign of $f'(x)$

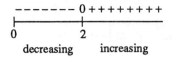

$$- - - - - - - 0 + + + + + + + +$$

0   2

   decreasing   increasing

$f(x)$

$f(2) = 0.75$ is a relative minimum

$f''(x) = \dfrac{2x(x^2 + 4)^2 - 4x(x^2 - 4)(x^2 + 4)}{(x^2 + 4)^4}$

$= \dfrac{2x(12 - x^2)}{(x^2 + 4)^3}$

(continued)

47. (continued)

sign of $f''(x)$

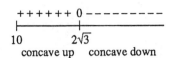

$$\begin{array}{ccc} +++++\,0 & --------- \\ \vdash\quad\quad\;\;\vdash & \\ 10 \quad\quad 2\sqrt{3} & \\ \text{concave up}\quad\text{concave down} \end{array}$$

$f(x)$

inflection point at $x = 2\sqrt{3}$
Horizontal asymptote: $y = 1$
y–intercept: $(0, 1)$

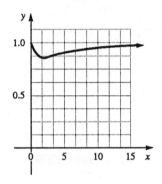

(b) $f(0) = 1$

(c) The level of oxygen is decreasing for 2 weeks.

(d) The level of oxygen begins to increase after 2 w

(e) The minimum level of oxygen is 0.75, occuring

(f) The maximum level of oxygen is 1.0, occurring

(g) The level of oxygen reaches its initial level asym
$f(100) \approx 0.990; f(200) \approx 0.995; f(1000) = 0.999.$

49. $C(x) = x^2 - 4x + 64 \ (x > 0)$

$$\bar{C}(x) = \frac{C(x)}{x} = x - 4 + \frac{64}{x}$$

(a) $\bar{C}'(x) = 1 - \dfrac{64}{x^2} = \dfrac{x^2 - 64}{x^2}$

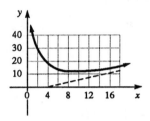

sign of $\bar{C}'(x)$

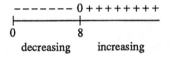

$$\begin{array}{ccc} --------\,0 & +++++++ \\ \vdash\quad\quad\;\;\vdash & \\ 0 \quad\quad\quad 8 & \\ \text{decreasing}\quad\text{increasing} \end{array}$$

$\bar{C}(x)$
$\bar{C}(8) = 12$ is a relative minimum.
$\bar{C}''(x) = \dfrac{128}{x^3} > 0 \Rightarrow \bar{C}(x)$ concave up

Oblique asymptote: $\bar{C}(x) = x - 4$

(b) $\bar{C}(8) = 12$ is a relative minimum.

(c) $x = 8$ minimizes the average cost per unit.

(d) $\bar{C}(8) = 12$

51. $R(x) = -0.01x^3 + 6x^2 \ (0 \le x \le 500)$

(a) $R'(x) = 0.03x^2 + 12x = 0.03x(400 - x)$

sign of $R'(x)$

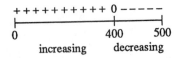

$R(x)$

$R''(x) = -0.06x + 12 = 0.06(200 - x)$

sign of $R''(x)$

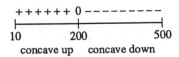

$R(x)$

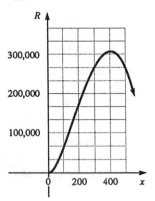

(b) $R(400) = 320,000$ is both a relative and an absol
$R(0) = 0$ is the absolute minimum.

(c) There is an inflection point at $x = 200$.

(continued)

51. *(continued)*

(d)

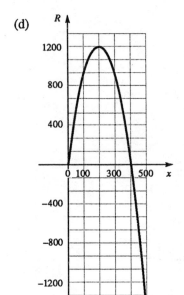

(e) $R'(x)$ has a relative maximum of 1200 at $x = 20$
-1500 at $x = 500$.

(f) $x = 200$

10–4 Optimization

1. $y = x^2 - 8x + 9$
$y' = 2x - 8 = 2(x - 4); \ y' = 0 \Rightarrow x = 4$
$y'' = 2; \ y''(4) > 0 \Rightarrow$ relative minimum;
$f(4) = -7$

3. $y = x^3 - 6x^2 + 7$
$y' = 3x^2 - 12x = 3x(x - 4); \ y' = 0 \Rightarrow x = 0, 4$
$y'' = 6x - 12; y''(0) = -12 \Rightarrow$ relative maximum; $f(0$
$\qquad\qquad y''(4) - 12 \Rightarrow$ relative minimum; $f(4)$

5. $f(x) = x^3 - 3x + 5$
$f'(x) = 3x^2 - 3 = 3(x^2 - 1); f'(x) = 0 \Rightarrow x = \pm 1$
$f''(x) = 6x; f''(-1) = -6 \Rightarrow$ relative maximum; $f(-1$
$f''(1) = 6 \Rightarrow$ relative minimum;
$f(1) = 3$

7. $f(x) = 2x^3 + 3x^2 - 12x + 1$
$f'(x) = 6x^2 + 6x - 12 = 6(x^2 + x - 2)$
$= 6(x + 2)(x - 1);$
$f'(x) = 0 \Rightarrow x = -2, 1$
$f''(x) = 12x + 6; f''(-2) = -18 \Rightarrow$ relative maximum;
$f''(1) = 18 \Rightarrow$ relative minimum; $f(1)$

9. $f(x) = x^3 - 3x^2 - 45x + 9$
$f'(x) = 3x^2 - 6x - 45 = 3(x^2 - 2x - 15)$
$= 3(x - 5)(x + 3);$
$f'(x) = 0 \Rightarrow x = -3, 5$
$f''(x) = 6x - 6; \ f''(-3) = -24 \Rightarrow$ relative maximum;
$f''(5) = 24 \Rightarrow$ relative minimum; $f($

11. $f(x) = x^4 - 8x^2 + 6$
$f'(x) = 4x^3 - 16x = 4x(x^2 - 4)$
$f'(x) = 0 \Rightarrow x = 0, \pm 2$
$f''(x) = 12x^2 - 16; f''(0) = -16 \Rightarrow$ relative maximum;
$f(0) = 6$
$f''(-2) = 32 \Rightarrow$ relative minimum;
$f(-2) = -10$
$f''(2) = 32 \Rightarrow$ relative minimum;
$f(2) = -10$

13. $y = f(x) = x + \dfrac{25}{x}$ (not defined at $x = 0$)
$y' = 1 - \dfrac{25}{x^2} = \dfrac{x^2 - 25}{x^2}$
$y' = 0 \Rightarrow x = \pm 5$
$y'' = \dfrac{50}{x^3}; y''(-5) = -\dfrac{2}{5} \Rightarrow$ relative maximum;
$y(-5) = -10$
$y''(5) = \dfrac{2}{5} \Rightarrow$ relative minimum;
$y(5) = 10$

15. $f(x) = x + \dfrac{27}{x^3}$ (not defined at $x = 0$)
$f'(x) = 1 - \dfrac{81}{x^4} = \dfrac{x^4 - 81}{x^4}$
$f' = 0 \Rightarrow x = \pm 3$
$f''(x) = \dfrac{324}{x^5}; f''(-3) = -\dfrac{4}{3} \Rightarrow$ relative maximum;
$f(-3) = -4$
$f''(3) = \dfrac{4}{3} \Rightarrow$ relative minimum;
$f(3) = 4$

17. $f(x) = x^2 - 16x + 9$
$f'(x) = 2x - 16 = 2(x - 8)$
$f'(x) = 0 \Rightarrow x = 8$
$f''(x) = 2 \Rightarrow f$ concave up; absolute minimum
of $f = f(8) = -55$
There is no absolute maximum of f.

19. $f(x) = -x^2 + 8x + 3$
$f'(x) = -2x + 8 = 2(4 - x)$
$f'(x) = 0 \Rightarrow x = 4$
$f''(x) = -2 \Rightarrow f$ concave down; absolute maximum
of $f = f(4) = 19$
There is no absolute minimum of f.

21. $f(x) = 4\sqrt{x} - x$

$$f'(x) = \frac{2}{\sqrt{x}} - 1 = \frac{2 - \sqrt{x}}{\sqrt{x}}$$

$$f'(x) = 0 \Rightarrow x = 4$$

$$f''(x) = -\frac{1}{x^{3/2}}; \; f''(x) < 0 \text{ for } x > 0 \Rightarrow f \text{ concave down}$$

There is no absolute minimum of f. For $x > 16, f(x)$

23. $f(x) = x^3 - 27x + 4(-4 \le x \le 4)$
$f'(x) = 3x^2 - 27 = 3(x^2 - 9)$
$f'(x) = 0 \Rightarrow x = \pm 3$
$f''(x) = 6x; \; f''(-3) = -18 \Rightarrow$ relative maximum;
$\qquad\qquad f(-3) = 58$
$\qquad f''(3) = 18 \Rightarrow$ relative minimum;
$\qquad\qquad f(3) = -50$
End points: $f(-4) = 48; f(4) = -40$
$\therefore$ absolute maximum of $f = f(-3) = 58$; absolute mi

25. $f(x) = 9 + 5x + \dfrac{125}{x} \quad (x > 0)$

$$f'(x) = 5 - \frac{125}{x^2} = \frac{5(x^2 - 25)}{x^2}$$

$$f'(x) = 0 \Rightarrow x = 5$$

$$f''(x) = \frac{250}{x^2} \Rightarrow f \text{ concave up for } x > 0;$$

$\qquad f(5) = 59$ is the absolute minimum of f.
There is no absolute maximum of f.

27. $f(x) = 2x + \dfrac{54}{x^3} \quad (x > 0)$

$$f'(x) = 2 - \frac{162}{x^4} = \frac{2(x^4 - 81)}{x^4}$$

$$f'(x) = 0 \Rightarrow x = 3$$

$$f''(x) = \frac{648}{x^5} \Rightarrow f \text{ concave up for } x > 0;$$

$\qquad f(3) = 8$ is the absolute minimum of f.
There is no absolute maximum of f.

29. $f(x) = 5(x - 2)^3 \quad (0 \le x \le 4)$
$\quad f'(x) = 15(x - 2)^2$
$\qquad f'(x) = 0 \Rightarrow x = 2$
$\quad f''(x) = 30(x - 2); f''(2) = 0 \Rightarrow$ test fails
Since $f'(x)$ does not change sign at $x = 2$, the point
$f(0) = -40; f(4) = 40$
$f(0)$ is the absolute minimum of f; $f(4)$ is the absolut

31. $f(x) = (x - 1)(x - 4)^2 \quad (0 \le x \le 6)$
$\quad f'(x) = 2(x - 1)(x - 4) + (x - 4)^2 = 3(x - 2)(x - 4)$
$\qquad f'(x) = 0 \Rightarrow x = 2, 4$
$\quad f''(x) = 3(x - 2) + 3(x - 4) = 6x - 18;$
$\qquad f''(2) = -6 \Rightarrow$ relative maximum
$\qquad f''(4) = 6 \Rightarrow$ relative minimum
$f(2) = 4; f(4) = 0$
End points: $f(0) = -16; f(6) = 20$
$\therefore$ absolute maximum $= f(6) = 20$; absolute minimu

33. $P = xy$, where $x + y = 24$
$P(x) = x(24 - x)$
$P'(x) = 24 - 2x = 2(12 - x)$
$P'(x) = 0 \Rightarrow x = 12$
$P(12) = 144$; at end points $P(0) = 0, P(24) = 0$.

35. $P = x^2 + y^2$, where $x + y = 600$
$P(x) = x^2 + (600 - x)^2$
$P'(x) = 2x - 2(600 - x) = 4x - 1200$
$P'(x) = 0 \Rightarrow x = 300, y = 300, P = 180,000$
$P''(x) = 4 \Rightarrow P(300)$ is a minimum.

37. $C(x) = x^2 - 80x + 3600 \quad (x > 0)$

Average cost per unit $= \bar{C}(x)$

$$= \frac{C(x)}{x}$$

$$= x - 80 + \frac{3600}{x}$$

(continued)

37. (continued)

$$\bar{C}'(x) = 1 - \frac{3600}{x^2} = \frac{x^2 - 3600}{x^2}$$

$$\bar{C}'(x) = 0 \Rightarrow x = 60$$

$$\bar{C}''(x) = \frac{7200}{x^3} > 0 \Rightarrow \bar{C} \text{ concave up.} \quad \therefore \bar{C}(60) \text{ is the}$$

39. $C(x) = 175,000x + \frac{6(2,362,500)}{x + 4}$ for 6 years

(a) $C'(x) = 175,000 - \frac{6(2,362,500)}{(x + 4)^2}$

$$= \frac{175,000(x + 4)^2 - 6(2,362,500)}{(x + 4)^2}$$

$C'(x) = 0 \neq (x + 4)^2 = 81 \Rightarrow x + 4 = 9 \Rightarrow$
$x = 5$ inches

$$C''(x) = \frac{12(2,362,500)}{(x + 4)^3} > 0 \Rightarrow C \text{ concave up.}$$

$\therefore$ $C(5)$ is the absolute minimum of C.

(b) $C(5) = (175,000)(5) + \frac{6(2,362,500)}{5 + 4}$

$= \$2,450,000$
Without insulation the 6–year cost is 6(590,625)
Savings $= \$3,543,750 - \$2,450,000$
$\qquad = \$1,093,750$

41. Area $= A = xy; 2x + y = 400$
$\quad A(x) = x(400 - 2x) = 400x - 2x^2$
$\quad A'(x) = 400 - 4x; A'(x) = 0 \Rightarrow x = 100$ feet
$\quad A''(x) = -4 \Rightarrow A$ concave down; $\therefore A(100)$ is
$\quad$ the absolute maximum of A.
$\quad y = 400 - 2(100) = 200$ feet

43. Length $= L = 4x + 2y; xy = 20,000$

$$L(x) = 4x + 2\left[\frac{20,000}{x}\right]$$

$$L'(x) = 4 - \frac{40,000}{x^2} = \frac{4x^2 - 40,000}{x^2}$$

$$L'(x) = 0 \Rightarrow x^2 = 10,000 \Rightarrow x = 100 \text{ feet;}$$

$$y = \frac{20,000}{100} = 200 \text{ feet.}$$

$$L''(x) = \frac{80,000}{x^3} > 0 \Rightarrow L \text{ concave up.} \quad \therefore L(100) \text{ is}$$

45. $A = xy; 2x + 2y = P$, a fixed number

$$A(x) = x\left[\frac{P - 2x}{2}\right] = \frac{P}{2}x - x^2$$

$$A'(x) = \frac{P}{2} - 2x$$

$$A'(x) = 0 \Rightarrow x = \frac{P}{4}; y = \frac{P}{4}.$$

$A''(x) = -2 \Rightarrow A$ concave down. $\therefore A\left[\dfrac{P}{4}\right]$ is the ab

Since $x = y$, the rectangle is a square.

47. $A = (400 + x)y; 400 + x + 2y = 1800$

$$A(x) = (400 + x)\left[700 - \frac{x}{2}\right]$$

$$A'(x) = -\frac{1}{2}(400 + x) + \left[700 - \frac{x}{2}\right] = 500 - x$$

$A'(x) = 0 \Rightarrow x = 500$ feet;
$\quad y = 700 - 250 = 450$ feet;
$\quad A(500) = (400 + 500)(450)$
$\quad = 405,000$ square feet
$A''(x) = -1 < 0 \Rightarrow A$ concave down. $\therefore A(500)$ is t

49. $R(x) = -4x^2 + 3200x;\ C(x) = 200x + 10,000$

(a) $P(x) = R(x) - C(x) = -4x^2 + 3000x - 10,000$

(b) $P'(x) = -8x + 3000$
$P'(x) = 0 \Rightarrow x = 375;\ P''(x) = -8 \Rightarrow P$ concave
down.
$\therefore\ P(375)$ is the absolute maximum of P.

(c) $P(375) = 552,500$

51. $R(x) = 20x;\ C(x) = 0.01x^2 + 0.2x + 40$

(a) $P(x) = -0.01x^2 + 19.8x - 40$

(b) $P'(x) = -0.02x + 19.8$
$P'(x) = 0 \Rightarrow x = 990;\ P''(x) = -0.02 \Rightarrow P$
concave down.
$\therefore\ P(990)$ is the absolute maximum of P.

(c) $P(990) = 9761$

53. $p = -5x + 3000;\ C(x) = 50x + 6000$

(a) $R(x) = xp = -5x^2 + 3000x$

(b) $P(x) = R(x) - C(x) = -5x^2 + 2950x - 6000$

(c) $P'(x) = -10x + 2950$
$P'(x) = 0 \Rightarrow x = 295;\ P''(x) = -10 \Rightarrow P$
concave down.
$\therefore\ P(295)$ is the absolute maximum of P.

(d) $P(295) = -5(295) + 3000 = \$1525$

(e) $P(295) = \$429,125$

55. $p = -2x + 2000;\ C(x) = 30x + 4000$

(a) $R(x) = xp = -2x^2 + 2000x$

(b) $P(x) = R(x) - C(x) = -2x^2 + 1970x - 4000$

(c) $P'(x) = -4x + 1970$
$P'(x) = 0 \Rightarrow x = 492.5;\ P''(x) = -4 \Rightarrow P$
concave down.
$\therefore\ P(492.5)$ is the absolute maximum of P.
Choose between $x = 492$ and $x = 493$.
$P(492) = \$481,112;\ P(493) = \$481,112.$ Choose

(d) $P(492) = -2(492) + 2000 = \$1016$

(e) $P(492) = \$481,112$

57. $R(x) = -5x^2 + 3000x;\ C(x) = 70x + 6000$
$P(x) = R(x) - C(x) = -5x^2 + 2930x - 6000$

(a) $P'(x) = -10x + 2930$
$P'(x) = 0 \Rightarrow x = 293$

(b) $p(293) = -5(293) + 3000 = \$1535$

(c) The unit price has increased by $10.

(d) 50% of the tax is passed on to the consumer.

59. $x = \dfrac{6400}{p^2}\quad (p > 0)$

$C(x) = 8x + 50\ (x > 0)$

$R(x) = xp = \dfrac{6400}{p}$

$P(x) = R(x) - C(x) = \dfrac{6400}{p} - 8x - 50$

$\qquad\qquad = \dfrac{6400}{80}\sqrt{x} - 8x - 50$

$\qquad\qquad = 80\sqrt{x} - 8x - 50$

(continued)

59. *(continued)*

(a) $P'(x) = \dfrac{40}{\sqrt{x}} - 8 = \dfrac{40 - 8\sqrt{x}}{\sqrt{x}}$

$P'(x) = 0 \Rightarrow 5 = \sqrt{x} \Rightarrow x = 25$ hundred units; P''

$\therefore$ P(25) is the absolute maximum of P.

(b) $p(25) = \dfrac{80}{\sqrt{25}} = 16$ thousand dollars

(c) $P(25) = (80)(5) - 8(25) - 50 = 150$ hundred thou

61. Let t = time in weeks from now, x = pounds of appl
$p(t) = 0.48 - 0.03t$
$x(t) = 120 + 10t$
$R(t) = x(t)p(t) = (120 + 10t)(0.48 - 0.03t)$
$R'(t) = -0.03(120 + 10t) + 10(0.48 - 0.03t)$
$\quad\quad = 1.2 - 0.6t$
$R'(t) = 0 \Rightarrow t = 2$ weeks; $R''(t) = -0.6 \Rightarrow R$ concav
$R(2) = (120 + 20)(0.48 - 0.06) = (140)(0.42)$
$\quad\quad = \$58.80/\text{tree}$

63. (a) Let x = number of \$50 increases in the fee.
number of attendees = $40 + 5x$
fee = $p(x) = 1000 - 50x$
$R(x) = (40 + 5x)(1000 - 50x)$

(b) $R'(x) = -50(40 + 5x) + 5(1000 - 50x)$
$\quad\quad = 3000 - 500x$
$R'(x) = 0 \Rightarrow x = 6$

(c) $R(6) = (40 + 30)(1000 - 300) = \$49,000$

65. $V(x) = (18 - 2x)^2 x \quad (0 \le x \le 9)$
$V'(x) = (18 - 2x)^2 - 4x(18 - 2x)$
$\quad\quad = (18 - 2x)(18 - 6x)$
$V'(x) = 0 \Rightarrow x = 9, 3$
$x = 9$ results in the minimum volume, $V = 0$.
$x = 3$ results in $V = 432$ cubic inches

67. $\text{Cost} = C = (2)(5)(6x) + (2)(10)(6y) + 20xy$
$\quad\quad\quad\quad\quad \text{ends} \quad\quad \text{sides} \quad\quad \text{bottom}$

$V = 6xy = 192; \; y = \dfrac{32}{x}$

$C(x) = 60x + 120\left[\dfrac{32}{x}\right] + 20(32)$

$C'(x) = 60 - \dfrac{3840}{x^2} = \dfrac{60(x^2 - 64)}{x^2}$

$C'(x) = 0 \Rightarrow x = 8; \; y = \dfrac{32}{8} = 4$

$C''(x) = \dfrac{7680}{x^3} > 0 \Rightarrow C$ concave up.

$\therefore \; x = 8$ results in the absolute minimum of C.

69. $H(P) = -0.01P^2 + 4P - P = -0.01P^2 + 3P$
$H'(P) = -0.02P + 3$
$H'(P) = 0 \Rightarrow P = 150$
$H(150) = 225 = $ maximum sustainable harvest.

71. $H(P) = 60\sqrt{P} - P$

$H'(P) = \dfrac{30}{\sqrt{P}} - 1 = \dfrac{30 - \sqrt{P}}{\sqrt{P}}$

$H'(P) = 0 \Rightarrow P = 900$

$H(900) = 60\sqrt{900} - 900 = 900 = $ maximum
sustainable harvest.

73. $H(P) = 1.10P - P = 0.10P$
$H'(P) = 0.10$
$H'(P) > 0$, and $H(P)$ is unbounded.

75. $C(x) = 600(20 - x) + 800\sqrt{9^2 + x^2}$
$\qquad\qquad$ onshore $\qquad$ under water

$$C'(x) = -600 + (800)\frac{x}{\sqrt{9^2 + x^2}}$$

$$C'(x) = 0 \Rightarrow 4x = 3\sqrt{9^2 + x^2};\ 16x^2 = 9(81 + x^2)$$

$$x^2 = \frac{729}{7};\ x = \frac{27}{\sqrt{7}} \approx 10.205 \text{ miles}$$

$$C''(x) = \frac{800\sqrt{81 + x^2} - \dfrac{800x^2}{\sqrt{81 + x^2}}}{81 + x^2} = \frac{64,800}{(81 + x^2)^{3/2}}$$

$C''(x) > 0 \Rightarrow C$ concave up. $\quad \therefore\ x = 27/\sqrt{7}$ yields th

10–5 Minimizing Inventory Cost

1. (a) $C(Q) = K\left[\dfrac{D}{Q}\right] + H\left[\dfrac{Q}{2}\right]$

$$= 100\left[\frac{25,000}{Q}\right] + 20\left[\frac{Q}{2}\right]$$

$$= \frac{2,500,000}{Q} + 10Q$$

(b) $C'(Q) = \dfrac{-2,500,000}{Q^2} + 10 = \dfrac{10Q^2 - 2,500,000}{Q^2}$

$\quad C'(Q) = 0 \Rightarrow Q = 500$, the order size that minimi

$\quad$ Note that $C''(Q) = \dfrac{5,000,000}{Q^3} > 0 \Rightarrow C$ concave u

3. (a) $C(Q) = K\left[\dfrac{D}{Q}\right] + H\left[\dfrac{Q}{2}\right]$

$$= 60\left[\frac{300,000}{Q}\right] + 0.05(500)\left[\frac{Q}{2}\right]$$

$$= \frac{18,000,000}{Q} + 12.5Q$$

(b) $C'(Q) = \dfrac{-18,000,000}{Q^2} + 12.5$

$$= \frac{12.5Q^2 - 18,000,000}{Q^2}$$

$\quad C'(Q) = 0 \Rightarrow Q = 1200$ units to minimize $C(Q)$

(c) $C(1200) = \dfrac{18,000,000}{1200} + 12.5(1200)$

$$= 15,000 + 15,000 = \$30,000/\text{year}$$

(d) $\dfrac{D}{Q} = \dfrac{300,000}{1200} = 250$ orders/year

5. (a) $H = \dfrac{7,000,000 + 1,000,000}{500,000} = \$16/\text{unit}$

(b) $C(Q) = K\left[\dfrac{D}{Q}\right] + H\left[\dfrac{Q}{2}\right]$

$$= 100\left[\frac{500,000}{Q}\right] + 16\left[\frac{Q}{2}\right]$$

$$= \frac{50,000,000}{Q} + 8Q$$

(c) $C'(Q) = \dfrac{-50,000,000}{Q^2} + 8$

$$= \frac{8Q^2 - 50,000,000}{Q^2}$$

$\quad C'(Q) = 0 \Rightarrow Q = 2500$ units to minimize $C(Q)$

(d) $C(2500) = \dfrac{50,000,000}{2500} + 8(2500)$

$$= 20,000 + 20,000 = \$40,000/\text{year}$$

(e) $\dfrac{D}{Q} = \dfrac{500,000}{2500} = 200$ orders/year

7. $C''(Q) = \dfrac{2KD}{Q^3} > 0 \Rightarrow C$ concave up. Hence, the rel
the graph of C vs. Q is concave up, the relative min

Review Exercises

1. $f(x) = x^2 - 6x + 8$

   (a) $f'(x) = 2x - 6$

   (b) $f''(x) = 2$

   (c) $f'(1) = 2(1) - 6 = -4$

   (d) $f''(1) = 2$

3. $f(x) = x^6 - 8x^4 + 7$

   (a) $f'(x) = 6x^5 - 32x^3$

   (b) $f''(x) = 30x^4 - 96x^2$

   (c) $f'(1) = 6(1) - 32(1) = -26$

   (d) $f''(1) = 30(1) - 96(1) = -66$

5. $S(t) = -2t^2 + 80t \quad (0 \le t \le 40)$

   (a) $S'(t) = -4t + 80$
   $S'(5) = -20 + 80 = 60$ feet/second
   $S'(10) = -40 + 80 = 40$ feet/second

   (b) Acceleration $= S''(t) = -4$
   $S''(5) = S''(10) = -4$ feet/second²

7. $f(x) = 2x^3 + 3x^2 - 60x + 8$

   (a) $f'(x) = 6x^2 + 6x - 60$
   $= 6(x^2 + x - 10)$

   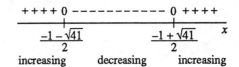

   $\approx 6(x - 2.70156)(x + 3.70156)$

   The critical values are $x = -\dfrac{1}{2} + \dfrac{\sqrt{41}}{2}$ and

   $x = -\dfrac{1}{2} - \dfrac{\sqrt{41}}{2}$

   (b) sign of $f'(x)$

   ```
   ++++ 0 ------------ 0 ++++
   ────┼───────────────┼──── x
     -1-√41          -1+√41
       2                2
   increasing   decreasing   increasing
   ```

   $f(x)$

   f is increasing over the intervals

   $-\infty < x < -\dfrac{1}{2} - \dfrac{\sqrt{41}}{2}$ and $-\dfrac{1}{2} + \dfrac{\sqrt{41}}{2} < x < \infty$

   (c) f is decreasing over the interval

   $-\dfrac{1}{2} - \dfrac{\sqrt{41}}{2} < x < -\dfrac{1}{2} + \dfrac{\sqrt{41}}{2}$

   (d) $f\left[-\dfrac{1}{2} - \dfrac{\sqrt{41}}{2}\right] \approx 169.764$ is a relative maximum

   $f\left[-\dfrac{1}{2} + \dfrac{\sqrt{41}}{2}\right] \approx -92.764$ is a relative minimum

   (e)

   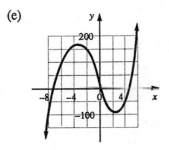

9. $f(x) = x^3 - 24x^2 + 6$

(a) $f'(x) = 3x^2 - 48x = 3x(x - 16)$
The critical values are $x = 0$ and $x = 16$.

(b) sign of $f'(x)$

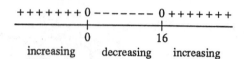

f is increasing over the intervals $-\infty < x < 0$ and $16 < x < \infty$

(c) f is decreasing over the interval $0 < x < 16$

(d) $f(0) = 6$ is a relative maximum
$f(16) = -2042$ is a relative minimum

(e)

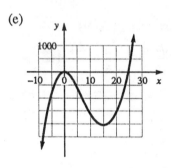

11. $f(x) = 3x^4 - 8x^3 + 6x^2 + 2$
$f'(x) = 12x^3 - 24x^2 + 12x$
$= 12x(x^2 - 2x + 1)$
$= 12x(x - 1)^2$

(continued)

11.   (continued)

sign of $f'(x)$

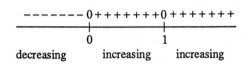

f(x)

$f''(x) = 36x^2 - 48x + 12$
$= 12(3x^2 - 4x + 1)$
$= 12(3x - 1)(x - 1)$

sign of $f''(x)$

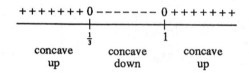

f(x)

$f(0) = 2$ is both a relative and an absolute minimum
inflection points occur at $x = \frac{1}{3}$ and $x = 1$

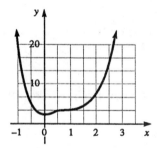

13. $f(x) = x^{6/5}$

$f'(x) = \dfrac{6}{5}x^{1/5}$

sign of $f'(x)$

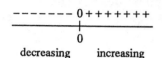

decreasing   increasing

$f(x)$

$f''(x) = \dfrac{6}{25}\dfrac{1}{x^{4/5}}$

sign of $f''(x)$

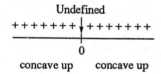

concave up   concave up

$f(x)$

$f(0) = 0$ is both a relative and an absolute minimum
There are no inflection points

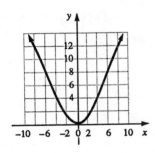

15. $f(x) = \dfrac{10}{x^2 + 4}$

$f'(x) = \dfrac{-20x}{(x^2 + 4)^2}$

sign of $f'(x)$

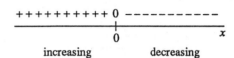

increasing   decreasing

$f(x)$

$f''(x) = \dfrac{-20(x^2 + 4)^2 + 80x^2(x^2 + 4)}{(x^2 + 4)^4} = \dfrac{20(3x^2 - 4)}{(x^2 + 4)^3}$

sign of $f''(x)$

$f(x)$

$f(0) = 2.5$ is a relative maximum
inflection points occur at $x = -\dfrac{2}{\sqrt{3}}$ and $x = \dfrac{2}{\sqrt{3}}$.

$y = 0$ is a horizontal asymptote.
$f(0) = 2.5$ is the absolute maximum.

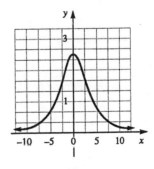

17. $C(x) = x^2 - 8x + 100$ $(x > 0)$

(a) $\bar{C}(x) = \dfrac{C(x)}{x} = x - 8 + \dfrac{100}{x}$

(b) $\bar{C}'(x) = 1 - \dfrac{100}{x^2} = \dfrac{x^2 - 100}{x^2} = \dfrac{(x - 10)(x + 10)}{x^2}$

sign of $\bar{C}'(x)$

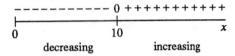

$\bar{C}(x)$

$\bar{C}''(x) = \dfrac{200}{x^3} > 0 \Rightarrow \bar{C}$ concave up.

$\bar{C}(10)$ is both a relative and an absolute minimum. $y = x - 8$ is an oblique asymptote; $x = 0$ is a vertical asymptote.

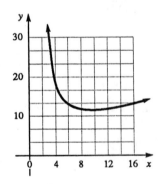

(c) $\bar{C}(10) = 12$ is both a relative and an absolute minimum.

(d) $x = 10$ minimizes the average cost per unit.

(e) $\bar{C}(10) = 12$

19. $f(x) = x^3 - 15x^2 + 48x + 4$

$f'(x) = 3x^2 - 30x + 48 = 3(x^2 - 10x + 16)$
$\qquad\qquad\qquad\qquad = 3(x - 2)(x - 8)$

$f'(x) = 0 \Rightarrow x = 2, 8$
$f''(x) = 6x - 30$
$f''(2) = -18 \Rightarrow f(2)$ is a relative maximum.
$f''(8) = 18 \Rightarrow f(8)$ is a relative minimum.
$f(2) = 48; f(8) = -60$

21. $f(x) = x + \dfrac{100}{x}$

$f'(x) = 1 - \dfrac{100}{x} = \dfrac{x^2 - 100}{x^2} = \dfrac{(x + 10)(x - 10)}{100}$

$f'(x) = 0 \Rightarrow x = -10, 10$
$f''(x) = \dfrac{200}{x^3}$
$f''(-10) < 0 \Rightarrow f(-10)$ is a relative maximum
$f''(10) > 0 \Rightarrow f(10)$ is a relative minimum
$f(-10) = -20; f(10) = 20$

23. $f(x) = 3\sqrt{x} - x$ $(x \geq 0)$

$f'(x) = \dfrac{3}{2\sqrt{x}} - 1 = \dfrac{3 - 2\sqrt{x}}{2\sqrt{x}}$

$f'(x) = 0 \Rightarrow x = \dfrac{9}{4}$

$f''(x) = -\dfrac{3}{4x^{3/2}} < 0 \Rightarrow f$ concave down

$\therefore \ f\left(\dfrac{9}{4}\right) = \dfrac{9}{4}$ is the absolute maximum of f.
There is no absolute minimum of f.

25. $f(x) = 7 + 3x + \dfrac{48}{x}$ $(x > 0)$

$f'(x) = 3 - \dfrac{48}{x^2} = \dfrac{3x^2 - 48}{x^2} = \dfrac{3(x^2 - 16)}{x^2}$

$f'(x) = 0 \Rightarrow x = 4$

$f''(x) = \dfrac{144}{x^3}; f''(4) > 0 \Rightarrow$ relative minimum

Since f is concave up for $x > 0$, $f(4) = 31$ is the absolute minimum of f. There is no absolute maximum of f.

27. $f(x) = 8 + 2x + \dfrac{128}{x^3}$ $(x > 0)$

$f'(x) = 2 - \dfrac{384}{x^4} = \dfrac{2x^4 - 348}{x^4} = \dfrac{2(x^4 - 192)}{x^4}$

$f'(x) = 0 \Rightarrow x = 2\sqrt[4]{12} \approx 3.7224$

$f''(x) = \dfrac{1536}{x^5} > 0 \Rightarrow f$ concave up for $x > 0$

$\therefore\ f(2\sqrt[4]{12}) \approx 17.9264$ is the absolute minimum of f. There is no absolute maximum of f.

29. $A = xy;\ 2x + 2y = 800$

$A = x(400 - x) = 400x - x^2$

$A'(x) = 400 - 2x = 2(200 - x)$

$A'(x) = 0 \Rightarrow x = 200,\ y = 200$

$A''(x) = -2 \Rightarrow A$ concave down. $\therefore\ x = 200$ and $y = 200$ maximize the area.

31. Let $x =$ number of \$1 decreases in the monthly rate

$R(x) = (20 - x)(5000 + 400x)$

(a) $R'(x) = 400(20 - x) - (5000 + 400x)$
$= 3000 - 800x$

$R'(x) = 0 \Rightarrow x = 3.75$

Calculate $R(3)$ and $R(4)$:

$R(3) = 17(6200) = 105,400$

$R(4) = 16(6600) = 105,600$

$\therefore$ let $x = 4$, or the monthly rate is \$16.

(b) $R(4) = \$105,600$

(c) 6600 subscribers

33.

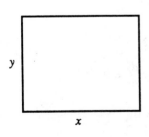

Cost $= C = 10(2x) + 10y + 20y$
$\underset{\text{cedar}}{\underbrace{\qquad\qquad}}$ chain link

(continued)

33. *(continued)*

Also, $xy = 800;\ y = \dfrac{800}{x}$

$C(x) = 20x + 30\left[\dfrac{800}{x}\right] = 20x + \dfrac{24,000}{x}$ $(x \neq 0)$

$C'(x) = 20 - \dfrac{24,000}{x^2} = \dfrac{20x^2 - 24,000}{x^2}$

$C'(x) = 0 \Rightarrow x^2 = 1200 \Rightarrow x = 20\sqrt{3} \approx 34.64$ feet

$y = \dfrac{800}{20\sqrt{3}} = \dfrac{40}{\sqrt{3}}$

≈ 23.09 feet

$C''(x) = \dfrac{48,000}{x^3};\ C''(20\sqrt{3}) > 0 \Rightarrow$ minimum of C.

CHAPTER
11

APPLICATIONS OF THE DERIVATIVE

11–1 Implicit Differentiation

1. $xy^3 + x^4 + y = 8$

$3xy^2 \dfrac{dy}{dx} + y^3 + 4x^3 + \dfrac{dy}{dx} = 0$

$(3xy^2 + 1)dx = -4x^3 - y^3$

$\dfrac{dy}{dx} = -\dfrac{4x^3 + y^3}{1 + 3xy^2}$

3. $xy^5 + x^3 + y^4 = 10$

$5xy^4\dfrac{dy}{dx} + y^5 + 3x^2 + 4y^2\dfrac{dy}{dx} = 0$

$(5xy^4 + 4y^3)\dfrac{dy}{dx} = -3x^2 - y^5$

$\dfrac{dy}{dx} = -\dfrac{3x^2 + y^5}{y^3(5xy + 4)}$

5. $x^3 + y^6 = 8xy$

$3x^2 + 6y^5\dfrac{dy}{dx} = 8x\dfrac{dy}{dx} + 8y$

$(6y^5 - 8x)\dfrac{dy}{dx} = 8y - 3x^2$

$\dfrac{dy}{dx} = \dfrac{8y - 3x^2}{6y^5 - 8x}$

7. $(x - y)(x^2 + y^2) = 9$

$(x - y)\left[2x + 2y\dfrac{dy}{dx}\right] + \left[1 - \dfrac{dy}{dx}\right](x^2 + y^2) = 0$

$(2xy - 2y^2 - x^2 - y^2)\dfrac{dy}{dx} = -2x^2 + 2xy - x^2 - y^2$

$\dfrac{dy}{dx} = \dfrac{-3x^2 + 2xy - y^2}{-x^2 + 2xy - 3y^2} = \dfrac{3x^2 - 2xy + y^2}{x^2 - 2xy + 3y^2}$

9. $(x + y)^4 = x^2 + y^2$

$4(x + y)^3\left[1 + \dfrac{dy}{dx}\right] = 2x + 2y\dfrac{dy}{dx}$

$[4(x + y)^3 - 2y]\dfrac{dy}{dx} = 2x - 4(x + y)^3$

$\dfrac{dy}{dx} = \dfrac{x - 2(x + y)^3}{2(x + y)^3 - y}$

11. $x^3 + xy^2 + y^4 = 21 \ (1,2)$

$3x^2 + 2xy\dfrac{dy}{dx} + y^2 + 4y^3\dfrac{dy}{dx} = 0$

$3 + 4\dfrac{dy}{dx} + 4 + 32\dfrac{dy}{dx} = 0$

$\dfrac{dy}{dx} = -\dfrac{7}{36}$

13. $x^2 - 3xy + 5y^2 = 15 \; (-2,1)$

$$2x - 3x\frac{dy}{dx} - 3y + 10y\frac{dy}{dx} = 0$$

$$-4 + 6\frac{dy}{dx} - 3y + 10\frac{dy}{dx} = 0$$

$$\frac{dy}{dx} = \frac{7}{16}$$

15. $4x - 2y^2 + 5y = -7 \; (-1,3)$

$$4 - 4y\frac{dy}{dx} + 5\frac{dy}{dx} = 0$$

$$4 - 12\frac{dy}{dx} + 5\frac{dy}{dx} = 0$$

$$\frac{dy}{dx} = \frac{4}{7}$$

17. $x^2 + y^2 = 64$

(a) $2x + 2y\frac{dy}{dx} = 0$

$$\frac{dy}{dx} = -\frac{x}{y}$$

(b) At $(4, 4\sqrt{3})$ $\dfrac{dy}{dx} = \dfrac{-4}{4\sqrt{3}} = -\dfrac{1}{\sqrt{3}}$

(c) $\dfrac{y - 4\sqrt{3}}{x - 4} = -\dfrac{1}{\sqrt{3}}$

$$y - 4\sqrt{3} = -\frac{1}{\sqrt{3}}x + \frac{4}{\sqrt{3}}$$

$$y = -\frac{1}{\sqrt{3}}x + \frac{16}{\sqrt{3}}$$

17. *(continued)*

(d)

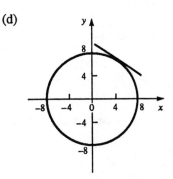

19. $y = 6x + 10$

$$\frac{dy}{dx} = 6; \; \frac{dx}{dy} = \frac{1}{6}$$

21. $y = 4x^2 - 3x + 7$

$$\frac{dy}{dx} = 8x - 3$$

$$\frac{dx}{dy} = \frac{1}{8x - 3}, \; x \neq \frac{3}{8}$$

23. $y = x^2 - 4x$

(a) $\dfrac{dy}{dx} = 2x - 4; \; \dfrac{dx}{dy} = \dfrac{1}{2x - 4}, \; x \neq 2$

(b) At $x = 3$; $\dfrac{dy}{dx} = \dfrac{1}{6 - 4} = \dfrac{1}{2}$

(c) $\dfrac{dy}{dx} = \dfrac{1}{2}$ is the rate of change of x with respect to

y at $x = 3$.

(continued)

25. $q = p^2 - 20p + 101$ $(0 \le p \le 10)$

(a) $\dfrac{dq}{dp} = 2p - 20;$ $\dfrac{dp}{dq} = \dfrac{1}{2p - 20},$ $p \ne 10$

(b) At $p = 7,$ $\dfrac{dp}{dq} = \dfrac{1}{14 - 20} = -\dfrac{1}{6}$

(c) $\dfrac{dp}{dq} = -\dfrac{1}{6}$ is the rate of change of p with respect to q at $p = 7.$

27. $S^3 - 2St + 4t^3 - 2t = 0$

$3S^2 \dfrac{dS}{dt} - 2S - 2t \dfrac{dS}{dt} + 12t^2 - 2 = 0$

$(3S^2 - 2t)\dfrac{dS}{dt} = 2S - 12t^2 + 2$

$\dfrac{dS}{dt} = \dfrac{2(S - 6t^2 + 1)}{3S^2 - 2t}$

29. $S^4 - 3St^2 + t^3 = 6$

$4S^3 \dfrac{dS}{dt} - 6St - 3t^2 \dfrac{dS}{dt} + 3t^2 = 0$

$(4S^3 - 3t^2)\dfrac{dS}{dt} = 3t(2S + t)$

$\dfrac{dS}{dt} = \dfrac{3t(2S + t)}{4S^3 - 3t^2}$

11–2 Related Rates

1. $y = 3x^2 + 4x;$ $\dfrac{dx}{dt} = -1,$ $x = 2,$ $\dfrac{dy}{dt} = ?$

$\dfrac{dy}{dt} = 6x\dfrac{dx}{dt} + 4\dfrac{dx}{dt} = (6 \cdot 2 + 4)(-1) = -16$

3. $y = 4x^2 + 9;$ $\dfrac{dy}{dt} = 4,$ $x = 2,$ $\dfrac{dx}{dt} = ?$

$\dfrac{dy}{dt} = 8x\dfrac{dx}{dt}$

$4 = 16\dfrac{dx}{dt};$ $\dfrac{dx}{dt} = \dfrac{1}{4}$

5. $2y^2 + 3x = 8;$ $\dfrac{dx}{dt} = 3,$ $x = 2,$ $y = 1,$ $\dfrac{dy}{dt} = ?$

$4y\dfrac{dy}{dt} + 3\dfrac{dx}{dt} = 0$

$4\dfrac{dy}{dt} + 9 = 0;$ $\dfrac{dy}{dt} = -\dfrac{9}{4}$

7. $3x^2 + 2y^2 = 30;$ $\dfrac{dx}{dt} = -2,$ $x = 2,$ $y = 3,$ $\dfrac{dy}{dt} = ?$

$6x\dfrac{dx}{dt} + 4y\dfrac{dy}{dt} = 0$

$12(-2) + 12\dfrac{dy}{dt} = 0;$ $\dfrac{dy}{dt} = 2$

9. $y^2 + xy = 6;$ $\dfrac{dx}{dt} = 1,$ $x = 5,$ $y = 1,$ $\dfrac{dy}{dt} = ?$

$2y\dfrac{dy}{dt} + x\dfrac{dy}{dt} + y\dfrac{dx}{dt} = 0$

$(2 + 5)\dfrac{dy}{dt} + 1 = 0;$ $\dfrac{dy}{dt} = -\dfrac{1}{7}$

11. $x^2 - xy + 4 = 0;$ $\dfrac{dx}{dt} = -1,$ $x = 2,$ $y = 6,$ $\dfrac{dy}{dt} = ?$

$2x\dfrac{dx}{dt} - x\dfrac{dy}{dt} - y\dfrac{dx}{dt} = 0$

$4(-1) - 2\dfrac{dy}{dt} - 6(-1) = 0;$ $\dfrac{dy}{dt} = 1$

13. $xy = 8;$ $\dfrac{dy}{dt} = 2$ at $(2,4),$ $\dfrac{dx}{dt} = ?$

$x\dfrac{dy}{dt} + y\dfrac{dx}{dt} = 0$

$2(2) + 4\dfrac{dx}{dt} = 0;$ $\dfrac{dx}{dt} = -1$ unit per second

15. $x^2 + y^2 = 25^2 = 625; \dfrac{dx}{dt} = 4, x = 20, \dfrac{dy}{dt} = ?$

at $x = 20$, $y = \sqrt{625 - 400} = \sqrt{225} = 15$.

$2x\dfrac{dx}{dt} + 2y\dfrac{dy}{dt} = 0$

$\dfrac{dy}{dt} = -\dfrac{x}{y}\dfrac{dx}{dt} = -\dfrac{4x}{y} = -\dfrac{4(20)}{15} = -\dfrac{16}{3}$ ft/second

17. $R = 800x - \dfrac{x^2}{10}$, $C = 40x + 5000$, $P = R - C$

$x = 2000$, $\dfrac{dx}{dt} = 100$

(a) $\dfrac{dR}{dt} = \left[800 - \dfrac{x}{5}\right]\dfrac{dx}{dt} = \left[800 - \dfrac{2000}{5}\right](100)$

$= 40{,}000$

(b) $\dfrac{dC}{dt} = 40\dfrac{dx}{dt} = 40(100) = 4000$

(c) $\dfrac{dP}{dt} = \dfrac{dR}{dt} - \dfrac{dC}{dt} = 40{,}000 - 4000 = 36{,}000$

19. $S = 2\pi r^2 + 2\pi rh$;
ends   lateral
   surface

$r = 20$, $h = 40$, $\dfrac{dr}{dt} = 0.40$, $\dfrac{dh}{dt} = 0.30$

$\dfrac{dS}{dt} = 4\pi r\dfrac{dr}{dt} + 2\pi r\dfrac{dh}{dt} + 2\pi h\dfrac{dr}{dt}$

$= 2\pi\left[2r\dfrac{dr}{dt} + r\dfrac{dh}{dt} + h\dfrac{dr}{dt}\right]$

$= 2\pi[40(0.40) + 20(0.30) + 40(0.40)]$
$= 76\pi$ cm²/minute
or approximately 238.8 cm²/minute

21. $V = lwh = (12)(10)h = 120h$

$\dfrac{dV}{dt} = 120\dfrac{dh}{dt}; \dfrac{dh}{dt} = \dfrac{1}{120}\dfrac{dV}{dt} = \dfrac{1}{120} \cdot 30 = \dfrac{1}{4}$ foot/day

23. (a) Let $x = $ the distance along I–95 to the motorist
and
$y = $ the distance from that state trooper to the motorist.

$y^2 = x^2 + 40^2$

$2y\dfrac{dy}{dx} = 2x\dfrac{dx}{dt}; \dfrac{y}{x}\dfrac{dy}{dt} = \dfrac{dx}{dt}$

When $y = 200$,

$x = \sqrt{200^2 - 40^2} = 40\sqrt{24} = 80\sqrt{6}$

$\dfrac{dx}{dt} = \dfrac{200}{80\sqrt{6}}(-80) = -\dfrac{100\sqrt{6}}{3}$ feet/second

≈ -81.65 feet/second

Speed $= \left|-81.65\right| = 81.65$ feet/second

(b) Miles/hour $= 81.65 \cdot \dfrac{3600}{5280} \approx 55.67$

25. $V = \dfrac{4}{3}\pi r^3$; $r = 20$, $\dfrac{dr}{dt} = 2$

$\dfrac{dV}{dt} = 4\pi r^2\dfrac{dr}{dt} = 4\pi(20^2)(2) = 3200\pi$

$\approx 10{,}053$ cm³/minute

27. $V = \pi r^2 h$; $r = 40$

$\dfrac{dV}{dt} = \pi r^2\dfrac{dh}{dt} = 1600\pi\dfrac{dh}{dt}$ (since r is a constant)

$\dfrac{dh}{dt} = \dfrac{1}{1600\pi}\left[\dfrac{dV}{dt}\right] = \dfrac{-4000}{1600\pi} = -\dfrac{5}{2\pi}$

≈ -0.796 foot/hour

29. $R = R_1 + R_2$; $\dfrac{dR_1}{dt} = 0.4$, $\dfrac{dR_2}{dt} = 0.6$

$\dfrac{dR}{dt} = \dfrac{dR_1}{dt} + \dfrac{dR_2}{dt} = 0.4 + 0.6 = 1.0$ ohm/second

31. $V = \frac{1}{3}\pi x^2 y;\ \frac{x}{1000} = \frac{y}{50},\ \frac{dV}{dt} = 200,\ y = 30$

$$V = \frac{\pi}{3}\left[\frac{1000}{50}y\right]y = \frac{400\pi}{3}y^3$$

$$\frac{dV}{dt} = 400\pi y^2\ \frac{dy}{dt}$$

$$200 = 400\pi(900)\ \frac{dy}{dt};$$

$$\frac{dy}{dt} = \frac{1}{1800\ \pi} \approx \frac{1}{1800(3.1416)}$$

$$\approx 1.768 \times 10^{-4}\ \text{foot/minute}$$

11–3 Elasticity of Demand

1. $q = -3p + 90$

$$E = \frac{p}{q} \cdot \frac{dq}{dp} = \left[\frac{p}{-3p + 90}\right](-3) = \frac{-3p}{-3 + 90} = \frac{-p}{30 - p}$$

3. $q = -p + 40$

$$E = \frac{p}{q} \cdot \frac{dq}{dp} = \left[\frac{p}{-p + 40}\right](-1) = \frac{-p}{-p + 40} = \frac{-p}{40 - p}$$

5. $q = 3(p - 4)^2$

$$E = \frac{p}{q} \cdot \frac{dq}{dp} = \frac{p}{3(p - 4)^2} \cdot 6(p - 4) = \frac{2p}{p - 4}$$

7. $q = 40 - 10\sqrt{p}$

$$E = \frac{p}{q} \cdot \frac{dq}{dp} = \frac{p}{40 - 10\sqrt{p}} \cdot \frac{-5}{\sqrt{p}} = \frac{\sqrt{p}}{2\sqrt{p} - 8}$$

9. $q = \sqrt{200 - p}$

$$E = \frac{p}{q} \cdot \frac{dq}{dp} = \frac{p}{\sqrt{200 - p}} \cdot \frac{-1}{2\sqrt{200 - p}} = \frac{-p}{2(200 - p)}$$

11. $q = \dfrac{p}{p - 10}$

$$E = \frac{p}{q} \cdot \frac{dq}{dp} = \frac{p(p - 10)}{p} \cdot \frac{p - 10 - p}{(p - 10)^2} = \frac{-10}{p - 10}$$

13. (a) $E = -1.5$ at $p = 10$

Since $E = \dfrac{\% \text{ change in quantity demanded}}{\% \text{ change in price}}$, the value of $E = -1.5$ means that if the price increases by 1% when $p = 10$, the quantity demanded will decrease by approximately 1.5%. If the price decreases by 1% when $p = 10$, the quantity demanded will increase by approximately 1.5%.

(b) Since $|E| > 1$, the demand is elastic.

15. (a) $E = -0.80$ at $p = 20$

Since $E = \dfrac{\% \text{ change in quantity demanded}}{\% \text{ change in price}}$, the value of $E = -0.80$ means that if the price increases by 1% when $p = 20$, the quantity demanded will decrease by approximately 0.80%. If the price decreases by 1% when $p = 20$, the quantity demanded will increase by approximately 0.80%.

(b) Since $|E| < 1$, the demand is inelastic.

17. $q = -2p + 130 \quad (0 \le p \le 65)$

(a) $E = \dfrac{p}{q} \cdot \dfrac{dq}{dp} = \left[\dfrac{p}{-2p + 130}\right](-2) = \dfrac{-p}{65 - p}$

(b) At $p = 50$, $E = \dfrac{-50}{65 - 50} = -\dfrac{10}{3} \approx -3.33$

(c) The value of $E = -3.33$ means that if the price increases by 1% when $p = 50$, the quantity demanded will decrease by approximately 3.33%. Also, if the price decreases by 1% when $p = 50$, the quantity demanded will increase by approximately 3.33%.

19. $q = \dfrac{p}{p-8}$ $(p > 8)$

(a) $E = \dfrac{p}{q} \cdot \dfrac{dq}{dp} = \dfrac{p(p-8)}{p} \cdot \dfrac{p-8-p}{(p-8)^2} = \dfrac{-8}{p-8}$

(b) At $p = 10$, $E = \dfrac{-8}{10-8} = -4$

(c) The value of $E = -4$ means that if the price increases by 1% when $p = 10$, the quantity demanded will decrease by approximately 4%. Also, if the price decreases by 1% when $p = 10$, the quantity demanded will increase by approximately 4%.

21. $q = 3(p-10)^2$ $(0 < p < 10)$

(a) $E = \dfrac{p}{q} \cdot \dfrac{dq}{dp} = \dfrac{p}{3(p-10)^2} \cdot 6(p-10) = \dfrac{2p}{p-10}$

(b) At $p = 4$, $E = \dfrac{2 \cdot 4}{4-10} = -\dfrac{4}{3} \approx -1.33$

(c) The value of $E = -1.33$ means that if the price increases by 1% when $p = 4$, the quantity demanded will decrease by approximately 1.33%. Also, if the price decreases by 1% when $p = 4$, the quantity demanded will increase by approximately 1.33%.

23. $q = \dfrac{10,000}{p}$ $(p > 0)$

(a) $E = \dfrac{p}{q} \cdot \dfrac{dq}{dp} = \dfrac{p^2}{10,000} \cdot \dfrac{-10,000}{p^2} = -1$

(b) $E = -1$ means that an increase of 1% in the price will lead to a decrease of approximately 1% in the quantity demanded; a decrease of 1% in the price will lead to an increase of approximately 1% in the quantity demanded.

(c,d) The demand has unit elasticity for all values of p.

25. $q = \sqrt{100-p}$ $(0 \le p \le 100)$

(a) $E = \dfrac{p}{q} \cdot \dfrac{dq}{dp} = \dfrac{p}{\sqrt{100-p}} \cdot \dfrac{-1}{2\sqrt{100-p}}$

$= \dfrac{-p}{2(100-p)}$

(b) The demand is elastic if $\left| E \right| > 1$.

$\left| E \right| > 1 \Rightarrow \dfrac{p}{2(100-p)} > 1 \Rightarrow$

$p > 2(100-p) \Rightarrow 3p > 200;$

$\dfrac{200}{3} < p < 100$

(c) The demand is inelastic if $\left| E \right| < 1$.

$\left| E \right| < 1 \Rightarrow \dfrac{p}{2(100-p)} < 1 \Rightarrow$

$p < 2(100-p) \Rightarrow 3p < 200;$

$0 \le p < \dfrac{200}{3}$

(d) The demand has unit elasticity when $p = \dfrac{200}{3}$.

11–4 Differentials

1. $y = x^3$
$dy = f'(x)dx = 3x^2 dx$

3. $y = f(x) = 8x^4 + 1$
$dy = f'(x)dx = 32x^3\, dx$

5. $y = 3x\sqrt{x} = 3x^{3/2}$

$dy = f'(x)dx = \dfrac{9}{2}\sqrt{x}\, dx$

7. $y = 8\left[1 + \dfrac{4}{x}\right]$

$dy = f'(x)dx = -\dfrac{32}{x^2}\, dx$

9. $y = \sqrt[3]{x}$

$dy = f'(x)dx = \dfrac{1}{3\sqrt[3]{x^2}}\, dx$

11. $y = x^2 - 6x + 5$; $x = 9$, $dx = 0.5$

$dy = f'(x)dx = (2x - 6)dx = (18 - 6)(0.5) = 6$

13. $y = f(x) = \dfrac{1}{(x - 9)^2}$; $x = 3$, $dx = 0.1$

$dy = f'(x)dx = \dfrac{-2}{(x - 9)^3}\, dx = \dfrac{-2(0.1)}{(3 - 9)^3} \approx 0.000926$

15. $y = x^3 - 8x + 5$; $x = 4$, $dx = 0.4$

$dy = f'(x)dx = (3x^2 - 8)dx = 40(0.4) = 16$

17. $\sqrt{81.68}$

Let $y = \sqrt{x}$; $dy = f'(x)dx = \dfrac{1}{2\sqrt{x}}\, dx$;

let $x = 81$, $dx = 0.68$

$dy = \dfrac{1}{2\sqrt{81}}\, (0.68) \approx 0.0378$; $\sqrt{81.68} \approx 9.03$

19. $\sqrt{65}$

Let $y = \sqrt{x}$; $dy = f'(x)dx = \dfrac{1}{2\sqrt{x}}\, dx$;

let $x = 64$, $dx = 1$

$dy = \dfrac{1}{2\sqrt{64}}\, (1) = \dfrac{1}{16} = 0.0625$

$\sqrt{65} \approx 8.063$

21. $\sqrt[4]{15.8}$

Let $y = x^{1/4}$; $dy = f'(x)dx = \dfrac{1}{4x^{3/4}}\, dx$;

let $x = 16$, $dx = -0.2$

$dy = \dfrac{-0.2}{4(16^{3/4})} = \dfrac{-0.2}{32} = -0.00625$

$\sqrt[4]{15.8} \approx 2 - 0.00625 \approx 1.9938$

23. $R(x) = -x^2 + 10x \quad (0 \le x \le 5)$

$dR = R'(x)dx = (-2x + 10)dx$; $x = 3.6$, $dx = 0.15$

$dR = (-7.2 + 10)(0.15) = 0.42$ million dollars

Actual change in $R = \Delta R = R(3.75) - R(3.6)$

$= 23.4375 - 23.04$

$= 0.3975$ million dollars

25. $p = \dfrac{6}{\sqrt{q}} \quad (q > 0)$

$dp = \dfrac{-3}{q^{3/2}}\, dq$; $q = 25$, $dq = 1$

$dp = \dfrac{-3}{125} = -\$0.024$

27. $C(x) = 0.15x^2 + 7000 \quad (0 \le x \le 500)$

$dC = C'(x)dx = 0.30x\, dx$; $x = 200$, $dx = 1$

$dC = 0.30(200)(1) = 60$ million dollars

Actual change in $C = \Delta C = C(201) - C(200)$

$= 13,060.15 - 13,000$

$= 60.15$ million dollars

Graphically, the slope of the tangent line at $x = 200$ is 60, and $60dx$ approximates Δy.

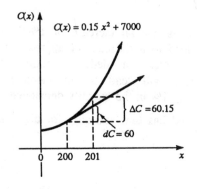

29. $C(x) = 100x + 50{,}000$

(a) $\bar{C}(x) = \dfrac{C(x)}{x} = 100 + \dfrac{50{,}000}{x}$

(b) $d\bar{C} = \bar{C}'(x)dx = \dfrac{-50{,}000}{x^2}\, dx;\; x = 200,\; dx = 1$

$d\bar{C} = \dfrac{-50{,}000}{200^2}(1) = -1.25$

31. $C(x) = 0.25x^2 + 8000\;\;(0 \le x \le 6000)$
$dC = C'(x)dx = 0.50x\, dx;\; x = 40,\; dx = 1$
$dC = 0.50(40)(1) = 20;$
$C(40) = 0.25(1600) + 8000 = 8400$

% change in $C \approx \dfrac{20}{8400}\, 100 \approx 0.238\%$

33. $C(x) = 500x^2$

a) $dC = C'(x)dx = 1000x\, dx;\; x = 20,\; dx = 2$
$dC = 1000(20)(2) = \$40{,}000$
$C(20) = 500(20)^2 = \$200{,}000$

% change in $C \approx \dfrac{40{,}000}{200{,}000}\, 100 = 20\%$

b) $dC = 1000(20)(0.05 \times 20) = \$20{,}000$

% change in $C \approx \dfrac{20{,}000}{200{,}000}\, 100 = 10\%$

Review Exercises

1. $xy^4 + x^3 + y = 43$

$4xy^3 \dfrac{dy}{dx} + y^4 + 3x^2 + \dfrac{dy}{dx} = 0$

$(4xy^3 + 1)\dfrac{dy}{dx} = -3x^2 - y^4$

$\dfrac{dy}{dx} = -\dfrac{3x^2 + y^4}{4xy^3 + 1}$

3. $(x - y)(x + y^3) = 7$

$(x - y)\left[1 + 3y^2\dfrac{dy}{dx}\right] + \left[1 - \dfrac{dy}{dx}\right](x + y^3) = 0$

$(3xy^2 - 3y^3 - x - y^3)\dfrac{dy}{dx} = -x + y - x - y^3$

$\dfrac{dy}{dx} = \dfrac{-2x + y - y^3}{-x + 3xy^2 - 4y^3} = \dfrac{2x - y + y^3}{x - 3xy^2 + 4y^3}$

5. $x^2 + y^2 = 20$

$2x + 2y\dfrac{dy}{dx} = 0;\; \dfrac{dy}{dx} = -\dfrac{x}{y}$

7. $y = 8x^2 + 6x + 4$

$\dfrac{dy}{dx} = 16x + 6 = 2(8x + 3)$

$\dfrac{dy}{dx} = \dfrac{1}{2(8x + 3)}$

9. $5x^2 + 3y^2 = 47;\; \dfrac{dy}{dt} = 2,\; x = 2,\; y = 3,\; \dfrac{dx}{dt} = ?$

$10x\dfrac{dx}{dt} + 6y\dfrac{dy}{dt} = 0$

$10(2)\dfrac{dx}{dt} + 6(3)(2) = 0$

$\dfrac{dx}{dt} = -\dfrac{18}{10} = -1.8$ unit/second

11. $V = \dfrac{4}{3}\pi r^3$

$\dfrac{dV}{dt} = 4\pi r^2 \dfrac{dr}{dt} = 4\pi(18^2)(1.5) = 1944\pi$ cm³/minute

≈ 6107 cm³/minute

13. $q = 4(p - 7)^2$

$E = \dfrac{p}{q} \cdot \dfrac{dq}{dp} = \dfrac{p}{4(p - 7)^2} \cdot 8(p - 7) = \dfrac{2p}{p - 7}$

15. $q = \dfrac{p}{p - 30}$

$$E = \frac{p}{q} \cdot \frac{dq}{dp} = \frac{p(p - 30)}{p} \cdot \frac{p - 30 - p}{(p - 30)^2} = \frac{-30}{(p - 30)}$$

17. $E = -1.4$ at $p = 20$

(a) The value of $E = -1.4$ means that an increase of 1% in the price when $p = 20$ leads to an approximate decrease of 1.4% in the quantity demanded. Also, a decrease of 1% in the price when $p = 20$ leads to an approximate increase of 1.4% in the quantity demanded.

(b) Since $|E| > 1$, the demand is elastic.

19. $y = x^3 - 4x^2 + 5$
$dy = f'(x)dx = (3x^2 - 8x)dx$

21. $y = 4x^2\sqrt{x} = 4x^{\frac{5}{2}}$

$dy = f'(x)dx = 10x^{\frac{3}{2}} dx$

23. $y = x^2 + 2x + 3$; $x = 4$, $dx = 0.05$
$dy = f'(x)dx = (2x + 2)dx = (8 + 2)(0.05) = 0.5$

25. $R(x) = -x^2 + 40x$ $(0 \le x \le 20)$

(a) $dR = R'(x)dx = (-2x + 40)dx$
$dR = (-20 + 40)(0.5) = 10$ million dollars

(b) $R(10) = -100 + 400 = 300$

% change in $R \approx \dfrac{10}{300} \, 100 = 3.33\%$

CHAPTER
12

DERIVATIVES OF EXPONENTIAL AND LOGARITHMIC FUNCTIONS

12–1 Derivatives of Exponential Functions

1. $y = e^{4x}; \; y' = 4e^{4x}$

3. $y = 4e^{-2x}; \; y' = -8e^{-2x}$

5. $f(x) = -2e^{-0.1x}; \; f'(x) = 0.2e^{-0.1x}$

7. $y = e^{x^5 - 7x}; \; y' = (5x^4 - 7)e^{x^5 - 7x}$

9. $y = e^{2x-5}; \; y' = 2e^{2x-5}$

11. $y = 4^{x^2 - 3x}; \; y' = (2x - 3)(\ln 4)4^{x^2 - 3x}$

13. $y = 4^{-0.02x} \; y' = -0.02(\ln 4)4^{-0.02x}$

15. $y = (x^5 - 4x^2)e^{x^4 - 7x};$

$y' = (x^5 - 4x^2)(4x^3 - 7)e^{x^4 - 7x} + (5x^4 - 8x)e^{x^4 - 7x}$

$= [x^2(x^3 - 4)(4x^3 - 7) + x(5x^3 - 8)]e^{x^4 - 7x}$

17. $f(x) = (x^4 + 8x)e^{x^3 + 5x};$

$f'(x) = (x^4 + 8x)(3x^2 + 5)e^{x^3 + 5x} + (4x^3 + 8)e^{x^3 + 5x}$

$[x(x^3 + 8)(3x^2 + 5) + 4x^3 + 8]e^{x^3 + 5x}$

19. $y = \dfrac{e^x + e^{-x}}{x}; \; y' = \dfrac{x(e^x + e^{-x}) - (e^x + e^{-x})}{x^2}$

21. $y = \dfrac{e^x - e^{-x}}{x}; \; y' = \dfrac{x(e^x + e^{-x}) - (e^x - e^{-x})}{x^2}$

23. $f(x) = \dfrac{e^{x^3 - 4x}}{x^2 - 3x}$

$f'(x) = \dfrac{(x^2 - 3x)(3x^2 - 4)e^{x^3 - 4}x - (2x - 3)e^{x^3 - 4x}}{(x^2 - 3x)^2}$

$= \dfrac{[(x^2 - 3x)(3x^2 - 4) - 2x + 3]e^{x^3 - 4x}}{(x^2 - 3x)^2}$

25. $y = e^{\sqrt{25 - x^2}}; \; y' = \dfrac{-x}{\sqrt{25 - x^2}} e^{\sqrt{25 - x^2}}$

27. $y = e^{\sqrt{x}}; y' = \dfrac{1}{2\sqrt{x}} e^{\sqrt{x}}$

29. $y = \dfrac{800}{1 + 40e^{-3x}}; \; y' = \dfrac{(120)(800)e^{-3x}}{(1 + 40e^{-3x})^2} = \dfrac{96{,}000e^{-3x}}{(1 + 40e^{-3x})^2}$

31. $y = \dfrac{1000}{5 - 80e^{-2x}}; \; y' = \dfrac{-160{,}000e^{-2x}}{(5 - 80e^{-2x})^2}$

33. $y = (4x + e^{-x})^2; \; y' = 2(4x + e^{-x})(4 - e^{-x})$

35. $f(x) = \sqrt{x + e^{-x}}; \; f'(x) = \dfrac{1 - e^{-x}}{2\sqrt{x + e^{-x}}}$

37. $y = \sqrt{x^2 + e^{-x^2}}; \; y' = \dfrac{2x - 2x\,e^{-x^2}}{2\sqrt{x^2 + e^{-x^2}}} = \dfrac{x(1 - e^{-x^2})}{\sqrt{x^2 + e^{-x^2}}}$

39. $y = e^x$ at $x = 0$; $y' = e^x = e^0 = 1$

Tangent line: $\dfrac{y - e^0}{x - 0} = 1$; $y - 1 = x$; $y = x + 1$

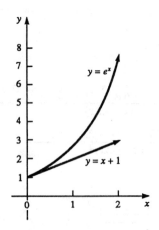

41. $S(x) = 10,000e^{0.10x}$

(a) $S'(x) = 1000e^{0.10x}$

(b) $S'(3) = 1000e^{0.3} \approx 1349.86$
  $S'(3)$ is the instantaneous rate of change of S with respect to x at $x = 3$.

(c) $S'(4) = 1000e^{0.4} \approx 1491.82$
  $S'(4)$ is the instantaneous rate of change of S with respect to x at $x = 4$.

(d) $S'(5) = 1000e^{0.5} \approx 1648.72$

(e) Assume that the annual rate of increase beyond the fifth year = $S'(5) = \$1648.72$. Then, the total amount at the en
  of the ninth year = $S(5) + (9 - 5)S'(5) = 10,000e^{0.5} + 4(1648.72) = 16,487.21 + 6594.88 = \$23,082.09$
  Compare with $S(9) = \$24,596.03$

(f)

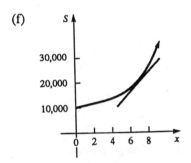

43. $P(x) = 1000e^{0.06x}$ $(x \geq 0)$

(a) $P'(x) = 60e^{0.06x}$; $P'(5) = 60e^{0.3} \approx 81$
  Since $P'(x) > 0$, the number of employees is increasing.

(b) $P'(10) = 60e^{0.6} \approx 109$. The number of
  employees is increasing.

(c) $P'(10) = 109$; $P(10) = 1822$
  Substitute into tangent line.
  $y - 1822 = 109(x - 109)$
  $\qquad x = 2149$

(d)

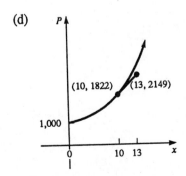

45. $N(x) = 50 - 50e^{-0.3x}$ $(x \geq 0)$

$N'(x) = 15e^{-0.3x}$; $N'(5) = 15e^{-1.5} \approx 3.35$
Since $N'(5) > 0$, daily production is increasing; the rate of increase is 3.35 items/day.

47. $y = 90 - 82e^{-0.30x}$

(a) $\dfrac{dy}{dx} = 24.6e^{-0.30x}$

(b) $\dfrac{dy}{dx}\bigg|_{x=1} = 24.6e^{-0.30} \approx 18.22$
This is the instantaneous rate of change of y with respect to x at $x = 1$.

(c) $\dfrac{dy}{dx}\bigg|_{x=6} = 24.6e^{-1.8} \approx 4.066$
This is the instantaneous rate of change of y with respect to x at $x = 6$.

(d) $\dfrac{dy}{dx}\bigg|_{x=25} = 24.6e^{-7.5} \approx 0.0136$
This is the instantaneous rate of change of y with respect to x at $x = 25$.

(e) See Figure. Note that the slope of the tangent line decreases rapidly as x increases. The graph of y vs. x is asymptotic to $y = 90$.

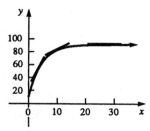

49. $N(t) = 10{,}000e^{0.05t}$

(a) $N'(t) = 500e^{0.05t}$
$N'(10) = 500e^{0.5} \approx 824.36$ flies/day

(b) $N'(20) = 500e \approx 1359.1$ flies/day

12–2 Derivatives of Logarithmic Functions

1. $y = 6 \ln x; \ y' = \dfrac{6}{x}$

3. $y = \ln 3x; \ y' = \dfrac{3}{3x} = \dfrac{1}{x}$

5. $f(x) = \ln(x^2 - 4x); \ f'(x) = \dfrac{2x - 4}{x^2 - 4x}$

7. $f(x) = \ln(x^2 - 4x + 1); \ f'(x) = \dfrac{2x - 4}{x^2 - 4x + 1}$

9. $y = x^4 + 3 \ln x; \ y' = 4x^3 + \dfrac{3}{x}$

11. $f(x) = \dfrac{1}{x^3} - 7 \ln x; \ f'(x) = -\dfrac{3}{x^4} - \dfrac{7}{x} = -\dfrac{3 + 7x^3}{x^4}$

13. $y = 5\sqrt{x} - 3 \ln x; \ y' = \dfrac{5}{2\sqrt{x}} - \dfrac{3}{x}$

15. $y = x^3 \ln(x^2 + 5); \ y' = \dfrac{2x^4}{x^2 + 5} + 3x^2 \ln(x^2 + 5)$

17. $y = \dfrac{\ln x^7}{x^9} = \dfrac{7 \ln x}{x^9};$

$y' = \dfrac{\dfrac{7x^9}{x} - 7 \cdot 9 \, x^8 \ln x}{x^{18}}$

$= \dfrac{7 - 63 \ln x}{x^{10}} = \dfrac{7(1 - 9 \ln x)}{x^{10}}$

19. $y = (x^2 + 1) \ln (x^2 + 5);$

$y' = \dfrac{2x(x^2 + 1)}{x^2 + 5} + 2x \ln (x^2 + 5)$

21. $y = \dfrac{x^2 + 6}{\ln (x^2 + 1)};$

$y' = \dfrac{2x \ln(x^2 + 1) - \dfrac{2x(x^2 + 6)}{x^2 + 1}}{[\ln (x^2 + 1)]^2}$

$= \dfrac{2x(x^2 + 1) \ln (x^2 + 1) - 2x(x^2 + 6)}{[x^2 + 1] \, [\ln(x^2 + 1)]^2}$

23. $y = [\ln(x^6 - 8x^3)]^{10};$

$y' = 10[\ln(x^6 - 8x^3)]^9 \cdot \dfrac{6x^5 - 24x^2}{x^6 - 8x^3}$

$= 60[\ln(x^6 - 8x^3)]^9 \left[\dfrac{x^3 - 4}{x^4 - 8x}\right]$

25. $y = \sqrt{\ln(x^3 + 7)};$

$y' = \dfrac{1}{2\sqrt{\ln(x^3 + 7)}} \cdot \dfrac{3x^2}{(x^3 + 7)}$

$= \dfrac{3x^2}{2(x^3 + 7)\sqrt{\ln(x^3 + 7)}}$

27. $y = \ln(\sqrt{x^3 + 7}) = \dfrac{1}{2} \ln(x^3 + 7);$

$y' = \dfrac{3x^2}{2(x^3 + 7)}$

29. $y = \dfrac{e^x}{\ln x}; \ y' = \dfrac{e^x \ln x - \dfrac{e^x}{x}}{(\ln x)^2} = \dfrac{e^x(x \ln x - 1)}{x(\ln x)^2}$

31. $y = \dfrac{x^3 e^x}{\ln x}; \ y' = \dfrac{(\ln x)(x^3 e^x + 3x^2 e^x) - x^2 e^x}{(\ln x)^2}$

$= \dfrac{x^2 e^x(x \ln x + 3 \ln x - 1)}{(\ln x)^2}$

33. $f(x) = [\ln(x^3 - 2x)]e^{x^4 - 8};$

$f'(x) = 4x^3[\ln(x^3 - 2x)]e^{x^4 - 8} + \left[\dfrac{3x^2 - 2}{x^3 - 2x}\right]e^{x^4 - 8}$

35. $y = \dfrac{\ln x^3}{\ln x^2} = \dfrac{3 \ln x}{2 \ln x} = \dfrac{3}{2}; \ y' = 0$

37. $y = \log_{10} 5x; \ y' = \dfrac{1}{x \ln 10}$

39. $y = \log_6(x^3 + 8x); \ y' = \dfrac{3x^2 + 8}{(x^3 + 8x) \ln 6}$

41. $y = \log_2(x^3 - 4x); \ y' = \dfrac{3x^2 - 4}{(x^3 - 4x) \ln 2}$

43. $y = \ln \dfrac{x^5}{\sqrt{x^3 + 7}} = 5 \ln x - \dfrac{1}{2} \ln(x^3 + 7);$

$\dfrac{dy}{dx} = \dfrac{5}{x} - \dfrac{1}{2}\left[\dfrac{3x^2}{x^3 + 7}\right] = \dfrac{7(x^3 + 10)}{2x(x^3 + 7)}$

45. $y = \ln \dfrac{(x^3 + 7)^5}{\sqrt{4 - 5x}} = 5 \ln(x^3 + 7) - \dfrac{1}{2} \ln(4 - 5x);$

$\dfrac{dy}{dx} = \dfrac{15x^2}{x^3 + 7} + \dfrac{5}{2(4 - 5x)}$

47. $y = \ln[(x^2 + 7)^4(x^6 - 5)^3]$
$= 4 \ln(x^2 + 7) + 3 \ln(x^6 - 5);$
$\dfrac{dy}{dx} = \dfrac{8x}{x^2 + 7} + \dfrac{18x^5}{x^6 - 5}$

49. $N(x) = 1000 \ln (0.5x)$

(a) $N'(x) = 1000 \dfrac{0.5}{0.5x} = \dfrac{1000}{x}$ items/day

(b) $N'(4) = \dfrac{1000}{4} = 250$

Thus, at $x = 4$, the number of items produced by the worker is increasing at a rate of 250 per day.

(c) $N'(8) = \dfrac{1000}{8} = 125$

Thus, at $x = 8$, the number of items produced by the worker is increasing at a rate of 125 per day.

51. $C(x) = 9000 + 10 \ln (x + 1)$

(a) $C'(x) = \dfrac{10}{x + 1}$

(b) $C'(10) = \dfrac{10}{11} \approx 0.909$

Thus, at $x = 10$, total cost is increasing at the rate of 0.909 per unit.

(c) $C'(20) = \dfrac{10}{21} \approx 0.476$

Thus, at $x = 20$, total cost is increasing at the rate of 0.476 per unit.

53. $t = \dfrac{\ln 3}{r}$

(a) $\dfrac{dt}{dr} = -\dfrac{1}{r^2} \ln 3 = -\dfrac{\ln 3}{r^2}$

(b) $\left.\dfrac{dt}{dr}\right|_{r = 0.08} = -\dfrac{\ln 3}{(0.08)^2} \approx -171.66$

Thus, a one percentage point increase in r (i.e. $\Delta r = 0.01$) reduces the tripling time by approximately $171.66(0.01) \approx 1.72$ years.

(c) $\left.\dfrac{dt}{dr}\right|_{r = 0.10} = -\dfrac{\ln 3}{(0.10)^2} \approx -109.86$

Thus, a one percentage point increase in r (i.e. $\Delta r = 0.01$) reduces the tripling time by approximately $109.86(0.01) \approx 1.10$ years.

(d) $\left.\dfrac{dt}{dr}\right|_{r = 0.20} = -\dfrac{\ln 3}{(0.20)^2} \approx -27.47$

When $r = 0.20$, a one percentage point increase in r reduces the tripling time by approximately $27.47\,(0.01) \approx 0.27$ years.

(e) $\left.\dfrac{dt}{dr}\right|_{r = 0.30} = -\dfrac{\ln 3}{(0.30)^2} \approx -12.21$

When $r = 0.30$, a one percentage point increase in r reduces the tripling time by approximately $12.21\,(0.01) \approx 0.12$ years.

Review Exercises

1. $y = e^{5x}$; $y' = 5e^{5x}$

3. $y = e^{8x+5}$; $y' = 8e^{8x+5}$

5. $y = (x^2 + 8x)e^{5x+9}$;
$y' = 5(x^2 + 8x)e^{5x+9} + (2x + 8)e^{5x+9}$
$= (5x^2 + 42x + 8)e^{5x+9}$

7. $y = \dfrac{(4e^x - e^{-x})}{x}$;
$y' = \dfrac{x(4e^x + e^{-x}) - (4e^x - e^{-x})}{x^2}$

9. $y = x^3e^{6x}$; $y' = 6x^3e^{6x} + 3x^2e^{6x}$
$= (2x + 1)(3x^2e^{6x})$

11. $y = \dfrac{600}{1 + 20e^{5x}}$; $y' = \dfrac{-100(600)e^{5x}}{(1 + 20e^{5x})^2} = \dfrac{-60,000e^{5x}}{(1 + 20e^{5x})^2}$

13. $y = (8x + e^x)^2$; $y' = 2(8x + e^x)(8 + e^x)$

15. $S(x) = 40,000e^{0.08x}$

(a) $S'(x) = 3200e^{0.08x}$

(b) $S'(3) = 3200e^{0.24} \approx 4068$
A small change in x, Δx, when $x = 3$, results in a change in S of approximately $4068\Delta x$.

(c) $S'(6) = 3200e^{0.48} \approx 5171.4$
A small change in x, Δx, when $x = 6$, results in a change in S of approximately $5171.4\Delta x$.

17. $f(x) = 90 - 50e^{-0.60x}$

(a) $f'(x) = 30e^{-0.60x}$

(b) $f'(3) = 30e^{-1.8} \approx 4.96$
A small change in x, Δx, when $x = 3$, results in a change in f of approximately $4.96\Delta x$.

(c) $f'(8) = 30e^{-4.8} \approx 0.247$
A small change in x, Δx, when $x = 8$, results in a change of f of approximately $0.247\Delta x$.

19. $y = -7 \ln x$; $y' = -\dfrac{7}{x}$

21. $y = \ln(x^3 - 6x^2)$; $y' = \dfrac{3x^2 - 12x}{x^3 - 6x^2} = \dfrac{3(x - 4)}{x(x - 6)}$

23. $y = x^2 \ln(x^3 + 7x)$;
$y' = \dfrac{x^2(3x^2 + 7)}{x^3 + 7x} + 2x \ln(x^3 + 7x)$

25. $y = [\ln(x^5 + 6x^2)]^8$;
$y' = 8[\ln(x^5 + 6x^2)]^7 \left[\dfrac{5x^4 + 12x}{x^5 + 6x^2} \right]$
$= 8[\ln(x^5 + 6x^2)]^7 \left[\dfrac{5x^3 + 12}{x^4 + 6x} \right]$

27. $y = e^{3x+5} \ln x$;
$y' = \dfrac{1}{x}e^{3x+5} + 3e^{3x+5} \ln x$
$= \left[\dfrac{1}{x} + 3 \ln x \right] e^{3x+5}$

29. $y = [\ln(x^5 - 6x^2)]e^{8x}$;
$y' = 8[\ln(x^5 - 6x^2)]e^{8x} + \left[\dfrac{5x^4 - 12x}{x^5 - 6x^2} \right] e^{8x}$
$= \left\{ 8[\ln(x^5 - 6x^2)] + \left[\dfrac{5x^3 - 12}{x^4 - 6x} \right] \right\} e^{8x}$

31. $C(x) = 20{,}000 + 40\ln(x + 3)$

(a) $C'(x) = \dfrac{40}{x + 3}$

(b) $C'(17) = \dfrac{40}{20} = 2$

A small change in x, Δx, when $x = 17$, will result in a change in C of approximately $2\Delta x$.

(c) $C'(22) = \dfrac{40}{25} = 1.6$

A small change in x, Δx, when $x = 22$, will result in a change in C of approximately $1.6\Delta x$.

CHAPTER
13
INTEGRATION

13–1 Antidifferentiation

1. $f(x) = 7x^6$, $F(x) = x^7$
$F'(x) = 7x^6 = f(x)$

3. $f(x) = x^2 - 8x + 5$, $F(x) = \frac{1}{3}x^3 - 4x^2 + 5x + 1$
$F'(x) = x^2 - 8x + 5 = f(x)$

5. $f(x) = e^{2x}$, $F(x) = \frac{1}{2}e^{2x}$
$F'(x) = e^{2x} = f(x)$

7. $f(x) = \frac{1}{x}$, $F(x) = \ln|x|$
$F'(x) = \frac{1}{x} = f(x)$

9. $\int x^{12}dx = \frac{1}{13}x^{13} + C$

11. $\int \frac{10}{\sqrt{x}}dx = 20\sqrt{x} + C$

13. $\int \sqrt[4]{x}\,dx = \frac{4}{5}x^{5/4} + C$

15. $\int \frac{1}{x^9}dx = -\frac{1}{8x^8} + C$

17. $\int \sqrt[3]{x^2}\,dx = \frac{3}{5}x^{5/3} + C$

19. $\int \frac{-4}{\sqrt[7]{x^2}}dx = -4 \cdot \frac{7}{5}x^{5/7} + C = -\frac{28}{5}x^{5/7} + C$

21. $\int \frac{-3}{\sqrt{x}}dx = -6\sqrt{x} + C$

23. $\int \dfrac{1}{\sqrt[5]{x^2}}\,dx = \dfrac{5}{3}x^{3/5} + C$

25. $\int e^{5x}\,dx = \dfrac{1}{5}e^{5x} + C$

27. $\int e^{-4x}\,dx = -\dfrac{1}{4}e^{-4x} + C$

29. $\int 8e^{2x}\,dx = 4e^{2x} + C$

31. $\int -\dfrac{4}{x}\,dx = -4 \ln|x| + C$

33. $\int \dfrac{300}{x}\,dx = 300 \ln|x| + C$

35. $\int \dfrac{1}{5x}\,dx = \dfrac{1}{5}\ln|x| + C$

37. $\int \dfrac{1}{x^6}\,dx = -\dfrac{1}{5}\dfrac{1}{x^5} + C = \dfrac{-1}{5x^5} + C$

39. $\int \dfrac{8}{x^7}\,dx = -\dfrac{8}{6}\dfrac{1}{x^6} + C = -\dfrac{4}{3x^6} + C$

41. $\int (4x^3 - 16)\,dx = x^4 - 16x + C$

43. $\int (x^5 - 7x)\,dx = \dfrac{x^6}{6} - \dfrac{7}{2}x^2 + C$

45. $\int \left[x^3 - \dfrac{4}{x^2} + 6\right]dx = \dfrac{x^4}{4} + \dfrac{4}{x} + 6x + C$

47. $\int (t^3 - 2t^2)\,dt = \dfrac{t^4}{4} - \dfrac{2}{3}t^3 + C$

49. $\int (y^6 - 5y^4 + 1)\,dy = \dfrac{y^7}{7} - y^5 + y + C$

51. $\int \left[4u^3 - \dfrac{8}{\sqrt{u}} + \dfrac{5}{u^2}\right]du = u^4 - 16\sqrt{u} - \dfrac{5}{u} + C$

53. $\int (7x^{-1} + e^{4x})\,dx = 7 \ln|x| + \dfrac{1}{4}e^{4x} + C$

55. $\int \left[\dfrac{9}{x} + e^{-x}\right]dx = 9 \ln|x| - e^{-x} + C$

57. $\int \left[\dfrac{4}{x} - e^{-x}\right]dx = 4 \ln|x| + \dfrac{1}{5}e^{-5x} + C$

59. $\int \left[\dfrac{4}{x^3} + \dfrac{2}{x}\right]dx = -\dfrac{2}{x^2} + 2 \ln|x| + C$

61. $\int \left[\dfrac{6}{x^4} - e^{4x}\right]dx = -\dfrac{2}{x^3} - \dfrac{1}{4}e^{4x} + C$

63. $\int (x^{-2} + e^{-x})\,dx = -\dfrac{1}{x} - e^{-x} + C$

65. $\int (x^{-1} - x^{-2} - e^x)dx = \ln|x| + \dfrac{1}{x} - e^x + C$

67. $\int \dfrac{x^6 - 4x}{x^3}dx = \int \left[x^3 - \dfrac{4}{x^2}\right]dx = \dfrac{x^4}{4} + \dfrac{4}{x} + C$

69. $\int (x - 1)^2 dx = \int (x^2 - 2x + 1)dx = \dfrac{x^3}{3} - x^2 + x + C$

71. $\int \dfrac{e^x - x}{xe^x}dx = \int \left[\dfrac{1}{x} - e^{-x}\right]dx = \ln|x| + e^{-x} + C$

73. $\int (x + 2)(x - 2)dx = \int (x^2 - 4)dx = \dfrac{x^3}{x} - 4x + C$

75. $\int (x - 3)(x + 2)dx = \int (x^2 - x - 6)dx$
$$= \dfrac{x^3}{3} - \dfrac{x^2}{2} - 6x + C$$

77. $\int \dfrac{x^2 - 4}{x + 2}dx = \int (x - 2)dx = \dfrac{x^2}{x} - 2x + C$

79. $C'(x) = 6x^2 + 4x - 5$, fixed cost = \$800
$C(x) = 2x^3 + 2x^2 - 5x + 800$

81. $V'(t) = 24{,}000t^2$, $V(0) = 1{,}000{,}000$
(a) $V(t) = 8{,}000t^3 + V(0) = 8{,}000t^3 + 1{,}000{,}000$
(b) $V(5) = 8{,}000(125) + 1{,}000{,}000 = \$2{,}000{,}000$

83. $V(t) = 6t^3$, $S(0) = 2$; $V(t) = S^1(t)$
$S(t) = 2t^3 + S(0) = 2t^3 + 2 = 2(t^3 + 1)$

85. $V(t) = 3t^2$, $S(0) = 6$; $V(t) = S^1(t)$
$S(t) = t^3 + S(0) = t^3 + 6$

87. $a(t) = 4t$, $v(0) = 30$; $a(t) = v^1(t)$
$v(t) = 2t^2 + v(0) = 2t^2 + 30$

89. $\dfrac{dA}{dt} = -40t^{-3}$ $(1 \le t \le 6)$

(a) $A(t) = \int (-40t^{-3})dt = \dfrac{20}{t^2} + C$

$A(1) = 19.5 \Rightarrow \dfrac{20}{1} + C = 19.5 \Rightarrow C = -0.5$

$A(t) = \dfrac{20}{t^2} - 0.5$

(b) $A(5) = \dfrac{20}{25} - 0.5 = 0.8 - 0.5 = 0.3\ cm^2$

13–2 The Definite Integral and Area Under a Curve

1. $\int_1^3 4x^3 dx = x^4 \Big|_1^3 = 81 - 1 = 80$

3. $\int_0^1 (4x + 1)dx = 2x^2 + x \Big|_0^1 = 2 + 1 - 0 = 3$

5. $\int_1^4 \dfrac{4}{x^3}dx = -\dfrac{2}{x^2}\Big|_1^4 = -\dfrac{2}{16} + \dfrac{2}{1} = \dfrac{15}{8} = 1.875$

7. $\int_1^5 \dfrac{10}{x}dx = 10 \ln|x| \Big|_1^5 = 10(\ln 5 - \ln 1)$
$$= 10 \ln 5 \approx 16.09$$

9. $\int_0^3 e^x dx = e^x \Big|_0^3 = e^3 - e^0 = e^3 - 1 \approx 19.086$

11. $\int_0^2 e^{-2x}dx = -\frac{1}{2}e^{-2x}\Big|_0^2 = -\frac{1}{2}(e^{-4} - e^0) = \frac{1}{2}(1 - e^{-4})$

≈ 0.4908

13. $f(x) = -x^2 + 4$, $0 \le x \le 1$, 4 subintervals

$\Delta x = \frac{1}{4}$, $x_i = 0 + i\Delta x$

i	x_i	$f(x_i)$
0	0	4.000
1	0.25	3.9375
2	0.50	3.7500
3	0.75	3.4375
4	1.00	3.0000

Lower approximation $= \Delta x \sum_{i=1}^{4} f(x_i) = 0.25(14.125)$

$= 3.53125$

Upper approximation $= \Delta x \sum_{i=0}^{3} f(x_i) = 0.25(15.125)$

$= 3.78125$

15. $f(x) = 2x^2$, $1 \le x \le 2$, 5 subintervals

$\Delta x = 0.2$, $x_i = 1 + i\Delta x$

i	x_i	$f(x_i)$
0	1.0	2.00
1	1.2	2.88
2	1.4	3.92
3	1.6	5.12
4	1.8	6.48
5	2.0	8.00

Lower approximation $= \Delta x \sum_{i=0}^{4} f(x_i) = (0.20)(20.4)$

$= 4.08$

Upper approximation $= \Delta x \sum_{i=1}^{5} f(x_i) = (0.20)(26.4)$

$= 5.28$

17. $f(x) = x$, $0 \le x \le 1$

$x_i = 0 + i\Delta x$; $f(x_i) = i\Delta x$

$A \approx \sum_{i=1}^{n} f(x_i)\Delta x = (\Delta x)^2 \sum_{i=1}^{n} i = (\Delta x)^2 \frac{n(n + 1)}{2}$

Let $\Delta x = \frac{1 - 0}{n} = n$

$A = \lim_{n \to \infty} \frac{1}{n^2} \cdot \frac{n(n + 1)}{2} = \lim_{n \to \infty} \frac{1}{2} + \frac{1}{2n} = \frac{1}{2}$

check: $\int_0^1 f(x)dx = \int_0^1 xdx = \frac{x^2}{2}\Big|_0^1 = \frac{1}{2}$

19. $f(x) = 3x^2$, $0 \le x \le 1$

(a) Let $x_i = i\Delta x$; $f(x_i) = 3i^2(\Delta x)^2$; $\Delta x = \frac{1}{n}$

$A = \lim_{n \to \infty} \sum_{i=1}^{n} f(x_i)\Delta x = \lim_{n \to \infty} \frac{3}{n^3} \sum_{i=1}^{n} i^2$

$= \lim_{n \to \infty} \frac{3}{n^3} \cdot \frac{n(n + 1)(2n + 1)}{6}$

$= \lim_{n \to \infty} \frac{1}{2n^3}(2n^3 + 3n^2 + n)$

$= \lim_{n \to \infty} \left[1 + \frac{3}{2n} + \frac{1}{2n^2}\right] = 1$

(b) $A = \int_0^1 3x^2dx = x^3\Big|_0^1 = 1$

21. $A = \int_1^2 2x^2dx = \frac{2}{3}x^3\Big|_1^2 = \frac{2}{3}(8 - 1) = \frac{14}{3} = \frac{14}{3} \approx 4.667$

The average of the lower and upper approximations found in Exercise 15 is 4.68 and in Exercise 16 is 4.67.

23. $A = \int_0^4 x^3dx = \frac{x^4}{4}\Big|_0^4 = \frac{1}{4}(256 - 0) = 64$

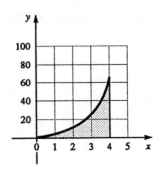

25. $A = \int_{1}^{3} \frac{1}{x^2} dx = -\frac{1}{x} \Big|_{1}^{3} = -\frac{1}{3} + 1 = \frac{2}{3}$

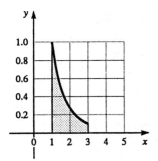

27. $A = \int_{0}^{3} (-2x + 8) dx = -x^2 + 8x \Big|_{0}^{3} = -9 + 24 = 15$

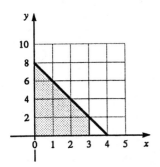

29. $A = \int_{-3}^{-1} (-2x) dx = -x^2 \Big|_{-3}^{-1} = -(1 - 9) = 8$

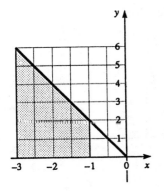

31. $A = \int_{0}^{5} (-x^2 + 25) dx = -\frac{x^3}{3} + 25x \Big|_{0}^{5} = -\frac{125}{3} + 125$
$$= \frac{250}{3} \approx 83.33$$

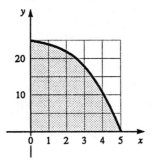

33. $A = \int_{0}^{4} (-5x^2 + 20x) dx = -\frac{5}{3}x^3 + 10x^2 \Big|_{0}^{4}$
$$= -\frac{5}{3} \cdot 64 + 160 = \frac{160}{3}$$
$$\approx 53.33$$

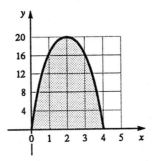

35. $A = \int_{0}^{4} (5x^2 - 6x + 12) dx = \frac{x^3}{3} - 3x^2 + 12x \Big|_{0}^{4}$
$$= \frac{64}{3} - 48 + 48 = \frac{64}{3} \approx 21.33$$

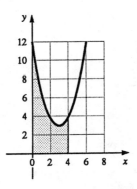

37. $A = \int_{-3}^{3}(x^2 - 2x + 8)dx = \frac{x^3}{3} - x^2 + 8x \Big|_{-3}^{3}$

$= \left[\frac{27}{3} - 9 + 24\right] - \left[-\frac{27}{3} - 9 - 24\right] = 66$

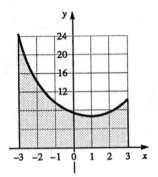

39. $A = -\int_{0}^{-3} e^{-x}dx = \int_{-3}^{0} e^{-x}dx = -e^{-x}\Big|_{-3}^{0} = -1 + e^3$

$= e^3 - 1 \approx 19.086$

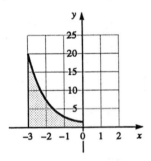

41. $A = \int_{-1}^{1} e^x dx = e^x \Big|_{-1}^{1} = e - \frac{1}{e} \approx 2.350$

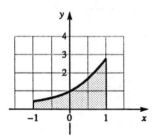

43. $A = \int_{1}^{5}\frac{1}{x}dx = \ln|x|\ \Big|_{1}^{5} = \ln 5 \approx 1.609$

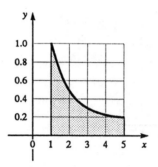

45. $A = \int_{2}^{4}\frac{1}{x^4}dx = -\frac{1}{3x^3}\Big|_{2}^{4} = -\frac{1}{3}\left[\frac{1}{64} - \frac{1}{8}\right] = \frac{7}{192}$

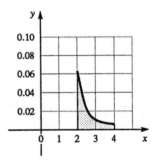

47. $A = \int_{0}^{1} x^5 dx = \frac{x^6}{6}\Big|_{0}^{1} = \frac{1}{6} - 0 = \frac{1}{6}$

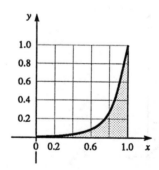

49. $y = x^2 - 4, \ 0 \le x \le 3$

For $0 \le x < 2, \ y < 0$; for $2 < x \le 3, \ y > 0$.

$$A = \left| \int_0^2 (x^2 - 4)dx \right| + \int_2^3 (x^2 - 4)dx$$

$$= -\left[\frac{x^3}{3} - 4x\right]\Big|_0^2 + \left[\frac{x^3}{3} - 4x\right]\Big|_2^3$$

$$= -\left[\frac{8}{3} - 8\right] + \left[\frac{27}{3} - 12\right] - \left[\frac{8}{3} - 8\right]$$

$$= 13 - \frac{16}{3} = \frac{23}{3} = 7\frac{2}{3}$$

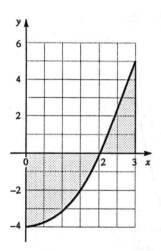

51. $y = -x, \ -4 \le x \le 4$

For $-4 \le x < 0, \ y > 0$; for $0 < x \le 4, \ y < 0$.

$$A = \int_{-4}^0 (-x)dx + \left| \int_0^4 (-x)dx \right|$$

$$= -\frac{x^2}{2}\Big|_{-4}^0 - \left[-\frac{x^2}{2}\right]\Big|_0^4$$

$$= 0 + \frac{16}{2} - \left[-\frac{16}{2}\right] + 0 = 16$$

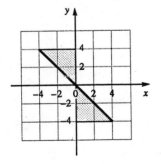

53. $y = -x^2 + 4, \ 0 \le x \le 3$

For $0 \le x < 2, \ y > 0$; for $0 < x \le 3, \ y < 0$.

$$A = \int_0^2 (-x^2 + 4)dx + \left| \int_2^3 (-x^2 + 4)dx \right|$$

$$= -\frac{x^3}{3} + 4x\Big|_0^2 - \left[-\frac{x^3}{3} + 4x\right]\Big|_2^3$$

$$= -\frac{8}{3} + 8 - 0 - \left[-\frac{27}{3} + 12\right] + \left[-\frac{8}{3} + 8\right]$$

$$= 13 - \frac{16}{3} = \frac{23}{3} = 7\frac{2}{3}$$

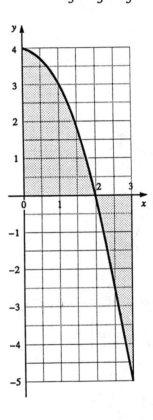

55. $y = -2x + 6,\ 0 \le x \le 6$
For $0 \le x < 3,\ y > 0;$ for $3 < x \le 6,\ y < 0.$

$$A = \int_0^3 (-2x + 6)\,dx + \left| \int_3^6 (-2x + 6)\,dx \right|$$

$$= -x^2 + 6x \Big|_0^3 - (-x^2 + 6x) \Big|_3^6$$

$$= -9 + 18 - 0 - (-36 + 36) + (-9 + 18)$$

$$= 18$$

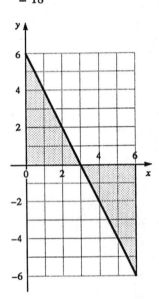

57. $y = -x^3,\ -2 \le x \le 4$
For $-2 \le x < 0,\ y > 0;$ for $0 < x \le 4,\ y < 0.$

$$A = \int_{-2}^0 (-x^3)\,dx + \left| \int_0^4 (-x^3)\,dx \right|$$

$$= -\frac{x^4}{4} \Big|_{-2}^0 - \left[-\frac{x^4}{4} \right] \Big|_0^4$$

$$= 0 + \frac{16}{4} - \left[-\frac{256}{4} \right] + 0$$

$$= 68$$

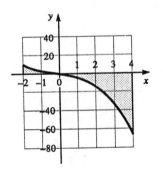

59. $y = x^2 - x - 2,\ 0 \le x \le 3$
For $0 \le x < 2,\ y < 0;$ for $2 < x \le 3,\ y > 0.$

$$A = \left| \int_0^2 (x^2 - x - 2)\,dx \right| + \int_2^3 (x^2 - x - 2)\,dx$$

$$= -\left[\frac{x^3}{3} - \frac{x^2}{2} - 2x \right] \Big|_0^2 + \left[\frac{x^3}{3} - \frac{x^2}{2} - 2x \right] \Big|_2^3$$

$$= -\left[\frac{8}{3} - \frac{4}{2} - 4 \right] + 0 + \left[\frac{27}{3} - \frac{9}{2} - 6 \right]$$

$$\quad - \left[\frac{8}{3} - \frac{4}{2} - 4 \right]$$

$$= -\frac{59}{6} + 15 = \frac{31}{6} = 5\frac{1}{6}$$

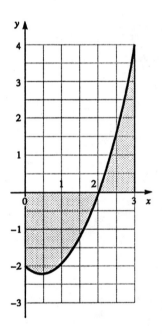

61. $y - 1 + \dfrac{4}{x}$, $1 \leq x \leq 6$

For $1 \leq x < 4$, $y > 0$; $4 < x \leq 6$, $y < 0$.

$$A = \int_1^4 \left[-1 + \frac{4}{x}\right] dx + \left| \int_4^6 \left[-1 + \frac{4}{x}\right] dx \right|$$

$$= (-x + 4 \ln x) \Big|_1^4 - \left[-x + 4 \ln x\right] \Big|_4^6$$

$$= (-4 + 4 \ln 4) - (-1 + 4 \ln 1) - (-6 + 4 \ln 6) + (-4 + 4 \ln 4)$$

$$= -1 + 8 \ln 4 - 4 \ln 6 = -1 + 4 \ln \frac{16}{6}$$

$$= -1 + 4 \ln \frac{8}{3} \approx 2.9233$$

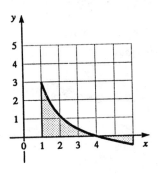

63. $y = -1 + \dfrac{9}{x^2}$, $1 \leq x \leq 6$

For $1 \leq x < 3$, $y > 0$; $3 < x \leq 6$, $y < 0$.

$$A = \int_1^3 \left[-1 + \frac{9}{x^2}\right] dx + \left| \int_3^6 \left[-1 + \frac{9}{x^2}\right] dx \right|$$

$$= \left[-x - \frac{9}{x}\right] \Big|_1^3 - \left[-x - \frac{9}{x}\right] \Big|_3^6$$

$$= (-3 - 3) - (-1 - 9) - \left[-6 - \frac{9}{6}\right] + \left[-3 - \frac{9}{3}\right]$$

$$= 5\frac{1}{2}$$

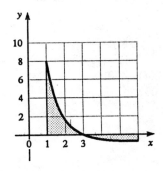

65. $y = -3x^2 + 12$, $0 \leq x \leq 6$

For $0 \leq x < 2$, $y > 0$; $2 < x \leq 6$, $y < 0$.

$$A = \int_0^2 (-3x^2 + 12) dx + \left| \int_2^6 (-3x^2 + 12) dx \right|$$

$$= (-x^3 + 12x) \Big|_0^2 - (-x^3 + 12x) \Big|_2^6$$

$$= (-8 + 24) - 0 - (-216 + 72) + (-8 + 24)$$

$$= 176$$

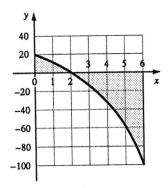

67. $C'(x) = 20x + 100$, $10 \leq x \leq 20$

$$\Delta C = \int_{10}^{20} (20x + 100) dx = 10x^2 + 100x \Big|_{10}^{20}$$

$$= 4000 + 2000 - (1000 + 1000)$$

$$= 4000$$

69. $R'(x) = -2x + 100$, $10 \leq x \leq 30$

$$\Delta R = \int_{10}^{30} (-2x + 100) dx = -x^2 + 100x \Big|_{10}^{30}$$

$$= -900 + 3000 - (-100 + 1000)$$

$$= 1200$$

71. $R'(x) = -8x + 400$; $5 \leq x \leq 40$

$$\Delta R = \int_5^{40} (-8x + 400) dx = -4x^2 + 400x \Big|_5^{40}$$

$$= -6400 + 1600 - (-100 + 2000)$$

$$= 7700$$

73. $v(t) = 6t^2 + 5t,\ 0 \le t \le 6$

$$s = \int_0^6 (6t^2 + 5t)dt = 2t^3 + \frac{5t^2}{2}\Big|_0^6$$

$$= 432 + 90 - 0 = 522 \rightarrow 522\text{ft}$$

75. $v(t) = 6t^5,\ 0 \le t \le 1$

$$s = \int_0^1 6t^5 dt = t^6\Big|_0^1$$
$$= 1 \rightarrow 1\text{ foot}$$

77. $L'(t) = 10t + 20$

(a) $L(5) = \int_0^5 (10t + 20)dt = 5t^2 + 20t\Big|_0^5$
$$= 125 + 100 - 0 = 225 \rightarrow 225\text{ cu. ft.}$$

(b) $L(10) = \int_0^{10} (10t + 20)dt = 5t^2 + 20t\Big|_0^{10}$
$$= 500 + 200 - 0 = 700 \rightarrow 700\ cu.\text{ ft.}$$

(c) $L(5) - L(4) = \int_4^5 (10t + 20)dt = 5t^2 + 20t$
$$= 125 + 100 - (80 + 80)$$
$$= 65 \rightarrow 65\text{ cu. ft.}$$

(d) $L(6) - L(5) = \int_5^6 (10t + 20)dt = 5t^2 + 20t$
$$= 180 + 120 - 225$$
$$= 75 \rightarrow 75\text{ cu. ft.}$$

79. $S'(t) = 100 - 100e^{-0.20t}$

(a) $S(5) = \int_0^5 (100 - 100e^{-0.20t})dt$
$$= (100t + 500e^{-0.20t})\Big|_0^5$$
$$= 500 + 500e^{-1} - 0 - 500$$
$$= \frac{500}{e} \approx 183.94$$

79. *(continued)*

(b) $S(10) = \int_0^{10} (100 - 100e^{-0.20t})dt$
$$= (100t + 500e^{-0.20t})\Big|_0^{10}$$
$$= 1000 + 500e^{-2} - 0 - 500$$
$$= 500 + \frac{500}{e^2} = 500(1 + e^{-2}) \approx 567.67$$

81. $J'(t) = 100e^{-0.05t}$

(a) $J(2) = \int_0^2 100e^{-0.05t}dt = -2000e^{-0.05t}\Big|_0^2$
$$= -2000^{-0.1} + 2000 = 2000(1 - e^{-0.1})$$
$$\approx 190 \rightarrow 190\text{ jobs}$$

(b) $J(3) = \int_0^3 100e^{-0.05t}dt = -2000e^{-0.05t}\Big|_0^3$
$$= -2000^{-0.15} + 2000 = 2000(1 - e^{-0.15})$$
$$\approx 279 \rightarrow 279\text{ jobs}$$

(c) $J(5) = \int_0^5 100e^{-0.05t}dt = -2000e^{-0.05t}\Big|_0^5$
$$= -2000e^{-0.25} + 2000 = 2000(1 - e^{-0.25})$$
$$\approx 442 \rightarrow 442\text{ jobs}$$

(d) $J(10) = \int_0^{10} 100e^{-0.05t}dt = -2000e^{-0.05t}\Big|_0^{10}$
$$= -2000e^{-0.05} + 2000$$
$$= 2000(1 - e^{-0.05}) \approx 787 \rightarrow 787\text{ jobs}$$

83. $\text{Tax} = \int_0^{18,400} 0.15dx = 0.15x\Big|_0^{18,400} = 0.15(18,400)$
$$= \$2760$$

85. $\text{Tax} = 0.15(18,550) + 0.28\int_{18,550}^{30,000} dx$
$$= 2782.50 + 0.28(11,450)$$
$$= 2782.50 + 3206$$
$$= \$5988.50$$

(continued)

87. Tax = 2782.50 + 0.28(44,900 − 18,550)

$$+ 0.33 \int_{44,900}^{50,000} dx.$$

= 10,160.50 + 0.33(5100) + 10,160.50 + 1683

= $11,843.50

89. Tax = $10,160.50 + 0.33 \int_{44,900}^{80,000} dx$

= 10,160.50 + 0.33(35,100)

= 10,160.50 + 11,583 = $21,743.50

13–3 Area Between Two Curves

1. $f(x) = 2x + 20$, $g(x) = x^2 + 5$; $1 \leq x \leq 3$

$$A = \int_1^3 [(2x + 20) - (x^2 + 5)]dx$$

$$= \int_1^3 (2x - x^2 + 15)dx = x^2 - \frac{x^3}{3} + 15x \Big|_1^3$$

$$= 9 - \frac{27}{3} + 45 - \left[1 - \frac{1}{3} + 15\right] = \frac{88}{3} = 29\frac{1}{3}$$

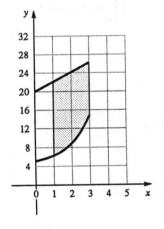

3. $f(x) = x^2$, $g(x) = x^3$; $0 \leq x \leq 1$

$$A = \int_0^1 (x^2 - x^3)dx = \frac{x^3}{3} - \frac{x^4}{4} \Big|_0^1 = \frac{1}{3} - \frac{1}{4} = \frac{1}{12}$$

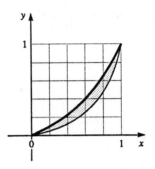

5. $f(x) = 3x^2 + 5$, $g(x) = x^2 + 5$; $0 \leq x \leq 2$

$$A = \int_0^2 [3x^2 + 5 - (x^2 + 5)]dx = \int_0^2 2x^2 dx = \frac{2}{3}x^3 \Big|_0^2$$

$$= \frac{16}{3} = 5\frac{1}{3}$$

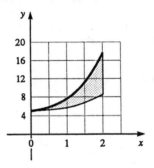

7. $f(x) = x$, $g(x) = x^5$; $0 \leq x \leq 1$

$$A = \int_0^1 (x - x^5)dx = \frac{x^2}{2} - \frac{x^6}{6} \Big|_0^1 = \frac{1}{2} - \frac{1}{6} = \frac{1}{3}$$

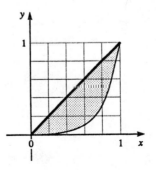

9. $f(x) = 2x + 5$, $g(x) = 1$; $0 \le x \le 3$

$$A = \int_0^3 [(2x + 5) - 1]dx = \int_0^3 (2x + 4)dx$$
$$= x^2 + 4x \Big|_0^3 = 9 + 12 = 21$$

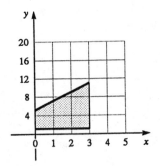

11. $y = -x + 4$, $y = 3/x$; $1 \le x \le 3$

$$A = \int_1^3 \left[(-x + 4) - \frac{3}{x}\right]dx = \int_1^3 \left[(-x + 4 - \frac{3}{x}\right]dx$$
$$= -\frac{x^2}{2} + 4x - 3 \ln x \Big|_1^3$$
$$= -\frac{9}{2} + 12 - 3 \ln 3 - \left[-\frac{1}{2} + 4 - 0\right]$$
$$= 4 - 3 \ln 3 \approx 0.7042$$

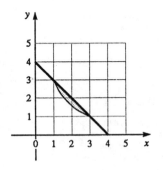

13. $f(x) = -x + 5$, $g(x) = 6/x$; $2 \le x \le 3$

$$A = \int_2^3 \left[-x + 5 - \frac{6}{x}\right]dx = -\frac{x^2}{2} + 5x - 6 \ln x \Big|_2^3$$
$$= -\frac{9}{2} + 15 - 6 \ln 3 - \left[-\frac{4}{2} + 10 - 6 \ln 2\right]$$
$$= \frac{5}{2} - 6 \ln \frac{3}{2} \approx 0.0672$$

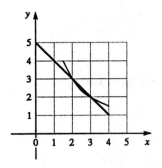

15. $f(x) = e^x$, $g(x) = 1$; $0 \le x \le 2$

$$A = \int_0^2 (e^x - 1)dx = e^x - x \Big|_0^2 = e^2 - 2 - 1 = e^2 - 3$$
$$= 4.389$$

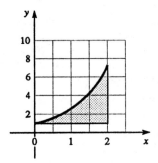

17. $y = -x, y = x; \ 0 \le x \le 2$

$$A = \int_0^2 [x - (-x)]dx = x^2 \Big|_0^2 = 4$$

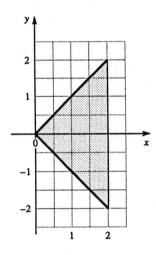

19. $y = x^4, y = x^3; \ 0 \le x \le 1$

$$A = \int_0^1 (x^3 - x^4)dx = \frac{x^4}{4} - \frac{x^5}{5}\Big|_0^1 = \frac{1}{20}$$

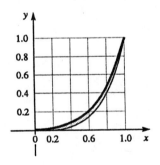

21. $y = e^x, y = e^{-x}; \ 0 \le x \le 1$

$$A = \int_0^1 (e^x - e^{-x})dx = e^x + e^{-x}\Big|_0^1 = e + e^{-1} - (1 + 1)$$

$$= e + e^{-1} \approx 1.0862$$

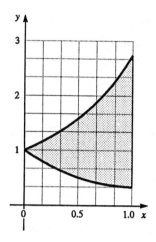

23. $f(x) = -x^2 + 6, g(x) = x; \ -3 \le x \le 2$

$$A = \int_{-3}^2 (-x^2 + 6 - x)dx = -\frac{x^3}{3} + 6x - \frac{x^2}{2}\Big|_{-3}^2$$

$$= -\frac{8}{3} + 12 - \frac{4}{2} - \left[\frac{27}{3} - 18 - \frac{9}{2}\right] = \frac{125}{6} = 20.833$$

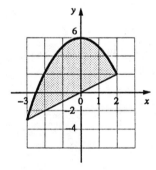

25. $y = x + 1$, $y = 2$; $0 \le x \le 4$

For $x < 1$, $x + 1 < 2$; for $x > 1$, $x + 1 > 2$.

$$A = \int_0^1 [2 - (x + 1)]dx + \int_1^4 (x + 1 - 2)dx$$

$$= x - \frac{x^2}{2}\Big|_0^1 + \frac{x^2}{2} - x\Big|_1^4$$

$$= 1 - \frac{1}{2} + \frac{16}{2} - 4 - \left[\frac{1}{2} - 1\right] = 5$$

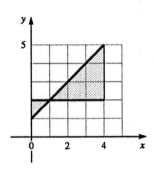

27. $f(x) = x$, $g(x) = x^2$; $0 \le x \le 3$

For $0 \le x < 1$, $x^2 < x$; for $1 < x \le 3$, $x < x^2$.

$$A = \int_0^1 (x - x^2)dx + \int_1^3 (x^2 - x)dx$$

$$= \frac{x^2}{2} - \frac{x^3}{3}\Big|_0^1 + \frac{x^3}{3} - \frac{x^2}{2}\Big|_1^3$$

$$= \frac{1}{2} - \frac{1}{3} + \frac{27}{3} - \frac{9}{2} - \left[\frac{1}{3} - \frac{1}{2}\right] = \frac{29}{6} \approx 4.833$$

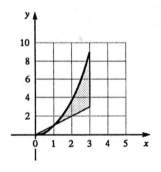

29. $f(x) = x$, $g(x) = x^3$; $-1 \le x \le 4$

For $-1 \le x < 0$, $x < x^3$; for $0 < x < 1$, $x^3 < x$; and for $1 < x \le 4$, $x < x^3$.

Use the symmetry of the graph about the origin.

$$A = 2\int_0^1 (x - x^3)dx + \int_1^4 (x^3 - x)dx$$

$$= 2\left[\frac{x^2}{2} - \frac{x^4}{4}\right]\Big|_0^1 + \frac{x^4}{4} - \frac{x^2}{2}\Big|_1^4$$

$$= 2\left[\frac{1}{2} - \frac{1}{4}\right] + \frac{256}{4} - \frac{16}{2} - \left[\frac{1}{4} - \frac{1}{2}\right] = \frac{227}{4} = 56.75$$

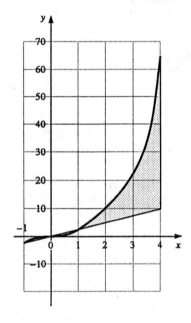

31. $y = -x^2 + 5, y = 1; \quad -2 \le x \le 6$
For $-2 \le x \le 2, 1 < -x^2 + 5; \quad$ for $2 < x \le 6, -x^2 + 5 < 1$.

$$A = \int_{-2}^{2}(-x^2 + 5 - 1)dx + \int_{2}^{6}[1 - (-x^2 + 5)]dx$$

$$= -\frac{x^3}{3} + 4x \Big|_{-2}^{2} + \frac{x^3}{3} - 4x \Big|_{2}^{6}$$

$$= -\frac{8}{3} + 8 - \left[\frac{8}{3} - 8\right] + \frac{216}{3} - 24 - \left[\frac{8}{3} - 8\right] = 64$$

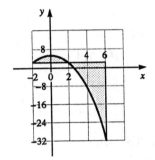

33. $f(x) = e^x, g(x) = e^{-x}; \quad -2 \le x \le 3$
For $-2 \le x < 0, e^x < e^{-x}; \quad$ for $0 < x \le 3, e^{-x} < e^x$.

$$A = \int_{-2}^{0}(e^{-x} - e^x)dx + \int_{0}^{3}(e^x - e^{-x})dx$$

$$= -e^{-x} - e^x \Big|_{-2}^{0} + e^x + e^{-x} \Big|_{0}^{3}$$

$$= -1 - - (-e^2 - e^{-2}) + e^3 + e^{-3} - (1 + 1)$$
$$= e^3 + e^2 + e^{-2} + e^{-3} - 4 \approx 23.66$$

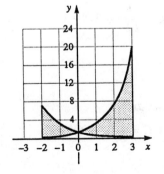

35. $f(x) = -x + 6, g(x) = 8/x; \quad 2 \le x \le 6$

$$-x + 6 = \frac{8}{x} \implies x^2 - 6x + 8 = 0 \implies$$

$$(x - 2)(x - 4) = 0; \quad x = 2,4$$

For $2 \le x < 4, \frac{8}{x} < -x + 6;$

for $4 < x \le x, -x + 6 < \frac{8}{x}.$

$$A = \int_{2}^{4}\left[-x + 6 - \frac{8}{x}\right]dx + \int_{4}^{6}\left[\frac{8}{x} - (x + 6)\right]dx$$

$$= -\frac{x^2}{2} + 6x - 8 \ln|x| \Big|_{2}^{4} + 8 \ln|x| + \frac{x^2}{2} - 6x \Big|_{4}^{6}$$

$$= -\frac{16}{2} + 24 - 8 \ln 4 - \left[-\frac{4}{2} + 12 - 8 \ln 2\right]$$

$$+ 8 \ln 6 \quad + \frac{36}{2} - 36 - \left[8 \ln 4 + \frac{16}{2} - 24\right]$$

$$= 4 + 8 \ln \frac{3}{4} \approx 1.699$$

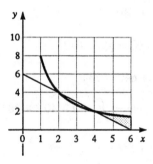

37. $p = D(q) = -0.5q + 20, p = S(q) = q + 8$
$\qquad -0.5q_E + 20 = q_E + 8 \implies 1.5q_E = 12; \quad q_E = 8$
$\qquad\qquad\qquad\qquad\qquad\qquad p_E = S(q_E) = 8 + 8 = 16$

$$CS = \int_{0}^{q_E}[D(q) - p_E]dq = \int_{0}^{8}(-0.5q + 20 - 16)dq$$

$$= -0.25q^2 + 4q \Big|_{0}^{8} = -16 + 32 = \$16$$

$$PS = \int_{0}^{q_E}[p_E - S(q)]dq = \int_{0}^{8}[16 - (q + 8)]dq$$

$$= 8q - \frac{q^2}{2} \Big|_{0}^{8} = 64 - 32 = \$32$$

39. $p = D(q) = (q - 30)^2, \; p = S(q) = q^2 + 300$

$$(q_E - 30)^2 = q_E^2 + 300 \;\Rightarrow\; q_E^2 - 60q_E$$

$$+ \; 900 = q_E^2 + 300$$

$$q_E = 10; \;\; p_E = q_E^2 + 300 = 400$$

$$CS = \int_0^{q_E}[D(q) - p_E]dq = \int_0^{10}[(q - 30)^2 - 400]dq$$

$$= \int_0^{10}(q^2 - 60q + 500)dq$$

$$= \frac{q^3}{3} - 30q^2 + 500q \Big|_0^{10} = \frac{1000}{3} - 3000 + 5000$$

$$= \frac{7000}{3} = \$2333\frac{1}{3}$$

$$PS = \int_0^{q_E}[p_E - S(q)]dq = \int_0^{10}[400 - (q^2 + 300)]dq$$

$$= 100q - \frac{q^3}{3}\Big|_0^{10} = 1000 - \frac{1000}{3} = \frac{2000}{3} = \$666\frac{2}{3}$$

41. $p = D(q) = (q - 40)^2, \; p = S(q) = q^2 + 640$

$$(q^E - 40)^2 = q_E^2 + 640$$

$$\Rightarrow q_E^2 - 80q^E + 1600 = q_E^2 + 640$$

$$q^E = 12; \;\; p^E = q_E^2 + 640 = 784$$

$$CS = \int_0^{q_E}[D(q) - p_E]dq = \int_0^{12}[(q - 40)^2 - 784]dq$$

$$= \int_0^{12}(q^2 - 80q + 816)dq$$

$$= \frac{q^3}{3} - 40q^2 + 816q\Big|_0^{12} = \frac{1728}{3} - 5760 + 9792$$

$$= \$4608$$

$$PS = \int_0^{q_E}[p_E - S(q)]dq = \int_0^{12}[784 - (q^2 + 640)]dq$$

$$= 144q - \frac{q^3}{3}\Big|_0^{12} = 1728 - \frac{1728}{3} = \$1152$$

43. $R'(t) = 20t + 5, \; C^1(t) = 4t + 2; \;\; 2 \le t \le 6$

$$P = \int_a^b[R^1(t) - C^1(t)]dt$$

$$= \int_2^6[(20t + 5) - (4t + 2)]dt = \int_2^6(16t + 3)dt$$

$$= 8t^2 + 3t\Big|_2^6 = 288 + 18 - (32 + 6) = \$268$$

45. $R^1(t) = 1000e^t + 1000, \; C^1(t) = 2t + 4; \;\; 0 \le t \le 8$

$$P = \int_0^8[R'(t) - C^1(t)]dt = \int_0^8(1000e^t + 996 - 2t)dt$$

$$= 1000e^t - t^2 + 996t\Big|_0^8$$

$$= 1000e^8 - 64 + 7968 - 1000$$

$$= 1000e^8 + 6904 \approx \$2{,}987{,}862$$

13–4 Integration by Substitution

1. $\displaystyle\int (x^3 - 7)^{10}\, 3x^2 dx;$

Let $u = x^3 - 7, \; du = 3x^2 dx$

$$\int u^{10}du = \frac{1}{11}u^{11} + C = \frac{1}{11}(x^3 - 7)^{11} + C$$

3. $\displaystyle\int (x^3 - 4x)^5 (3x^2 - 4)dx;$

Let $u = x^3 - 4x, \; du = (3x^2 - 4)dx$

$$\int u^5 du = \frac{1}{6}u^6 + C = \frac{1}{6}(x^3 - 4x)^6 + C$$

5. $\displaystyle\int (x^4 - 8)^9 x^3 dx;$

Let $u = x^4 - 8, \; du = 4x^3 dx$

$$\frac{1}{4}\int u^9 du = \frac{1}{40}u^{10} + C = \frac{1}{40}(x^4 - 8)^{10} + C$$

7. $\displaystyle\int (x - 3)^9 dx;$

Let $u = x - 3, \; du = dx$

$$\int u^9 du = \frac{1}{10}u^{10} + C = \frac{1}{10}(x - 3)^{10} + C$$

9. $\displaystyle\int (x^4 + 6)^5 5x^3 dx;$

Let $u = x^4 + 6, \; du\; 4x^3 dx$

$$\frac{5}{4}\int u^5 du = \frac{5}{4} \cdot \frac{u^6}{6} + C = \frac{5}{24}(x^4 + 6)^6 + C$$

11. $\int (x^6 + 9)^{10} 2x^5 dx;$

Let $u = x^6 + 9$, $du = 6x^5 dx$

$\frac{2}{6} \int u^{10} du = \frac{1}{3} \cdot \frac{u^{11}}{11} + C = \frac{1}{33}(x^6 + 9)^{11} + C$

13. $\int \frac{x^2}{(x^3 - 5)^{10}} dx;$

Let $u = (x^3 - 5)$, $du = 3x^2 dx$

$\frac{1}{3} \int \frac{du}{u^{10}} = -\frac{1}{3 \cdot 9} \cdot \frac{1}{u^9} + C = -\frac{1}{27(x^3 - 9)^9} + C$

15. $\int \frac{x^3}{\sqrt{x^4 - 6}} dx;$

Let $u = x^4 - 6$, $du = 4x^3 dx$

$\frac{1}{4} \int \frac{du}{\sqrt{u}} = \frac{1}{4} \cdot 2\sqrt{u} + C = \frac{1}{2}\sqrt{x^4 - 6} + C$

17. $\int \frac{dx}{(3x - 5)^4};$

Let $u = 3x - 5$, $du = 3dx$

$\frac{1}{3} \int \frac{du}{u^4} = -\frac{1}{3 \cdot 3} \cdot \frac{1}{u^3} + C = -\frac{1}{9(3x - 5)^3} + C$

19. $\int_2^4 (\sqrt{x^2 + 9})x dx;$

Let $u = x^2 + 9$, $du = 2xdx$; $x = 2 \Rightarrow u = 13$, $x = 4 \Rightarrow u = 25$

$\frac{1}{2}\int_{13}^{25} \sqrt{u} \, du = \frac{1}{2} \cdot \frac{2}{3}u^{3/2} \Big|_{13}^{25} = \frac{1}{3}(125 - 13^{3/2}) \approx 26.043$

21. $\int_3^5 \frac{xdx}{(x^2 - 5)^2};$

$\int_3^5 \frac{xdx}{(x^2 - 5)^2};$

Let $u = x^2 - 5$, $du = 2xdx$; $x = 3 \Rightarrow u = 4$, $x = 5 \Rightarrow u = 20$

$\frac{1}{2}\int_4^{20} \frac{du}{u^2} = -\frac{1}{2} \cdot \frac{1}{u} \Big|_4^{20} = -\frac{1}{2}\left[\frac{1}{20} - \frac{1}{4}\right] = \frac{1}{10}$

23. $A = \int_5^7 (x - 5)^3 dx = \frac{1}{4}(x - 5)^4 \Big|_5^7 = \frac{1}{4}(16 - 0) = 4$

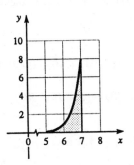

25. $A = \int_0^5 \frac{dx}{(x - 6)^2};$

Let $u = x - 6$, $du = dx$; $x = 0 \Rightarrow u = -6$, $x = 5 \Rightarrow u = -1$

$\int_{-6}^{-1} \frac{du}{u^2} = -\frac{1}{u} \Big|_{-6}^{-1} = -\left[-1 + \frac{1}{6}\right] = \frac{5}{6}$

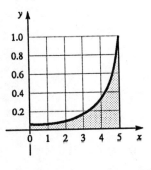

27. $\int (x^3 - 5)^{10} x dx;$

The substitution $u = x^3 - 5$, $du = 3x^2 dx$ does not work, because we need to multiply xdx by $3x$ in order to obtain $du = 3x^2 dx$. However, we cannot do this because $3x$ is not a constant.

29. $B'(t) = 200t + 50;$ t in hours. Find B at the end of 2 days.

$$B(48) = \int_0^{48} (200t + 50)dt = 100t^2 + 50t \Big|_0^{48}$$

$$= 230{,}400 + 2400 = 232{,}800 \text{ barrels}$$

31. $P'(x) = 500{,}000x + 100{,}000;$ $P(2) = 1{,}000{,}000$

$$\begin{aligned} P(x) &= \int (500{,}000x + 100{,}000)dx \\ &= 250{,}000x^2 + 100{,}000x + C \\ P(2) &= 1{,}000{,}000 = 250{,}000(4) + 100{,}000(2) + C \\ C &= -200{,}000 \\ \therefore\ P(x) &= 250{,}000x^2 + 100{,}000x - 200{,}000 \end{aligned}$$

13–5 Integrals Involving Exponential and Logarithmic Functions

1. $\int e^x dx = e^x + C$

3. $\int e^{4x} dx;$

Let $u = 4x$, $du = 4\ dx$

$\dfrac{1}{4}\int e^u du = \dfrac{1}{4}e^u + C = \dfrac{1}{4}e^{4x} + C$

5. $\int e^{x/2} dx;$

Let $u = \dfrac{x}{2}$, $du = \dfrac{1}{2}dx$

$2\int e^u du = 2e^u + C = 2e^{x/2} + C$

7. $\int e^{x^3 - 5} x^2 dx;$

Let $u = x^3 - 5$, $du = 3x^2 dx$

$\dfrac{1}{3}\int e^u\ du = \dfrac{1}{3}e^u + C = \dfrac{1}{3}e^{x^3 - 5} + C$

9. $\int x^3 e^{x^4 + 6} dx;$

Let $u = x^4 + 6;$ $du = 4x^3 dx$

$\dfrac{1}{4}\int e^u du = \dfrac{1}{4}e^u + C = \dfrac{1}{4}e^{x^4 + 6} + C$

11. $\int (x + 1)e^{x^2 + 2x} dx;$

Let $u = x^2 + 2x;$ $du = 22(x + 1)dx$

$\dfrac{1}{2}\int e^u du = \dfrac{1}{2}e^u + C = \dfrac{1}{2}e^{x^2 + 2x} + C$

13. $\int 5\dfrac{dx}{x} = 5\int \dfrac{dx}{x} = 5\ln|x| + C$

15. $\int \dfrac{x}{x^2 + 1} dx;$

Let $u = x^2 + 1;$ $du = 2xdx$

$\dfrac{1}{2}\int \dfrac{du}{u} = \dfrac{1}{2}\ln|u| + C = \dfrac{1}{2}\ln(x^2 + 1) + C$

$$= \ln\sqrt{x^2 + 1} + C$$

17. $\int \dfrac{6x^2 dx}{x^3 + 4};$

Let $u = x^3 + 4$, $du = 3x^2 dx$

$2\int \dfrac{du}{-u} = 2\ln|u| + C = 2\ln|x^3 + 4| + C$

$$= \ln(x^3 + 4)^2 + C$$

19. $\displaystyle\int \frac{-xdx}{5x^2 - 6};$

Let $u = 5x^2 - 6$, $du = 10xdx$

$$-\frac{1}{10}\int \frac{du}{u} = -\frac{1}{10}\ln|u| + C = -\frac{1}{10}\ln|5x^2 - 6| + C$$

21. $\displaystyle\int_0^4 e^xdx = e^x\Big|_0^4 = e^4 - 1 \approx 53.598$

23. $\displaystyle\int_0^1 xe^{-x^2}dx;$

Let $u = x^2$, $du = 2xdx$; $x = 0 \Rightarrow u = 0$, $x = 1$
$$\Rightarrow u = 1$$

$$\frac{1}{2}\int_0^1 e^{-u}du = -\frac{1}{2}e^{-u}\Big|_0^1 = -\frac{1}{2}(e^{-1} - 1)$$

$$= \frac{1}{2}(1 - e^{-1}) \approx 0.3161$$

25. $\displaystyle\int_1^4 \frac{1}{x}dx = \ln|x|\,\Big|_1^4 = \ln 4 \approx 1.3863$

27. $\displaystyle\int_1^6 -\frac{6}{x}dx = -6\ln|x|\,\Big|_1^6 = -6\ln 6 \approx -10.7506$

29. $\displaystyle\int \frac{(\ln x)^2}{x}dx;$ $(x > 0)$

Let $u = \ln x$, $du = \frac{1}{x}dx$

$$\int u^2du = \frac{1}{3}u^3 + C = \frac{1}{3}(\ln x)^3 + C$$

31. $\displaystyle\int \frac{(\ln 3x)^4}{x}dx;$ $(x > 0)$

Let $u = \ln 3x$, $du = \frac{1}{x}dx$

$$\int u^4du = \frac{1}{5}u^5 + C = \frac{1}{5}(\ln 3x)^5 + C$$

33. $\displaystyle\int \frac{x - 2}{x^2 - 4x + 1}dx;$

Let $u = x^2 - 4x + 1$, $du = 2(x - 2)dx$

$$\frac{1}{2}\int \frac{du}{u} = \frac{1}{2}\ln|u| + C = \frac{1}{2}\ln|x^2 - 4x + 1| + C$$

35. $\displaystyle\int \frac{1}{x \ln x}dx;$ $(x > 0)$

Let $u = \ln x$, $du = \frac{1}{x}dx$

$$\int \frac{1}{u}du = \ln|u| + C = \ln|\ln x| + C$$

37. $\displaystyle\int \frac{1}{x \ln x^2}dx;$ $(x \neq 0)$

Let $u = \ln x^2$, $du = \frac{2x}{x^2}dx = \frac{2}{x}dx$

$$\frac{1}{2}\int \frac{du}{u} = \frac{1}{2}\ln|u| + C = \frac{1}{2}\ln|\ln x^2| + C$$

39. $\displaystyle A = \int_0^1 e^{-2x}dx = -\frac{1}{2}e^{-2x}\Big|_0^1 = -\frac{1}{2}(e^{-2} - 1)$

$$= \frac{1}{2}(1 - e^{-2}) \approx 0.4323$$

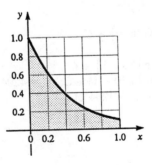

41. $A = \left| \int_{1}^{4} \frac{1}{x - 9} dx \right| = \left| \ln|x - 9| \Big|_{1}^{4} \right|$

$= \left| \ln|-5| - \ln|-8| \right| = \left| \ln \frac{5}{8} \right| \approx |-0.470|$

$= 0.470$

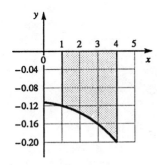

43. $R^1(t) = 100t + 10e^{-t}$

$\int_{1}^{10} R^1(t)\, dt = \int_{1}^{10} (100t + 10e^{-t})\, dt = 50t^2 - 10e^{-t} \Big|_{1}^{10}$

$= 50(100) - 10e^{-10} - (50 - 10e^{-1})$

$= 4950 + 10e^{-1} - 10e^{-10} \approx \$4953.68$

45. $B^1(x) = \frac{10,000}{5x + 1}$; $B(2) = 9000$

$B(x) = \int B^1(x)dx = 10,000 \int \frac{dx}{5x + 1}$

Let $u = 5x + 1$, $du = 5dx$

$\frac{10,000}{5} \int \frac{du}{u} = 2000 \ln|u| + C$

$= 2000 \ln|5x + 1| + C$

$B(2) = 9000 = 2000 \ln(10 + 1) + C$; $C \approx 4204.21$

$B(x) = 2000 \ln(5x + 1) + 4204.21$

13–6 Integration by Parts

1. $\int xe^{3x}dx$;

Let $u = x$ and $dv = e^{3x}dx$

$du = dx$, $\quad v = \frac{1}{3}e^{3x}$

$\int u\, dv = uv - \int v\, du = \frac{1}{3}xe^{3x} - \frac{1}{3}\int e^{3x}dx$

$= \frac{1}{3}xe^{3x} - \frac{1}{9}e^{3x} + C$

3. $\int \frac{x}{\sqrt{x - 3}}dx$;

Let $u = x$ and $dv = \frac{dx}{\sqrt{x - 3}}$

$du = dx$, $\quad v = 2\sqrt{x - 3}$

$\int u\, dv = uv - \int v\, du = 2x\sqrt{x - 3} - 2\int \sqrt{x - 3}\, dx$

$= 2x\sqrt{x - 3} - \frac{4}{3}(x - 3)^{3/2} + C$

5. $\int \frac{x}{e^x}dx = \int xe^{-x}dx$;

Let $u = x$ and $dv = e^{-x}dx$

$du = dx$, $\quad v = -e^{-x}$

$\int u\, dv = -xe^{-x} + \int e^{-x}dx$

$= -xe^{-x} - e^{-x} + C = -e^{-x}(x + 1) + C$

7. $\int \frac{x}{(x - 4)^3}dx$;

Let $u = x$ and $dv = \frac{dx}{(x - 4)^3}$

$du = dx$, $\quad v = -\frac{1}{2(x - 4)^2}$

$\int u\, dv = -\frac{x}{2(x - 4)^2} + \frac{1}{2}\int \frac{dx}{(x - 4)^2}$

$= -\frac{x}{2(x - 4)^2} - \frac{1}{2(x - 4)} + C$

9. $\displaystyle\int \frac{x}{e^4{}^x}dx = \int xe^{-4x}dx;$

Let $u = x$ and $dv = e^{-4x}dx$

$du = dx,\qquad v = -\dfrac{1}{4}e^{-4x}$

$\displaystyle\int u\,dv = -\frac{1}{4}e^{-4x} + \frac{1}{4}\int e^{-4x}dx$

$\qquad = -\dfrac{1}{4}xe^{-4x} - \dfrac{1}{16}e^{-4x} + C$

$\qquad = -\dfrac{1}{4}e^{-4x}\left[x + \dfrac{1}{4}\right] + C$

11. $\displaystyle\int x^3 \ln x\,dx;$

Let $u = \ln x$ and $dv = x^3 dx$

$du = \dfrac{1}{x}dx,\qquad v = \dfrac{1}{4}x^4$

$\displaystyle\int u\,dv = \frac{1}{4}x^4 \ln x - \frac{1}{4}\int \frac{x^4}{x}dx$

$\qquad = \dfrac{1}{4}x^4 \ln x - \dfrac{1}{16}x^4 + C$

$\qquad = \dfrac{1}{4}x^4\left[\ln x - \dfrac{1}{4}\right] + C$

13. $\displaystyle\int x(x - 9)^5 dx;$

Let $u = x$ and $dv = (x - 9)^5 dx$

$du = dx,\qquad v = \dfrac{1}{6}(x - 9)^6$

$\displaystyle\int u\,dv = \frac{x(x - 9)^6}{6} - \frac{1}{6}\int (x - 9)^6 dx$

$\qquad = \dfrac{x(x - 9)^6}{6} - \dfrac{1}{42}(x - 9)^7 + C$

15. $\displaystyle\int \frac{x}{(6 + 5x)^2} dx$

Let $u = x$ and $dv = \dfrac{1}{(6 + 5x)^2} dx$

$du = dx,\qquad v = \displaystyle\int \frac{dx}{(6 + 5x)^2} = \frac{1}{5}\int \frac{5dx}{(6 + 5x)^2}$

$\qquad\qquad\qquad = \dfrac{-1}{5(6 + 5x)}$

$\displaystyle\int u\,dv = x\left[-\frac{1}{5(6 + 5x)}\right] - \left[-\frac{1}{5(6 + 5x)}\right]dx$

$\qquad = -\dfrac{x}{5(6 + 5x)} + \dfrac{1}{25}\int \dfrac{5dx}{6 + 5x}$

$\qquad = -\dfrac{x}{5(6 + 5x)}\Bigg] + \dfrac{1}{25}\ln |6 + 5x| + C$

17. $\displaystyle\int \frac{x}{\sqrt{5 + 3x}}dx;$

Let $u = x$ and $dv = \dfrac{dx}{\sqrt{5 + 3x}}$

$du = dx,\qquad v = \dfrac{2}{3}\sqrt{5 + 3x}$

$\displaystyle\int u\,dv = \frac{2}{3}x\sqrt{5 + 3x} - \frac{2}{3}\int \sqrt{5 + 3x}\,dx$

$\qquad = \dfrac{2}{3}x\sqrt{5 + 3x} - \dfrac{4}{27}(5 + 3x)^{3/2} + C$

19. $\displaystyle\int 3x\sqrt{3x + 7}\,dx;$

Let $u = x$ and $dv = 3\sqrt{3x + 7}\,dx$

$du = dx,\qquad v = \dfrac{2}{3}(3x + 7)^{3/2}$

$\displaystyle\int u\,dv = \frac{2}{3}x(3x + 7)^{3/2} - \frac{2}{3}\int (3x + 7)^{3/2}dx$

$\qquad = \dfrac{2}{3}x(3x + 7)^{3/2} - \dfrac{4}{45}(3x + 7)^{5/2} + C$

21. $\int x \ln 9x\,dx$

Let $u = \ln 9x$ and $dv = x\,dx$

$$du = \frac{1}{x}dx, \qquad v = \frac{x^2}{2}$$

$$\int u\,dv = \frac{x^2}{2}\ln 9x - \frac{1}{2}\int \frac{x^2}{x}dx$$

$$= \frac{x^2}{2}\ln 9x - \frac{1}{4}x^2 + C$$

23. $\int x^3 e^x\,dx$;

Let $u = x^3$ and $dv = e^x\,dx$

$$du = 3x^2\,dx, \quad v = e^x$$

$$\int u\,dv = x^3 e^x - 3\int x^2 e^x\,dx$$

Evaluate $\int x^2 e^x\,dx$ by parts;

Let $u = x^2$ and $dv = e^x\,dx$

$$du = 2x\,dx \quad v = e^x$$

$$\int u\,dv = x^2 e^x - 2\int xe^x\,dx$$

Evaluate $\int xe^x\,dx$ by parts:

Let $u = x$ and $dv = e^x\,dx$

$$du = dx, \qquad v = e^x$$

$$\int u\,dv = xe^x - \int e^x\,dx = xe^x - e^x$$

The complete integral is

$$x^3 e^x - 3[x^2 e^x - 2(xe^x - e^x)] + C$$
$$= x^3 e^x - 3x^2 e^x + 6xe^x - 6e^x + C$$
$$= e^x(x^3 - 3x^2 + 6x - 6) + C$$

25. Show that $\int x^n e^x\,dx = x^n e^x - n\int x^{n-1}e^x\,dx$,

n integer > 0.

Let $u = x^n$ and $dv = e^x\,dx$

$$du = nx^{n-1}, \quad v = e^x$$

$$\int x^n e^x\,dx = x^n e^x - n\int x^{n-1}e^x\,dx$$

27. $\int (\ln x)^2\,dx$;

Let $u = (\ln x)^2$ and $dv = dx$

$$du = \frac{2\ln x}{x}dx, \qquad v = x$$

$$\int u\,dv = x(\ln x)^2 - 2\int \frac{x\ln x}{x}dx$$

Evaluate $\int \ln x\,dx$ by parts:

Let $u = \ln x$ and $dv = dx$

$$du = \frac{1}{x}dx, \qquad v = x$$

$$\int u\,dv = x\ln x - \int \frac{x}{x}dx = x\ln x - x$$

Complete integral is $x(\ln x)^2 - 2x\ln x + 2x + C$

29. Show that $\int (\ln x)^n\,dx = x(\ln x)^n - n\int (\ln x)^{n-1}\,dx$,

$n =$ integer > 0

Let $u = (\ln x)^n$ and $dv = dx$

$$du = n(\ln x)^{n-1}\frac{dx}{x}, \; v = x$$

$$\int u\,dv = x(\ln x)^n - n\int (\ln x)^{n-1}\,dx$$

31. $f(x) = xe^x,\ 0 \le x \le 1$;

$$A = \int_0^2 xe^x\,dx = e^x(x-1)\Big|_0^2 = e^2(1) - 1(-1)$$

$$= e^2 + 1 \approx 8.3891$$

33. $f(x) = x^2 e^{x^2},\ 0 \le x \le 1$;

$$A = \int_0^1 x^2 e^{x^2}\,dx$$

Let $u = x$ and $dv = xe^{x^2}\,dx$

$$du = dx, \qquad v = \frac{1}{2}e^{x^2}$$

$$\int u\,dv = \frac{1}{2}xe^{x^2} - \frac{1}{2}\int e^{x^2}\,dx$$

$$A = \int_0^1 x^3 e^{x^2}\,dx$$

(continued)

33. *(continued)*

Let $u = x^2$ and $dv = xe^{x^2}dx$

$$du = 2xdx, \quad v = \frac{1}{2}e^{x^2}$$

$$\int u\,dv = \frac{1}{2}x^2e^{x^2} - \int xe^{x^2}dx$$

$$= \frac{1}{2}x^2e^{x^2} - \frac{1}{2}e^{x^2}$$

$$A = \frac{e^{x^2}}{2}(x^2 - 1)\Big|_0^1 = \frac{e}{2}(0) - \frac{1}{2}(-1) = \frac{1}{2}$$

35. $\int x^2e^x dx;$

Let $u = x^2$ and $dv = e^x dx$

$$du = 2x \quad v = e^x$$

$$\int u\,dv = x^2e^x - 2\int xe^x dx$$

Evaluate $\int xe^x dx$ by parts:

Let $u = x$ and $dv = e^x dx$

$$du = dx, \quad v = e^x$$

$$\int u\,dv = xe^x - \int e^x dx = xe^x - e^x$$

Complete integral: $x^2e^x - 2xe^x + 2e^x + C$
$$e^x(x^2 - 2x + 2) + C$$

37. $\int \frac{1}{x + 6}dx = \ln|x + 6| + C$

39. $\int \frac{x}{2x + 9}dx = \frac{1}{2}\int \left[1 - \frac{9}{2x + 9}\right]dx$

$$= \frac{1}{2}\left[x - \frac{9}{2}\ln|2x + 9|\right] + C$$

41. $P'(t) = 2000te^{0.2t}$
$$= 2000[5te^{0.2t} - 25e^{0.2t}] + C$$
Assume $P(0) = 0$
$$0 = 2000[0 - 25] + C \Rightarrow C = 50,000$$
$$P(t) = 10,000e^{0.2t}(t - 5) + 50,000$$
$$P(6) = 10,000e^{1.2}(1) + 50,000 \approx \$83,201.17$$

43. $D'(t) = 10,000te^{-t}$

$$D(t) = 10,000\int te^{-t}dt = 100\,(-te^{-t} - e^{-t}) + C$$
$$D(0) = 0 = 10,000(0 - 1) + C; \quad C = 100$$
$$D(t) = 10,000[1 - e^{-t}(t + 1)]$$
$$D(2) = 10,000[1 - e^{-t}(3)] \approx 5940 \text{ decrease in cells}$$

45. $R'(x) = x^3\ln x \ (x > 1)$

$$R(x) = \int x^3 \ln x\,dx$$

Let $u = \ln x$ and $dv = x^3 dx$

$$du = \frac{1}{x}dx, \quad v = \frac{x^4}{4}$$

$$R(x) = \frac{x^4}{4}\ln x - \int \frac{x^3}{4}dx$$

$$= \frac{x^4}{4}\ln x - \frac{1}{16}x^4 + C$$

$$R(1) = 80 = \frac{1}{4}(0) - \frac{1}{16} + C \Rightarrow C = 80\frac{1}{16}$$

$$R(x) = \frac{x^4}{4}\left[\ln x - \frac{1}{4}\right] + \frac{1281}{16}$$

13–7 Using Tables of Integrals

1. $\int \frac{dx}{\sqrt{81 + x^2}} = \ln\left[x + \sqrt{81 + x^2}\right] + C$

3. $\int_0^8 \frac{dx}{\sqrt{x^2 + 36}} = \ln(x + \sqrt{x^2 + 36})\Big|_0^8$

$$= \ln(8 + \sqrt{100}) - \ln(\sqrt{36})$$
$$= \ln 18 - \ln 6 = \ln 3 \approx 1.0986$$

5. $\int \frac{-7dx}{81 - x^2} = \frac{-7}{2 \cdot 9}\ln\left|\frac{9 + x}{9 - x}\right| + C$

$$= \frac{7}{18}\ln\left|\frac{9 - x}{9 + x}\right| + C$$

7. $\displaystyle\int \frac{3dx}{x^2(4 - 7x)} = 3\left[-\frac{1}{4x} - \frac{7}{16}\ln\left|\frac{4 - 7x}{x}\right|\right] + C$

9. $\displaystyle\int_0^1 x^3 e^x dx = x^3 e^x - 3x^2 e^x + 6xe^x - 6e^x\Big|_0^1$

$= e - 3e + 6e - 6e + 6$

$= 6 - 2e \approx 0.5634$

11. $\displaystyle\int \ln 5x\, dx = x \ln|5x| - x + C$

13. $\displaystyle\int \frac{dx}{x\sqrt{5 - 2x}} = \frac{1}{\sqrt{5}}\ln\left|\frac{\sqrt{5 - 2x} - \sqrt{5}}{\sqrt{5 - 2x} + \sqrt{5}}\right| + C$

15. $\displaystyle\int \frac{dx}{\sqrt{x^2 - 100}} = \ln\left|x + \sqrt{x^2 + 100}\right| + C$

17. $\displaystyle\int \frac{dx}{x\sqrt{25 - x^2}} = -\frac{1}{5}\ln\left|\frac{5 + \sqrt{25 - x^2}}{x}\right| + C$

19. $\displaystyle\int (36 - x^2)^{-3/2}dx = \frac{x}{36(\sqrt{36 - x^2})} + C$

21. $\displaystyle\int (\ln x)^2 dx = x(\ln|x|)^2 - 2x \ln|x| + 2x + C$

23. $\displaystyle\int x^2 \ln x dx = \frac{x^3}{3}\left[\ln|x| - \frac{1}{3}\right] + C$

25. $\displaystyle\int (x^2 + 25)^{-3/2}dx = \frac{1}{25}\left[\frac{x}{\sqrt{x^2 + 25}}\right] + C$

27. $\displaystyle\int (x + 1) \ln (x + 1)dx$

$= \frac{(x + 1)^2}{2}\ln|x + 1| - \frac{(x + 1)^2}{4} + C$

29. $\displaystyle\int \frac{4}{x(3 + 7x)}dx = \frac{4}{3}\ln\left|\frac{x}{3 + 7x}\right| + C$

31. $v(t) = \dfrac{1}{64 - t^2}\ (0 \le t \le 6)$

$S(6) = \displaystyle\int_0^6 \frac{1}{64 - t^2}dt = \frac{1}{8}\ln\left|\frac{8 + t}{\sqrt{64 - t^2}}\right|\Big|_0^6$

$= \frac{1}{8}\left[\ln\frac{14}{\sqrt{64 - 36}} - \ln\frac{8}{\sqrt{64}}\right] = \frac{1}{8}\ln\sqrt{7}$

$= \frac{1}{16}\ln 7 \approx 0.1216$ foot

33. $R'(t) = \dfrac{200}{\sqrt{t^2 + 1}}\ (t \ge 0)$

$R(10) = \displaystyle\int_0^{10} \frac{200}{\sqrt{t^2 + 1}}dt = 200 \ln\left|t + \sqrt{t^2 + 1}\right|\Big|_0^{10}$

$= 200[\ln(10 + \sqrt{101}) - \ln 1]$

≈ 599.64 gallons

Review Exercises

1. $\displaystyle\int x^4 dx = \frac{x^5}{5} + C$

3. $\displaystyle\int \frac{20}{x}dx = 20 \ln|x| + C$

5. $\displaystyle\int e^{-x}dx = -e^{-x} + C$

7. $\int 7dx = 7x + C$

9. $\int (5x^4 + 6x^2 - 8x + 4)dx = x^5 + 2x^3 - 4x^2 + 4x + C$

11. $\int (4x^{-1} + e^{6x})dx = 4 \ln|x| + \frac{1}{6}e^{6x} + C$

13. $\int \left[\frac{4 - x^6}{x^2}\right] dx = \int \frac{4}{x^2}dx - \int x^4 dx$

$= -\frac{4}{x} - \frac{x^5}{5} + C$

15. $\int (x - 5)(x + 5)dx = \int (x^2 - 25)dx = \frac{x^3}{3} - 25x + C$

17. $C^1(x) = 9x^2 + 4x - 1,\ C(0) = 5000$

$C(x) = \int (9x^2 + 4x - 1)dx = 3x^3 + 2x^2 - x + A$

$C(0) = 5000 = A$

$C(x) = 3x^3 + 2x^2 - x + 5000$

19. $\int_1^4 (8x + 2)dx = 4x^2 + 2x + C \Big|_1^4$

$= 4(16) + 8 - 4 - 2 = 66$

21. $\int_2^5 \frac{20}{x}dx = 20 \ln|x| + C \Big|_2^5 = 20 \ln 5 - 20 \ln 2$

$= 20 \ln \frac{5}{2} \approx 18.3$

23. $\int_1^4 \frac{40}{\sqrt{x}}dx = \frac{40x^{1/2}}{\frac{1}{2}} \Big|_1^4 = 80\sqrt{x} \Big|_1^4 = 80(2 - 1) = 80$

25. $A = \int_0^6 (-x^2 + 36)dx = -\frac{x^3}{3} + 36x \Big|_0^6 = -\frac{216}{3} + 216$

$= 144$

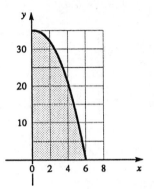

27. $A = \int_0^2 (x^4 + 10)dx = \frac{x^5}{5} + 10x \Big|_0^2 = \frac{32}{5} + 20 = \frac{132}{5}$

$= 26.4$

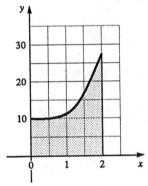

29. $f(x) = x^2 - 4, 0 \le x \le 4$;

For $0 \le x < 2$, $x^2 - 4 < 0$; for $2 < x \le 4$, $x^2 - 4 > 0$

$$A = \left| \int_0^2 (x^2 - 4)dx \right| + \int_2^4 (x^2 - 4)dx$$

$$= -\left[\frac{x^3}{3} - 4x \right] \Big|_0^2 + \frac{x^3}{3} - 4x \Big|_2^4$$

$$= -\left[\frac{8}{3} - 8 \right] + \frac{64}{3} - 16 - \left[\frac{8}{3} - 8 \right]$$

$$= 16$$

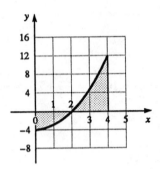

31. $f(x) = -x^2 + 16, g(x) = x^2 - 16, -4 \le x \le 4$

$$A = \int_{-4}^4 [(-x^2 + 16) - (x^2 - 16)]dx$$

$$= \int_{-4}^4 (-2x^2 + 32)dx$$

$$= -\frac{2}{3}x^3 + 32x \Big|_{-4}^4$$

$$= -\frac{2}{3}(64) + 128 - \left[\frac{2}{3}(64) - 128 \right]$$

$$= \frac{512}{3} = 170\frac{2}{3}$$

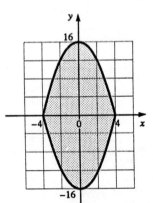

33. $p = D(q) = -4q + 56, p = S(q) = 2q + 8$

$-4q_E + 56 = 2q_E + 8 \Rightarrow q_E = 8$

$p_E = 2q_E + 8 = 24$

$$CS = \int_0^{q_E} [D(q) - p_E]dq = \int_0^8 [(-4q + 56) - 24]dq$$

$$= -2q^2 + 32q \Big|_0^8 = -128 + 256 = 128$$

$$PS = \int_0^{q_E} [p_E - S(q)]dq = \int_0^8 [24 - (2q + 8)]dq$$

$$= 16q - q^2 \Big|_0^8 = 128 - 64 = 64$$

35. $\int (x^4 + 6x)^5 (4x^3 + 6)dx$;

Let $u = x^4 + 6x$, $du = (4x^3 + 6)dx$

$$\int u^5 du = \frac{1}{6}u^6 + C = \frac{1}{6}(x^4 + 6x)^6 + C$$

37. $\int e^{x^2+6}x\,dx$;

Let $u = x^2 + 6$, $du = 2x\,dx$

$$\frac{1}{2}\int e^u du = \frac{1}{2}e^u + C = \frac{1}{2}e^{x^2+6} + C$$

39. $\int \frac{x}{x^2 + 4}dx$;

Let $u = x^2 + 4$, $du = 2x\,dx$

$$\frac{1}{2}\int \frac{du}{u} = \frac{1}{2}\ln|u| + C = \frac{1}{2}\ln(x^2 + 4) + C$$

41. $\int \frac{(\ln 5x)^3}{x}dx$;

Let $u = \ln 5x$, $du = \frac{dx}{x}$

$$\int u^3 du = \frac{u^4}{4} + C = \frac{(\ln 5x)^4}{4} + C$$

43. $\int (x^2 + 10x - 7)^6(x + 5)dx$;

Let $u = x^2 + 10x - 7$, $du = (2x + 10)dx$

$\frac{1}{2}\int u^6 du = \frac{u^7}{14} + C = \frac{1}{14}(x^2 + 10x - 7)^7 + C$

45. $\int x\sqrt{x - 8}\,dx$;

Let $u = x$ and $dv = \sqrt{x - 8}\,dx$

$\quad du = dx \qquad v = \frac{2}{3}(x - 8)^{3/2}$

$\int u\,dv = \frac{2}{3}x(x - 8)^{3/2} - \frac{2}{3}\int (x - 8)^{3/2}dx$

$\qquad = \frac{2}{3}x(x - 8)^{3/2} - \frac{4}{15}(x - 8)^{5/2} + C$

47. $\int \frac{x}{8 + 5x}dx = \frac{1}{5}\int \left[1 - \frac{8}{5x + 8}\right]dx$

$\qquad = \frac{x}{5} - \frac{8}{25}\ln|5x + 8| + C$

49. $\int \frac{1}{7 - x}dx = -\ln|7 - x| + C$

51. $\int \frac{1}{x\sqrt{7 - 2x}}dx = \frac{1}{\sqrt{7}}\ln\left|\frac{\sqrt{7 - 2x} - \sqrt{7}}{\sqrt{7 - 2x} + \sqrt{7}}\right| + C$

53. $\int (x^2 + 16)^{-3/2}dx = \frac{1}{16}\left[\frac{1}{\sqrt{x^2 + 16}}\right] + C$

55. $\int x^2 e^{8x}\,dx = \frac{e^{8x}}{8}\left[x^2 - \frac{1}{4}x + \frac{1}{32}\right] + C$

57. $P'(t) = 0.25te^t$;

$P(t) = 0.25\int te^t dt = 0.25e^t(t - 1) + C$

Let $P(0) = 0 = 0.25(-1) + C \Rightarrow C = 0.25$

$P(t) = 0.25e^t(t - 1) + 0.25$

$P(8) = 0.25e^8(8 - 1) + 0.25 \approx 5217$

CHAPTER

14

FURTHER TOPICS OF INTEGRATION

14–1 Continuous Money Flow

1. $f(x) = 2000, 0 \leq x \leq 5, r = 0.09$

(a) Total money flow $= \displaystyle\int_0^5 2000dx = 2000x \Big|_0^5$

$\qquad\qquad\qquad\qquad = \$10,000$

(b) Present value $= \displaystyle\int_0^5 2000e^{-rx}\, dx$

$\qquad\qquad = 2000 \displaystyle\int_0^5 e^{-0.09x}\, dx$

$\qquad\qquad = -\dfrac{2000}{0.09}e^{-0.09x} \Big|_0^5$

$\qquad\qquad = -\dfrac{2000}{0.09}(e^{-0.45} - 1)$

$\qquad\qquad \approx \$8052.71$

(c) Future value $= e^{rt} \displaystyle\int_0^t 2000e^{-rx}\, dx$

$\qquad\qquad = e^{0.09(5)}(\text{Present value})$

$\qquad\qquad = e^{0.45}(8052.71)$

$\qquad\qquad \approx \$12,629.16$

3. $f(x) = 60,000$

(a) Total money flow $= \displaystyle\int_0^5 60,000\, dx = 60,000x \Big|_0^5$

$\qquad\qquad\qquad\qquad = \$300,000$

(b) Present value $= \displaystyle\int_0^5 60,000e^{-0.09x}\, dx$

$\qquad\qquad = -\dfrac{60,000}{0.09}e^{-0.09x} \Big|_0^5$

$\qquad\qquad = -\dfrac{60,000}{0.09}(e^{-0.45} - 1)$

$\qquad\qquad \approx \$241,581.23$

(c) Future value $= e^{0.09(5)}(\text{Present value})$
$\qquad\qquad = e^{0.45}(241,581.23)$
$\qquad\qquad \approx \$378,874.79$

5. $f(x) = 6000x$

(a) Total money flow $= \displaystyle\int_0^5 6000x\, dx$

$$= 3000x^2 \Big|_0^5 = \$75,000$$

(b) Present value $= \displaystyle\int_0^5 6000x\, e^{-0.09x}\, dx$

Let $u = -0.09x$, $du = -0.09dx$
$x = 0 \Rightarrow u = 0$, $x = 5 \Rightarrow u = -0.45$

$$\frac{6000}{(0.09)^2} \int_0^{-0.45} u\, e^u\, du = \frac{6000}{0.0081} e^u(u - 1) \Big|_0^{-0.45}$$

$$= \frac{6000}{0.0081} \left[e^{-0.45}(-1.45) - e^0(-1) \right]$$

$$\approx \$55,880.87$$

(c) Future value $= e^{0.09(5)}$(Present value)
$= e^{0.45}(55,880.87) \approx \$87,638.66$

7. $f(x) = 0.8x$

(a) Total money flow $= \displaystyle\int_0^5 0.8x\, dx = 0.4x^2 \Big|_0^5$

$$= \$10$$

(b) Present value $= \displaystyle\int_0^5 0.8x\, e^{-0.09x}\, dx$

$$= \frac{0.8}{0.0081} \left[e^{-0.45}(-1.45) - e^0(-1) \right]$$

(using the method of Exercise 5)
$\approx \$7.45$

(c) Future value $= e^{0.09(5)}$ (Present value)
$= e^{0.45}(7.45) \approx \$11.69$

9. $f(x) = 3000e^{0.05x}$

(a) Total money flow $= \displaystyle\int_0^5 3000e^{0.05x}\, dx$

$$= \frac{3000}{0.05} e^{0.05x} \Big|_0^5$$

$$= 60,000(e^{0.25} - 1)$$

$$\approx \$17,041.53$$

(b) Present value $= \displaystyle\int_0^5 3000e^{0.5x}\, e^{-0.09x}\, dx$

$$= \frac{3000}{-0.04} e^{-0.04x} \Big|_0^5$$

$$= -75,000(e^{-0.2} - 1)$$

$$\approx \$13,595.19$$

(c) Future value $= e^{0.09(5)}$ (Present value)
$= e^{0.45}(13,595.19) \approx \$21,321.51$

11. $f(x) = 40,000\, e^{-0.04x}$

(a) Total money flow $= \displaystyle\int_0^5 40,000e^{-0.04x}\, dx$

$$= \frac{40,000}{-0.04} e^{-0.04x} \Big|_0^5$$

$$= -1,000,000(e^{-0.2} - 1)$$

$$\approx \$181,269.25$$

(b) Present value $= \displaystyle\int_0^5 40,000e^{-0.04x}\, e^{-0.09x}\, dx$

$$= \frac{40,000}{-0.13} e^{-0.13x} \Big|_0^5$$

$$= \frac{40,000}{-0.13} (e^{-0.65} - 1)$$

$$\approx \$147,062.84$$

(c) Future value $= e^{0.09(5)}$ (Present value)
$= e^{0.45}(147,062.84) \approx \$230,640.44$

13. $f(x) = 0.20x + 1000$

(a) Total money flow $= \int_0^5 (0.20x + 1000)dx$

$= 0.10x^2 + 1000x \Big|_0^5 = 2.5 + 5000 = \$5002.50$

(b) Present value $= \int_0^5 (0.20x + 1000)e^{-0.09x}\, dx$

$= 0.20 \int_0^5 xe^{-0.09x}\, dx + 1000 \int_0^5 e^{-0.09x}\, dx$

$= \dfrac{0.20}{0.0081}\left[e^{-0.45}(-1.45) - e^0(-1) \right]$

$\quad - \dfrac{1000}{0.09}(e^{-0.45} - 1)$

(using the method of Exercise 5)
$\approx 1.8627 + 4026.35 \approx \$4028.22$

(c) Future value $= e^{0.09(5)}$ (Present value)
$= e^{-0.45}(4028.22) \approx \$6317.50$

15. $f(x) = 2500 - 250x$

(a) Total money flow $= \int_0^5 (2500 - 250x)dx$

$= 2500x - 125x^2 \Big|_0^5 = 12,500 - 3125 = \$9375$

(b) Present value $= \int_0^5 (2500 - 250x)e^{-0.09x}\, dx$

$= 2500 \int_0^5 e^{-0.09x}\, dx - 250 \int_0^5 x\, e^{-0.09x}\, dx$

$= \dfrac{2500}{-0.09} e^{-0.09x} \Big|_0^5 - \dfrac{250}{0.0081}\left[e^{-0.09x}(-0.09x - 1) \right]\Big|_0^5$

$= -\dfrac{2500}{0.09}(e^{-0.45} - 1)$

$\quad + \dfrac{250}{0.0081}\left[e^{-0.45}(0.45 + 1) - 1 \right]$

$\approx 10,065.89 - 2328.37 = \$7737.51$

(c) Future value $= e^{0.09(5)}$(Present value)
$= e^{0.45}(7737.51) \approx \$12,134.83$

17. $f(x) = 10,000;\ 0 \le x \le 7$

(a) Total money flow $= \int_0^7 10,000dx = 10,000x \Big|_0^7$

$= \$70,000$

(b) Present value $= \int_0^7 10,000e^{-0.10x}dx$

$= \dfrac{10,000}{-0.10} e^{-0.01x} \Big|_0^7 = -100,000(e^{-0.7} - 1)$

$\approx \$50,341.47$

(c) Future value $= e^{0.10(7)}$ (Present value)
$= e^{0.70}(50,341.47) \approx \$101,375.27$

19. $f(x) = 10,000,000x;\ \ 0 \le x \le 4$

(a) Total money flow $= \int_0^4 10,000,000x\, dx$

$= 5,000,000x^2 \Big|_0^4 = \$80,000,000$

(b) Present value $= \int_0^4 (10,000,000x)e^{-0.12x}\, dx$

$= \dfrac{10,000,000}{(-0.12)^2} \int_0^{-0.48} u\, e^u\, du,$ where $u = -0.12x$

$= \dfrac{10,000,000}{0.0144} e^u(u - 1) \Big]_0^{-0.48}$

$= \dfrac{10,000,000}{0.0144}[e^{-0.48}(-1.48) + 1] \approx \$58,472,625$

Future value $= e^{0.12(4)}$ (Present value)
$= e^{0.48}(58,472,625) \approx \$94,496,112$

14-2 Improper Integrals

1. $\displaystyle\int_1^\infty \dfrac{dx}{x^2} = \lim_{b \to \infty} -\dfrac{1}{x}\Big|_0^b = 0 + 1 = 1$

3. $\int_6^\infty \frac{dx}{x} = \lim_{b\to\infty} \ln x \Big|_6^b$

Since $\lim_{b\to\infty} \ln b$ does not exist, the integral diverges.

5. $\int_3^\infty \frac{dx}{x+1} = \lim_{b\to\infty} \ln(x+1)\Big|_3^b$

Since $\lim_{b\to\infty} \ln(b+1)$ does not exist, the integral diverges.

7. $\int_0^\infty e^{-2x}\,dx = \lim_{b\to\infty}\left[-\frac{1}{2}\right]e^{-2x}\Big|_0^b$

$= \left[-\frac{1}{2}\right]\left[\lim_{b\to\infty} e^{-b} - 1\right] = \left[-\frac{1}{2}\right](0-1) = \frac{1}{2}$

9. $\int_0^\infty xe^{-x}\,dx = \lim_{b\to\infty}\int_0^b xe^{-x}\,dx$

Integrate by parts, with $u = x$ and $dv = e^{-x}\,dx$

$\int u\,dv = -xe^{-x} + \int e^{-x}\,dx = -e^{-x}(x+1)$

$\lim_{b\to\infty} -e^{-x}(x+1)\Big|_0^b = -\lim_{b\to\infty} be^{-b} + \lim_{b\to\infty} e^{-b} + 1$

The first limit can be shown to equal 0. Let

$g(x) = \frac{e^x}{x^2}$ for $x > 0$.

$g'(x) = \frac{x^2 e^x - 2xe^x}{x^4} = \frac{xe^x - 2e^x}{x^3} = \frac{e^x(x-2)}{x^3} > 0$ if

$x > 2$. Set up the inequality $0 \le \frac{x}{e^x} \le \frac{x}{x^2} = \frac{1}{x}$. Take

the $\lim_{x\to\infty}$ to get $0 \le \lim_{x\to\infty} \frac{x}{e^x} \le 0 \Rightarrow \lim_{x\to\infty} xe^{-x} = 0$.

Hence $\int_0^\infty xe^{-x}\,dx = \lim_{b\to\infty} e^{-b} + 1 = 0 + 1 = 1$

11. $\int_{-\infty}^0 e^{2x}\,dx = \frac{1}{2}\lim_{a\to-\infty} e^{2x}\Big|_a^0 = \frac{1}{2}(1-0) = \frac{1}{2}$

13. $\int_4^\infty x^{-1/2}\,dx = \lim_{b\to\infty} 2\sqrt{x}\Big|_4^b$

Since $\lim_{b\to\infty} \sqrt{b}$ does not exist, the integral diverges.

15. $\int_1^\infty \frac{dx}{(x+2)^2} = \lim_{b\to\infty} -\frac{1}{(x+2)}\Big|_1^b = 0 + \frac{1}{3} = \frac{1}{3}$

17. $\int_0^\infty xe^{-x^2}\,dx = \lim_{b\to\infty}\left[-\frac{1}{2}e^{-x^2}\right]\Big|_0^b = -\frac{1}{2}(0-1) = \frac{1}{2}$

19. $\int_{-\infty}^\infty xe^{-x^2}\,dx = \int_{-\infty}^0 xe^{-x^2}\,dx + \int_0^\infty xe^{-x^2}\,dx$

$= \lim_{a\to-\infty}\left[-\frac{1}{2}e^{-x^2}\right]\Big|_a^0 + \lim_{b\to\infty}\left[-\frac{1}{2}e^{-x^2}\right]\Big|_0^b$

$= -\frac{1}{2}(1-0) - \frac{1}{2}(0-1) = 0$

21. $\int_1^\infty \frac{dx}{\sqrt{x+6}} = \lim_{b\to\infty} 2\sqrt{x+6}\Big|_1^b$

Since $\lim_{b\to\infty} \sqrt{b+6}$ does not exist, the integral diverges.

23. $\int_0^\infty x^3\,dx = \lim_{b\to\infty} \frac{x^4}{4}\Big|_0^b$

Since $\lim_{b\to\infty} b^4$ does not exist, the integral diverges.

25. $A = \int_0^\infty e^{-x}\,dx = \lim_{b\to\infty} -e^{-x}\Big|_0^b = 0 + 1 = 1$

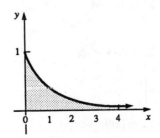

27. $A = \int_4^\infty \dfrac{1}{(x-3)^2}\, dx = \lim_{b \to \infty} \left. -\dfrac{1}{x-3} \right|_4^b$

$= 0 + \dfrac{1}{1} = 1$

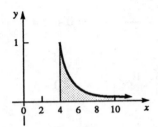

29. $A = \int_3^\infty \dfrac{1}{(x-1)^3}\, dx = \lim_{b \to \infty} \left. -\dfrac{1}{2(x-1)^2} \right|_3^b$

$= 0 + \dfrac{1}{8} = \dfrac{1}{8}$

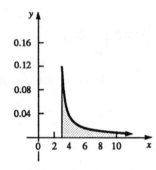

31. Present value $= \int_0^\infty 40,000\, e^{-0.09x}\, dx$

$= \lim_{b \to \infty} \left. -\dfrac{40,000}{0.09} e^{-0.09x} \right|_0^b$

$= \dfrac{-40,000}{0.09}(0 - 1) \approx \$444,444.44$

33. Present value $= \int_0^\infty 20,000 e^{-0.10x}\, dx$

$= \lim_{b \to \infty} \left. -\dfrac{20,000}{0.10} e^{-0.10x} \right|_0^b$

$= -200,000(0 - 1) = \$200,000$

35. Total amount $= \int_0^\infty 8000 e^{-0.04t}\, dt$

$= \lim_{b \to \infty} \left. -\dfrac{8000}{0.04} e^{-0.04t} \right|_0^b$

$= -200,000(0 - 1) = 200,000$ units

37. Total amount $= \int_0^\infty \dfrac{600}{(1+t)^3}\, dt$

$= \lim_{b \to \infty} \left. -\dfrac{600}{2(1+t)^2} \right|_0^b = -300(0 - 1)$

$= 300$ gallons

14-3 Probability Distributions

1. Let $f(x)$ be the probability density function.

$f(x) = \begin{cases} 0.05 & \text{if } 0 \le x \le 20 \\ 0 & \text{otherwise} \end{cases}$

(a) $P(12 \le x \le 20) = \int_{12}^{20} f(x)dx = \int_{12}^{20} 0.05\, dx$

$\qquad = \left. 0.05x \right|_{12}^{20} = 0.40$

(b) $P(0 \le x \le 5) = \int_0^5 0.05\, dx = \left. 0.05x \right|_0^5 = 0.25$

(c) $P(5 \le x \le 10) = \int_5^{10} 0.05\, dx = \left. 0.05x \right|_5^{10} = 0.25$

3. (a) $f(x) = \begin{cases} 0.20 & \text{if } 2 \le x \le 7 \\ 0 & \text{otherwise} \end{cases}$

(b) $P(2 \le x \le 3) = \int_2^3 f(x)dx = \int_2^3 0.20\, dx$

$\qquad = \left. 0.20x \right|_2^3 = 0.20$

(c) $P(4 \le x \le 6) = \int_4^6 0.20\, dx = \left. 0.20x \right|_4^6 = 0.40$

5. $f(x) = 5e^{-5x}$ $(x \geq 0)$

(a) $P(0 \leq x \leq 0.1) = \int_0^{0.1} f(x)dx = \int_0^{0.1} 5e^{-5x}dx$

$= -\frac{5}{5}e^{-5x}\Big|_0^{0.1} = -(e^{-0.5} - e^0)$

$= (1 - e^{-0.5}) \approx 0.3935$

(b) $P(0.2 \leq x \leq 0.6) = \int_{0.2}^{0.6} 5e^{-5x}\,dx = -\frac{5}{5}e^{-5x}\Big|_{0.2}^{0.6}$

$= -(e^{-3} - e^{-1}) = (e^{-1} - e^{-3}) \approx 0.3181$

(c) $P(0.4 \leq x < \infty) = \int_{0.4}^{\infty} 5e^{-5x}$

$= \lim_{b \to \infty} (-1)e^{-5x}\Big|_{0.4}^{b}$

$= -(0 - e^{-2}) \approx 0.1353$

7. $f(x) = 2e^{-2x}$ $(x \geq 0)$

(a) $P(0.5 \leq x \leq 1) = \int_{0.5}^{1} 2e^{-2x}\,dx = -e^{-2x}\Big|_{0.5}^{1}$

$= -(e^{-2} - e^{-1}) = (e^{-1} - e^{-2}) \approx 0.2325$

(b) $P(0 \leq x \leq 1.5) = \int_0^{1.5} 2e^{-2x}\,dx = -e^{-2x}\Big|_0^{1.5}$

$= -(e^{-3} - e^0) = 1 - e^{-3} \approx 0.9502$

(c) $P(x \geq 3) = \int_3^{\infty} 2e^{-2x}\,dx = \lim_{b \to \infty} -e^{-2x}\Big|_3^{b}$

$= -(0 - e^{-6}) \approx 0.00248$

9. $f(x) = 0.5e^{-0.5}x$ $(x \geq 0)$

(a) $R(t) = P(x > t) = \int_t^{\infty} f(x)dx = \int_t^{\infty} 0.5e^{-0.5x}\,dx$

$= \lim_{b \to \infty} (-e^{-0.5x})\Big|_t^{b} = -(0 - e^{-0.5t}) = e^{-0.5t}$

(b) $R(1) = e^{-0.5(1)} \approx 0.6065$
$R(1)$ is the probability that the circuitry will have a lifetime of at least 1 year.

(c) $R(4) = e^{-0.5(4)} \approx 0.1353$
$R(4)$ is the probability that the circuitry will have a lifetime of at least 4 years.

14-4 Applications: Average Value and Volume

1. $f(x) = -3x + 15, 0 \leq x \leq 5$

Average value of $f = \frac{1}{5-0}\int_0^5 f(x)dx$

$= \frac{1}{5}\int_0^5 (-3x + 15)dx = \frac{1}{5}\left[-\frac{3}{2}x^2 + 15x\right]\Big|_0^5$

$= \frac{1}{5}\left[-\frac{75}{2} + 75\right] = 7.5$

3. $f(x) = \sqrt{x}, 1 \leq x \leq 4$

Average value of $f = \frac{1}{4-1}\int_1^4 \sqrt{x}\,dx = \frac{1}{3}\cdot\frac{2}{3}x^{3/2}\Big|_1^4$

$= \frac{2}{9}(8 - 1) = \frac{14}{9} \approx 1.556$

5. $f(x) = -3x^2 + 6, 0 \leq x \leq 1$

Average value of $f = \frac{1}{1-0}\int_0^1 (-3x^2 + 6)dx$

$= -x^3 + 6x\Big|_0^1 = -1 + 6 = 5$

7. $f(x) = -2x + 10, 0 \leq x \leq 5$

Average value of $f = \dfrac{1}{5-0}\displaystyle\int_0^5 (-2x + 10)dx$

$= \dfrac{1}{5}(-x^2 + 10x)\Big|_0^5 = \dfrac{1}{5}(-25 + 50) = 5$

The area of the rectangle of height 5 over

$0 \leq x \leq 5$ is the same as that of $\displaystyle\int_0^5 f(x)dx.$

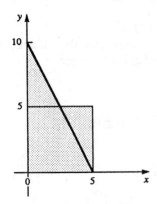

9. $R(x) = -5x^2 + 40, 0 \leq x \leq 2$

Average value of $R = \dfrac{1}{2-0}\displaystyle\int_0^2 (-5x^2 + 40)dx$

$= \dfrac{1}{2}\left[-\dfrac{5}{3}x^3 + 40x \right]\Big|_0^2 = \dfrac{1}{2}\left[-\dfrac{40}{3} + 80 \right] = \dfrac{100}{3} = 33\dfrac{1}{3}$

The area of the rectangle of height $33\dfrac{1}{3}$ over

$0 \leq x \leq 2$ equals $\displaystyle\int_0^2 f(x)dx.$

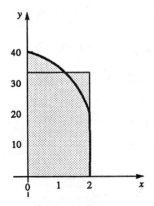

11. $y = f(t) = 150e^{-0.02t} \ (t \geq 0)$

Average value of y over $0 \leq t \leq 5$

$= \dfrac{1}{5-0}\displaystyle\int_0^5 150e^{-0.02t}dt = \dfrac{-150}{5(0.02)}e^{-0.02t}\Big|_0^5$

$= -1500(e^{-0.1} - e^0) \approx 142.74°\text{F}.$

13. $S(t) = 10,000e^{0.08t}$ (continuous compounding)

Average value of S over $0 \leq t \leq 5$

$= \dfrac{1}{5-0}\displaystyle\int_0^5 10,000e^{0.08t}dt = \dfrac{10,000}{5(0.08)}e^{0.08t}\Big|_0^5$

$= 25,000(e^{0.40} - e^0) \approx \$12,295.62$

15. $y = f(x) = mx + b, 0 \leq x \leq -b/m$

Average value of $f = \dfrac{1}{-b/m - 0}\displaystyle\int_0^{-b/m} (mx + b)dx$

$= \dfrac{-m}{b}\left[\dfrac{mx^2}{2} + bx \right]\Big|_0^{-b/m} = \dfrac{-m}{b}\left[\dfrac{b^2}{2m} - \dfrac{b^2}{m} \right] = \dfrac{b}{2}$

17. $y = -4x + 8, 0 \leq x \leq 2$

$V = \pi\displaystyle\int_0^2 [f(x)]^2dx = \pi\displaystyle\int_0^2 (-4x + 8)^2dx$

$= \dfrac{\pi}{(-4)(3)}(-4x + 8)^3\Big|_0^2 = \dfrac{-\pi}{12}(0 - 512)$

$= \dfrac{128\pi}{3} \approx 134.0$

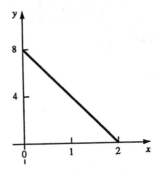

19. $y = -6x^2 + 6,\ 0 \le x \le 1$

$$V = \pi \int_0^1 (-6x^2 + 6)^2 dx = \pi \int (36x^4 - 72x^2 + 36)dx$$

$$= 36\pi \left[\frac{x^5}{5} - \frac{2x^3}{3} + x \right] \Big|_0^1 = 36\pi \left[\frac{1}{5} - \frac{2}{3} + 1 \right]$$

$$= \frac{96\pi}{5} \approx 60.32$$

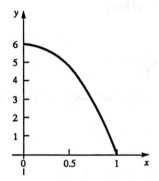

21. $y = e^{-x},\ 0 \le x \le 1$

$$V = \pi \int_0^1 (e^{-x})^2 dx = \frac{-\pi}{2} e^{-2x} \Big|_0^1 = -\frac{\pi}{2}(e^{-2} - e^0)$$

$$= \frac{\pi}{2}(1 - e^{-2}) \approx 1.358$$

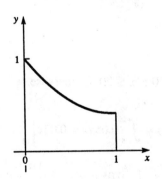

23. $y = \sqrt{x},\ 1 \le x \le 4$

$$V = \pi \int_1^4 (\sqrt{x})^2 dx = \pi \frac{x^2}{2} \Big|_1^4 = \pi \left[\frac{16}{2} - \frac{1}{2} \right]$$

$$= \frac{15\pi}{2} \approx 23.56$$

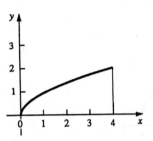

25. $y = \sqrt{4 - x^2},\ -2 \le x \le 2$

$$V = \pi \int_{-2}^2 (\sqrt{4 - x^2})^2 dx = \pi \int_{-2}^2 (4 - x^2)dx$$

$$= \pi \left[4x - \frac{x^3}{3} \right] \Big|_{-2}^2 = \pi \left[\left(8 - \frac{8}{3} \right) - \left(-8 + \frac{8}{3} \right) \right]$$

$$= \frac{32\pi}{3} \approx 33.51$$

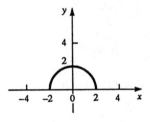

27. $y = \sqrt{16 - x^2}, -4 \le x \le 4$

$$V = \pi \int_{-4}^{4} (\sqrt{16 - x^2})^2 dx = \pi \int_{-4}^{4} (16 - x^2)dx$$

$$= \pi \left[16x - \frac{x^3}{3} \right] \Big|_{-4}^{4} = \pi \left[\left(64 - \frac{64}{3} \right) - \left(-6 + \frac{64}{3} \right) \right]$$

$$= \frac{256}{3}\pi \approx 268.08$$

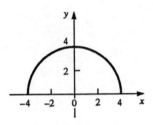

Review Exercises

1. $f(x) = 5000, 0 \le x \le 4$

(a) Total money flow $= \displaystyle\int_{0}^{4} 5000 dx = 5000x \Big|_{0}^{4}$

$= \$20,000$

(b) Present value $= \displaystyle\int_{0}^{4} 5000e^{-0.08x} \, dx$

$$= -\frac{5000}{0.08} e^{-0.08x} \Big|_{0}^{4} = -\frac{5000}{0.08} (e^{-0.32} - e^0)$$

$$= \frac{5000}{0.08} (1 - e^{-0.32}) \approx \$17,115.69$$

(c) Future value $= e^{0.08(4)}$ (Present value)
$= e^{0.32}(17,115.69) \approx \$23,570.49$

3. $f(x) = 2000e^{0.10x}, 0 \le x \le 4$

(a) Total money flow $= \displaystyle\int_{0}^{4} 2000e^{0.10x} dx$

$$= \frac{2000}{0.10} e^{0.10x} \Big|_{0}^{4} = 20,000(e^{0.40} - e^0) \approx \$9836.49$$

(continued)

3. *(continued)*

(b) Present value $= \displaystyle\int_{0}^{4} 2000e^{0.10x}e^{-0.08x}dx$

$$= \int_{0}^{4} 2000e^{0.02x}dx = \frac{2000}{0.02} e^{0.02x} \Big|_{0}^{4}$$

$$= 100,000(e^{0.08} - e^0) \approx \$8328.71$$

(c) Future value $= e^{0.08(4)}(8328.71)$
$= e^{0.32}(8328.71) \approx \$11,469.70$

5. $\displaystyle\int_{1}^{\infty} \frac{1}{x}dx = \lim_{b \to \infty} \ln x \Big|_{1}^{b}$

Since $\lim_{b \to \infty} \ln b$ does not exist, the integral diverges.

7. $\displaystyle\int_{0}^{\infty} e^{-x}dx = \lim_{b \to \infty} -e^{-x} \Big|_{0}^{b} = -(0 - 1) = 1$

9. $\displaystyle\int_{1}^{\infty} \frac{1}{(x + 4)^2}dx = \lim_{b \to \infty} \frac{-1}{x + 4} \Big|_{1}^{b} = -\left[0 - \frac{1}{5} \right] = \frac{1}{5}$

11. Present value $= \displaystyle\int_{0}^{\infty} 100,000e^{-0.08x}dx$

$$= \lim_{b \to \infty} \frac{100,000}{-0.08} e^{-0.08x} \Big|_{0}^{b} = -1,250,000(0 - 1)$$

$$= \$1,250,000$$

13. $f(x) = \dfrac{1}{20} = 0.05$ if $0 \le x \le 20$, 0 otherwise.

(a) $P(15 \le x \le 20) = \displaystyle\int_{15}^{20} 0.05 dx = 0.05x \Big|_{15}^{20} = 0.25$

(b) $P(0 \le x \le 12) = \displaystyle\int_{0}^{12} 0.05 dx = 0.05x \Big|_{0}^{12} = 0.60$

(c) $P(5 \le x \le 17) = \displaystyle\int_{5}^{17} 0.05 dx = 0.05x \Big|_{5}^{17} = 0.60$

(d) $P(6 \le x \le 15) = \displaystyle\int_{6}^{15} 0.05 dx = 0.05x \Big|_{6}^{15} = 0.45$

15. $f(x) = 0.10e^{-0.10x}$ $(x \geq 0)$

(a) $P(0 \leq x \leq 5) = \int_0^5 0.10e^{-0.10x}dx = -e^{-0.10x} \Big|_0^5$

$= -(e^{-0.50} - e^0) = 1 - e^{-0.50} \approx 0.3935$

(b) $P(0 \leq x \leq 12) = -e^{-0.10x} \Big|_0^{12} = -(e^{-1.20} - e^0)$

$= (1 - e^{-1.20}) \approx 0.6988$

(c) $P(8 \leq x < \infty) = \lim_{b \to \infty} (-e^{-0.10x}) \Big|_8^b = -(0 - e^{-0.80})$

$= e^{-0.80} \approx 0.4493$

(d) $P(10 \leq x < \infty) = \lim_{b \to \infty} (-e^{-0.10x}) \Big|_{10}^b$

$= -(0 - e^{-1.0}) = e^{-1} \approx 0.3679$

17. $f(x) = -5x + 20, 0 \leq x \leq 4$

Average value of $f = \dfrac{1}{4 - 0} \int_0^4 (-5x + 20)dx$

$= \dfrac{1}{4}\left[-\dfrac{5}{2}x^2 + 20x \right] \Big|_0^4 = \dfrac{1}{4}(-40 + 80) = 10$

19. $R(x) = -6x^2 + 240x, 0 \leq x \leq 20$

Average value of $R = \dfrac{1}{20 - 0} \int_0^{20} (-6x^2 + 240x)dx$

$= \dfrac{1}{20}(-2x^3 + 120x^2) \Big|_0^{20} = \dfrac{1}{20}(-16,000 + 48,000)$

$= \$1600$

21. $y = 6x^2, 0 \leq x \leq 4$

$V = \pi \int_0^4 (6x^2)^2 dx = 36\pi \int_0^4 x^4 dx = \dfrac{36\pi}{5}x^5 \Big|_0^4$

$= \dfrac{36\pi(1024)}{5} = \dfrac{36,864\pi}{5} \approx 23,162.33$

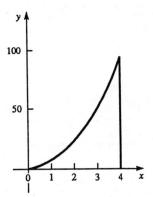

23. $y = -x^2 + 4, 0 \leq x \leq 2$

$V = \pi \int_0^2 (-x^2 + 4)^2 dx = \pi \int_0^2 (x^4 - 8x^2 + 16)^2 dx$

$= \pi\left[\dfrac{x^5}{5} - \dfrac{8}{3}x^3 + 16x \right] \Big|_0^2 = \pi\left[\dfrac{32}{5} - \dfrac{64}{3} + 32 \right]$

$= \dfrac{256\pi}{15} \approx 53.617$

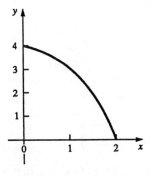

CHAPTER
15
FUNCTIONS OF SEVERAL VARIABLES

15–1 Functions of Several Variables

1. $f(x, y) = 3x^2 - 4y + 6xy + y^2 + 5$;

    (a) $f(0, 1) = 0 - 4 + 0 + 1 + 5 = 2$

    (b) $f(2, 5) = 12 - 20 + 60 + 25 + 5 = 82$

    (c) $f(1, 0) = 3 - 0 + 0 + 0 + 5 = 8$

3. $z = x^3 - \dfrac{y^2}{4}$;

    $z(2, 1) = 8 - \dfrac{1}{4} = \dfrac{31}{4} = 7\dfrac{3}{4}$

5. $z = f(x_1, x_2, x_3) = x_1^2 - 4x_1x_3 + x_3^2 + x_2x_3$;

    (a) z is a function of 3 variables.

    (b) $f(2, 1, -3) = 2^2 - 4(2)(-3) + (-3)^2 + 1(-3)$
        $= 4 + 24 + 9 - 3 = 34$

    (c) $f(-2, -1, 4) = (-2)^2 - 4(-2)(4) + 4^2 + (-1)(4)$
        $= 4 + 32 + 16 - 4 = 48$

7. $g(x_1, x_2, x_3, x_4) = x_1^2 + x_2x_3 + x_3^2x_4^3 + 5$;

    (a) g is a function of 4 variables.

    (b) $g(1, -1, 2, -3) = 1^2 + (-1)(2) + 2^2(-3)^3 + 5$
        $= 1 - 2 - 108 + 5 = -104$

    (c) $g(-1, 0, -3, 4) = (-1)^2 + 0 + (-3)^2(4^3) + 5$
        $= 1 + 576 + 5 = 582$

9. $g(x, y) = 4x^2 + \dfrac{5x^3}{y} + \dfrac{8}{x - 2}$;

    The domain of g is the set of all ordered pairs (x, y) for which the function is defined, namely, all ordered pairs (x, y) such that $x \neq 2$ and $y \neq 0$.

11. $f(x, y) = \dfrac{(x^3 - 4xy + y^2)}{\sqrt{x - 3}}$;

    The domain of f is the set of all ordered pairs (x, y) for which the function is defined, namely, all ordered pairs (x, y) such that $x > 3$.

13. $f(x, y) = \ln\left[\dfrac{x}{y}\right];$

The domain of f is the set of all ordered pairs (x, y) for which the function is defined, namely, all ordered pairs (x, y) such that x and y have the same sign, $x \neq 0$, and $y \neq 0$.

15. $(-1, 5, 2)$

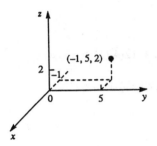

17. $(3, 2, -4)$

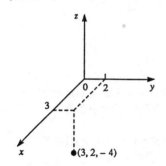

19. $(-2, 1, 8)$

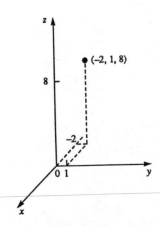

21. (a) $R(x_1, x_2) = x_1 p_1 + x_2 p_2$
$= x_1(90 - 4x_1 - 3x_2) + x_2(60 - 2x_1 - 5x_2)$
$= 90x_1 - 4x_1{}^2 - 5x_1 x_2 + 60x_2 - 5x_2{}^2$

  (b) $R(3, 5) = 90(3) - 4(9) - 5(15) + 60(5) - 5(25)$
$= 334$

23. $C(x, y) = x^2 + 4y^2 - 2xy + 8x + 6y + 3;$
$C(3, 5) = 9 + 4(25) - 2(15) + 24 + 30 + 3$
$= 136$
(or \$136,000)

25. $F(e, i, c) = \sqrt{\dfrac{1 + e}{c(1 + i)}};$

  (a) If $e = 0.15$, $i = 0.10$, and $c = 0.03$

$F(0.15, 0.10, 0.03) = \sqrt{\dfrac{1 + 0.15}{0.03(1 + 0.10)}}$
≈ 5.9033, optimum R–value

  (b) $F(0.10, 0.12, 0.03) = \sqrt{\dfrac{1 + 0.10}{0.03(1 + 0.12)}}$
≈ 5.7217, optimum R–value

15–2 Partial Derivatives

1. $f(x,y) = 3x^2 + 4y^3 + 6xy - x^2 y^3 + 5;$

  (a) $f_x = 6x + 6y - 2xy^3$

  (b) $f_y = 12y^2 + 6x - 3x^2 y^2$

  (c) $f_x(1, -1) = 6 + 6(-1) - 2(1)(-1)^3 = 2$

  (d) $f_y(2, 1) = 12(1) + 6(2) - 3(4)(1) = 12$

3. $z = x^3 + y^5 - 8xy + 2x^3y^2 + 11$;

   (a) $\dfrac{\partial z}{\partial x} = 3x^2 - 8y + 6x^2y^2$

   (b) $\dfrac{\partial z}{\partial y} = 5y^4 - 8x + 4x^3y$

5. $f(x,y) = 8x^3 - 2y^2 - 7x^5y^8 + 8xy^2 + y + 6$

   (a) $f_x(x,y) = 24x^2 - 35x^4y^8 + 8y^2$

   (b) $f_y(x,y) = -4y - 56x^5y^7 + 16xy + 1$

   (c) $f_x(1,2) = 24(1) - 35(1)(256) + 8(4) = -8904$

   (d) $f_y(1,2) = -4(2) - 56(1)(128) + 16(1)(2) + 1$
   $= -7143$

7. $z = 4x^6 - 8x^3 - 7x + 6xy + 8y + x^3y^5$

   (a) $\dfrac{\partial z}{\partial x} = 24x^5 - 24x^2 - 7 + 6y + 3x^2y^5$

   (b) $\dfrac{\partial z}{\partial y} = 6x + 8 + 5x^3y^4$

   (c) $\dfrac{\partial^2 z}{\partial y^2} = 120x^4 - 48x + 6xy^5$

   (d) $\dfrac{\partial^2 z}{\partial y^2} = 20x^3y^3$

   (e) $\dfrac{\partial^2 z}{\partial x \partial y} = 6 + 15x^2y^4$

   (f) $\dfrac{\partial^2 z}{\partial y \partial x} = 6 + 15x^2y^4$

9. $f(x,y) = 1000 - x^3 - y^2 + 4x^3y^6 + 8y$;

   (a) $f_x = -3x^2 + 12x^2y^6$

   (b) $f_y = -2y + 24x^3y^5 + 8$

   (c) $f_{xx} = -6x + 24xy^6$

   (d) $f_{yy} = -2 + 120x^3y^4$

   (e) $f_{xy} = 72x^2y^5$

   (f) $f_{yx} = 72x^2y^5$

   (g) $f_x(2,-1) = -3(4) + 12(4)(1) = 36$

   (h) $f_{yy}(1,3) = -2 + 120(1)(81) = 9718$

11. $z = x^3e^{x^2+y^2}$;

   (a) $\dfrac{\partial z}{\partial x} = x^2e^{x^2+y^2} + (2x)(x^3e^{x^2+y^2}) = x^2e^{x^2+y^2}(3 + 2x^2)$

   (b) $\dfrac{\partial z}{\partial y} = (2y)x^3e^{x^2+y^2} = 2x^3ye^{x^2+y^2}$

13. $f(x,y) = (x^3 + y^2)e^{2x+3y+5}$;

   (a) $f_x(x,y) = 3x^2e^{2x+3y+5} + 2(x^3 + y^2)e^{2x+3y+5}$
   $= e^{2x+3y+5}(3x^2 + 2x^3 + 2y^2)$

   (b) $f_y(x,y) = 2ye^{2x+3y+5} + 3(x^3 + y^2)e^{2x+3y+5}$
   $= e^{2x+3y+5}(3x^3 + 2y + 3y^2)$

   (c) $f_x(0,1) = e^8(2) = 2e^8$

15. $C(x,y) = 40x + 200y + 10xy + 300$;
$$C(50,90) = 40(50) + 200(90) + 10(50)(90) + 300$$
$$= \$65,300$$

(a) $C_x(x,y) = 40 + 10y$
   $C_y(50,90) = 40 + 900 = \$940 =$ the cost of producing one more washer

(b) $C_y(x,y) = 200 + 10x$
   $C_y(50,90) = 200 + 500 = \$700 =$ the cost of producing one more dryer

17. $P(x,y) = 100x^2 + 4y^2 + 2x + 5y + 100,000$;
$$P(90,1000) = 100(90^2) + 4(1000^2) + 2(90)$$
$$+ 5(1000) + 100,000$$
$$= \$4,915,180$$

(a) $P_x{'}(x,y) = 200x + 2$
   $P_x{'}(90,1000) = 200(90) + 2 = \$18,002 \approx$ the increase in $P(x,y)$ if x is increased by 1.

(b) $P_y{'}(x,y) = 8y + 5$
   $P_y{'}(90,1000) = 8(1000) + 5 = \$8005 \approx$ the increase in $P(x,y)$ if y is increased by 1.

15–3 Relative Maxima and Minima (Functions of Two Variables)

1. $z = x^2 + 3y^2 - 10x + 48y + 86$;

$$\frac{\partial z}{\partial x} = 2x - 10; \; 2x - 10 = 0 \Rightarrow x = 5$$

$$\frac{\partial z}{\partial y} = 6y + 48; \; 6y + 48 = 0 \Rightarrow y = -8$$

Critical point: (5,–8)

3. $f(x,y) = -x^2 - y^2 + 4x + 8y + xy + 56$

$$\frac{\partial f}{\partial x} = -2x + 4 + y$$

$$\frac{\partial f}{\partial x} = -2y + 8 + x$$

Set $\frac{\partial f}{\partial x} = 0$ and $\frac{\partial f}{\partial y} = 0$, and solve for x and y.

$$\begin{array}{l} 2x - y = 4 \\ -x + 2y = 8 \end{array} \rightarrow \begin{array}{l} 2x - y = 4 \\ \underline{-2x - 4y = 16} \\ 3y = 20 \end{array} \Rightarrow y = \frac{20}{3}, x = \frac{16}{3}$$

Critical point: $\left[\frac{16}{3}, \frac{20}{3}\right]$

5. $z = 3x^2 + 2y^2 + 24x - 36y + 50$

$$\frac{\partial z}{\partial x} = 6x + 24; \frac{\partial z}{\partial x} = 0 \Rightarrow x = -4$$

$$\frac{\partial x}{\partial y} = 4y - 36; \frac{\partial z}{\partial y} = 0 \Rightarrow y = 9$$

Critical point: (–4, 9)

7. $f(x,y) = -x^2 - 5y^2 + 30x + 20y + 3xy + 8$;

$$\frac{\partial f}{\partial x} = -2x + 30 + 3y$$

$$\frac{\partial f}{\partial y} = -10y + 20 + 3x$$

Set $\frac{\partial f}{\partial x} = 0$ and $\frac{\partial f}{\partial y} = 0$, and solve for x and y.

$$\begin{array}{l} 2x - 3y = 30 \\ -3x + 10y = 20 \end{array} \rightarrow \begin{array}{l} 6x - 9y = 90 \\ \underline{-6x + 20y = 40} \\ 1y = 130 \end{array} \Rightarrow y = \frac{130}{11}, y = \frac{360}{11}$$

Critical point: $\left[\frac{360}{11}, \frac{130}{11}\right]$

9. $f(x,y) = 2x^3 + x^2y + 5x - 36y + 90$;

$\dfrac{\partial f}{\partial x} = 6x^2 + 2xy + 5$

$\dfrac{\partial f}{\partial y} = x^2 - 36$

Set $\dfrac{\partial f}{\partial x} = 0$ and $\dfrac{\partial f}{\partial x} = 0$, and solve for x and y.

$\dfrac{\partial f}{\partial y} = 0 \Rightarrow x^2 = 36 \Rightarrow x = \pm 6$

Let $x = 6$ in $\dfrac{\partial f}{\partial x} = 0$:

$6(36) + 12y + 5 = 0 \Rightarrow y = -\dfrac{221}{12}$

Let $x = -6$ in $\dfrac{\partial f}{\partial x} = 0$:

$6(36) - 12y + 5 = 0 \Rightarrow y = \dfrac{221}{12}$

Critical points: $\left[6, -\dfrac{221}{12}\right]$, $\left[-6, \dfrac{221}{12}\right]$

11. $z = -x^2 + y^3 - 12y + 8x + 80$;

$\dfrac{\partial z}{\partial x} = -2x + 8; \dfrac{\partial z}{\partial x} = 0 \Rightarrow x = 4$

$\dfrac{\partial z}{\partial y} = 3y^2 - 12; \dfrac{\partial z}{\partial y} = 0 \Rightarrow y = \pm 2$

Critical points: $(4, -2)$, $(4, +2)$

13. $z = f(x,y) = x^2 + 2y^2 - 8x - 20y + 18$;

$\dfrac{\partial z}{\partial x} = 2x - 8; \dfrac{\partial z}{\partial x} = 0 \Rightarrow x = 4$

$\dfrac{\partial z}{\partial y} = 4y - 20; \dfrac{\partial z}{\partial y} = 0 \Rightarrow y = 5$

$A = \dfrac{\partial^2 z}{\partial x^2}\bigg|_{(4,5)} = 2; B = \dfrac{\partial^2 z}{\partial y^2}\bigg|_{(4,5)} = 4$;

$C = \dfrac{\partial^2 z}{\partial x \partial y}\bigg|_{(4,5)} = 0$

$AB - C^2 = (2)(4) - 0 = 8$

Since $AB - C^2 > 0$ and $A > 0$, $f(4,5) = -48$ is a relative minimum.

15. $f(x,y) = 9x - 50y + x^2 + 5y^2 + 100$;

$\dfrac{\partial f}{\partial x} = 9 + 2x; \dfrac{\partial f}{\partial x} = 0 \Rightarrow x = -\dfrac{9}{2}$

$\dfrac{\partial f}{\partial y} = -50 + 10y; \dfrac{\partial f}{\partial y} = 0 \Rightarrow y = 5$

$A = \dfrac{\partial^2 f}{\partial x^2}\bigg|_{(-\frac{9}{2}, 5)} = 2; B = \dfrac{\partial^2 f}{\partial y^2}\bigg|_{(-\frac{9}{2}, 5)} = 10$;

$C = \dfrac{\partial^2 f}{\partial x \partial y}\bigg|_{(-\frac{9}{2}, 5)} = 0$

$AB - C^2 = 20$

Since $AB - C^2 > 0$ and $A > 0$, $f\left[-\dfrac{9}{2}, 5\right] = -\dfrac{181}{4}$ is a relative minimum.

17. $f(x,y) = 1000 + 80x + 100y - 2x^2 - y^2$;

$\dfrac{\partial f}{\partial x} = 80 - 4x; \dfrac{\partial f}{\partial x} = 0 \Rightarrow x = 20$

$\dfrac{\partial f}{\partial y} = 100 - 2y; \dfrac{\partial f}{\partial y} = 0 \Rightarrow y = 50$

$A = \dfrac{\partial^2 f}{\partial x^2}\bigg|_{(20,50)} = -4; B = \dfrac{\partial^2 f}{\partial y^2}\bigg|_{(20,50)} = -2$;

$C = \dfrac{\partial^2 f}{\partial x \partial y}\bigg|_{(20,50)} = 0; AB - C^2 = 8$

Since $AB - C^2 > 0$ and $A < 0$, $f(20,50) = 4300$ is a relative maximum.

19. $z = f(x,y) = 3x^2 + 4y^2 + 2xy - 30x - 32y + 50$;

$\dfrac{\partial z}{\partial x} = 6x + 2y - 30$

$\dfrac{\partial z}{\partial y} = 8y + 2x - 32$

Set $\dfrac{\partial z}{\partial x} = 0$ and $\dfrac{\partial z}{\partial y} = 0$, and solve for x and y.

$\begin{array}{l} 3x + y = 15 \\ x + 4y = 16 \end{array} \rightarrow \begin{array}{l} 3x + y = 15 \\ \underline{3x + 12y = 48} \\ 11y = 33 \Rightarrow y = 3, x = 4 \end{array}$

$A = \dfrac{\partial^2 z}{\partial x^2}(4,3) = 6; B = \dfrac{\partial^2 z}{\partial y^2}\bigg|_{(4,3)} = 8$;

$C = \dfrac{\partial^2 z}{\partial x \partial y}\bigg|_{(4,3)} = 0; AB - C^2 = 48$

Since $AB - C^2 > 0$ and $A > 0$, $f(4,3) = -58$ is a relative minimum.

21. $z = f(x,y) = 200 - 2x^2 - 6y^2 + 2xy + 32x + 28y;$

$\dfrac{\partial z}{\partial x} = -4x + 2y + 32;\ \dfrac{\partial z}{\partial y} = -12y + 2x + 28$

Set $\dfrac{\partial z}{\partial x} = 0$ and $\dfrac{\partial z}{\partial y} = 0$, and solve for x and y.

$\begin{aligned} 2x - \ y &= 16 \\ -2x + 12y &= 28 \\ \hline 11y &= 44 \Rightarrow y = 4,\ x = 10 \end{aligned}$

$A = \dfrac{\partial^2 z}{\partial x^2}\bigg|_{(10,4)} = -4;\ B = \dfrac{\partial^2 z}{\partial y^2}\bigg|_{(10,4)} = -12;$

$C = \dfrac{\partial^2 z}{\partial x \partial y}\bigg|_{(10,4)} = 2;\ AB - C^2 = 48 - 4 = 44$

Since $AB - C^2 > 0$ and $A < 0$, $f(10,4) = 416$ is a relative maximum.

23. $z = f(x,y) = 2x^3 + x^2y + 8x - 25y + 800;$

$\dfrac{\partial z}{\partial x} = 6x^2 + 2xy + 8;\ \dfrac{\partial z}{\partial y} = x^2 - 25$

$\dfrac{\partial z}{\partial y} = 0 \Rightarrow x = \pm 5;$

$\dfrac{\partial z}{\partial x}\bigg|_{-5} = 150 - 10y + 8 = 158 - 10y$

$\dfrac{\partial z}{\partial x}\bigg|_{-5} = 0 \Rightarrow y = 15.8$

$\dfrac{\partial z}{\partial x}\bigg|_{5} = 150 + 10y + 8 = 158 + 10y;$

$\dfrac{\partial z}{\partial x}\bigg|_{5} = 0 \Rightarrow y = -15.8$

Critical points: $(-5, 15.8),\ (5, -15.8)$
$f_{xx}(x,y) = 12x + 2y;\ f_{yy}(x,y) = 0$
$f_{xy}(x,y) = 2x$
At $(-5, 15.8)$,
$AB - C^2 = f_{xx}f_{yy} - f^2_{xy}$(evaluated at $(-5, 15.8)$)
$\quad\quad = [12(-5) + 2(15.8)][0] - [2(-5)]^2 = -100$
Since $AB - C^2 < 0$ at $(-5, 15.8)$, the point is a saddle point.
At $(5, -15.8)$, $AB - C^2 = 0 - 100 = -100$.
Hence, $(5, -15.8)$ is a saddle point.
Thus, $(5, -15.8)$ and $(-5, 15.8)$ are saddle points.

25. $Z = f(x,y) = -x^2 + y^3 - 24y + 10x + 95;$

$\dfrac{\partial z}{\partial x} = -2x + 10;\ \dfrac{\partial z}{\partial x} = 0 \Rightarrow x = 5$

$\dfrac{\partial z}{\partial y} = 3y^2 - 24;\ \dfrac{\partial z}{\partial y} = 0 \Rightarrow y = \pm 2\sqrt{2}$

Critical points: $(5, -2\sqrt{2}),\ (5, 2\sqrt{2})$
$f_{xx}(x,y) = -2;\ f_{yy}(x,y) = 6y;\ f_{xy} = 0$
At $(5, -2\sqrt{2})$,

$AB - C^2 = f_{xx}f_{yy} - f^2_{xy}$ (evaluated at $(5, -2\sqrt{2})$)
$\quad\quad\quad + (-2)(-12\sqrt{2}) - 0$
$\quad\quad = 24\sqrt{2}$

Since $AB - C^2 > 0$ and $A < 0$, $f(5, -2\sqrt{2}) \approx 165.25$ is a relative maximum.
At $(5, 2\sqrt{2})$, $AB - C^2 = (-2)(12\sqrt{2}) - 0 = -24\sqrt{2}$
Since $AB - C^2 < 0$, $(5, 2\sqrt{2})$ is a saddle point.

27. $f(x,y) = 500 + x^2 - 2y^2 - 18x + 16y;$

$\dfrac{\partial f}{\partial x} = 2x - 18;\ \dfrac{\partial f}{\partial x} = 0 \Rightarrow x = 9$

$\dfrac{\partial f}{\partial y} = -4y + 16;\ \dfrac{\partial f}{\partial y} = 0 \Rightarrow y = 4$

$A + f_{xx}(9,4) = 2;\ B = f_{yy}(9,4) = -4;\ C = f_{xy}(9,4) = 0$
$AB - C^2 = (2)(-4) - 0 = -8$

Since $AB - C^2 < 0$, $(9,4)$ is a saddle point. There are no relative maxima or minima.

29. $P(x,y) = 1{,}000{,}000 + 1600x + 2000y - 4x^2 - 2y^2$

(a) $\dfrac{\partial P}{\partial x} = 1600 - 8x;\ \dfrac{\partial P}{\partial x} = 0 \Rightarrow x = 200$

$\dfrac{\partial P}{\partial y} = 2000 - 4y;\ \dfrac{\partial P}{\partial y} = 0 \Rightarrow y = 500$

$A = P_{xx}(200,500) = -8;\ B = P_{yy}(200,500) = -4;$
$C = P_{xy}(200,500) = 0;$
$AB - C^2 = (-8)(-4) - 0 = 32$
Since $AB - C^2 > 0$ and $A < 0$, $P(200,500)$ is a relative maximum.

(b) $P(200,500) = \$1{,}660{,}000$

31. $z(x,y) = 1000x + 1600y + 2xy - 5x^2 - 2y^2$

(a) $\dfrac{\partial z}{\partial x} = 1000 + 2y - 10x;\ \dfrac{\partial z}{\partial y} = 1600 + 2x - 4y$

Set $\dfrac{\partial z}{\partial x} = 0$ and $\dfrac{\partial z}{\partial y} = 0$, and solve for x and y.

$$
\begin{array}{rl}
10x - 2y &= 1000 \\
-x + 2y &= 800 \\
\hline
9x\quad\ &= 1800 \Rightarrow x = 200,\ y = 500
\end{array}
$$

$A = z_{xx}(200,500) = -10;\ B + z_{yy}(200,500) = -4;$
$C = z_{xy}(200,500) = 2;\ AB - C^2 = 40 - 4 = 36$
Since $AB - C^2 > 0$ and $A < 0$, $z(200,500)$ is a relative maximum.

(b) $z(200,500) = 500,000$ maximum output

33. $x_1 = 200 - 20p_1 + p_2$
     $x_2 = 300 - 15p_2 + 2p_1$

(a) $R(p_1,p_2) = x_1 p_1 + x_2 p_2$
     $= 200p_1 - 20p_1{}^2 + p_1 p_2 + 300p_2$
     $\quad - 15p_2{}^2 + 2p_1 p_2$
     $= 200p_1 - 20p_1{}^2 + 3p_1 p_2 + 300p_2$
     $\quad - 15p_2{}^2$

(b) $\dfrac{\partial R}{\partial p_1} = 200 - 40p_1 + 3p_2$

     $\dfrac{\partial R}{\partial p_2} = 3p_1 + 300 - 30p_2$

Set $\dfrac{\partial R}{\partial p_1} = 0$ and $\dfrac{\partial R}{\partial p_2} = 0$, and solve for p_1 and p_2.

$$
\begin{array}{ll}
40p_1 - 3p_2 = 200 & 400p_1 - 30p_2 = 2000 \\
-3p_1 + 30p_2 = 300 \rightarrow & \underline{-3p_1 + 30p_2 = 300} \\
& 397p_1 \qquad = 2300 \Rightarrow
\end{array}
$$

$$p_1 = \frac{2300}{397},\ p_2 = \frac{4200}{397}$$

(continued)

33.   *(continued)*

$A = Rp_1p_1\left[\dfrac{2300}{397}, \dfrac{4200}{397}\right] = -40$

$B = Rp_2p_2\left[\dfrac{2300}{397}, \dfrac{4200}{397}\right] = -30$

$C = Rp_1p_2\left[\dfrac{2300}{397}, \dfrac{4200}{397}\right] = 3$

$AB - C^2 = 1200 - 9 = 1191$
Since $AB - C^2 > 0$ and $A < 0$, $R\left[\dfrac{2300}{397}, \dfrac{4200}{397}\right]$
is a relative maximum.

(c) $R\left[\dfrac{2300}{397}, \dfrac{4200}{397}\right] \approx \$2166.25$

(d) $x_1 = 200 - 20\left[\dfrac{2300}{397}\right] + \dfrac{4200}{397} \approx 94.71$ cases

$x_2 = 300 - 15\left[\dfrac{4200}{397}\right] + 2\left[\dfrac{2300}{397}\right]$

≈ 152.90 cases

35. $S(w,l,h) = 2wl + 2wh + 2hl;$

$V = wlh = 200;\ w = \dfrac{200}{lh}$

$S(l,h) = \dfrac{400}{h} + \dfrac{400}{l} + 2hl$

$\dfrac{\partial S}{\partial l} = -\dfrac{400}{l^2} + 2h;\ \dfrac{\partial S}{\partial h} = -\dfrac{400}{h^2} + 2l$

Set $\dfrac{\partial S}{\partial l} = 0$ and $\dfrac{\partial S}{\partial h} = 0$, and solve for h and l.

$h = \dfrac{200}{l^2};\ l - \dfrac{200}{200^2}l^4 = 0;\ l(l^3 - 200) = 0;\ l = \sqrt[3]{200};$

$h = \dfrac{200}{200^{\frac{2}{3}}} = \sqrt[3]{200};\ w = \dfrac{200}{200^{\frac{2}{3}}} = \sqrt[3]{200}$

$S_{ll}S_{hh} - S^2{}_{lh} = \dfrac{800}{l^3}\dfrac{800}{h^3} - 4 > 0.$ Also, $S_{ll} > 0.$

$\therefore\ S(\sqrt[3]{200}, \sqrt[3]{200})$ is a relative minimum.

37. $C(w,l,h) = 10wl + 4(2wh) + 4(2hl)$

$V = wlh = 10,000; \ w = \dfrac{10,000}{lh}$

$C(l,h) = \dfrac{100,000}{h} + \dfrac{80,000}{l} + 8hl$

$\dfrac{\partial C}{\partial l} = \dfrac{-80,000}{l^2} + 8h; \ \dfrac{\partial C}{\partial h} = \dfrac{-100,000}{h^2} + 8l$

Set $\dfrac{\partial C}{\partial l} = 0$ and $\dfrac{\partial C}{\partial h} = 0$, and solve for h and l.

$h = \dfrac{10,000}{l^2}; \ l = \dfrac{12,500}{(10,000)^2}l^4; \ l^3 = 8000; \ l = 20$ feet;

$h = \dfrac{10,000}{400} = 25$ feet; $w = \dfrac{10,000}{(20)(25)} = 20$ feet

$C_{ll} = \dfrac{160,000}{l^3}; \ C_{hh} = \dfrac{200,000}{h^3}; \ C_{lh} = 8$

$C_{ll}C_{hh} - C^2_{lh} > 0$ and $Cll > 0$.

$\therefore \ C(20,20,25)$ is a relative minimum.

15–4 Application: The Method of Least Squares

1.

x	y	x^2	xy	$2.5 + 2.5x$	$y - (2.5 + 2.5x)$
2	8	4	16	7.5	0.5
3	10	9	30	10	0
2	7	4	14	7.5	0.5
5	15	25	75	0	0
12	40	42	135		

$m = \dfrac{\Sigma x_i y_i - n\bar{x}\bar{y}}{\Sigma x_i^2 - n\bar{x}^2} = \dfrac{135 - 4\left[\dfrac{12}{4}\right]\left[\dfrac{40}{4}\right]}{42 - 4\left[\dfrac{12}{4}\right]^2} = 2.5$

$b = \bar{y} - m\bar{x} = \dfrac{40}{4} - (2.5)\left[\dfrac{12}{4}\right] = 2.5$

(a) Regression line: $y = 2.5 + 2.5x$

(continued)

1. *(continued)*

(b)

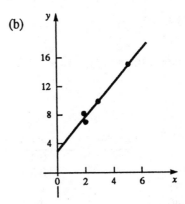

(c) Residuals: –0.5, 0, 0.5, 0

(d) Sum of squares error $= \displaystyle\sum_{i=1}^{4} [y_i - (2.5 + 2.5x_i)]^2$

$= 0.5$

3.

x	y	x^2	xy	$2 + 1.4x$	$y - (2 + 1.4x)$
4	8	16	32	7.6	0.4
2	5	4	10	4.8	0.2
8	14	64	112	13.2	0.8
6	9	36	54	10.4	–1.4
20	36	120	208		

$n = 4, \ \bar{x} = \dfrac{20}{4} = 5, \ \bar{y} = \dfrac{36}{4} = 9$

$m = \dfrac{\Sigma x_i y_i - n\bar{x}\bar{y}}{\Sigma x_i^2 - n\bar{x}^2} = \dfrac{208 - 4(5)(9)}{120 - 4(25)} = 1.4$

$b = \bar{y} - m\bar{x} = 9 - 1.4(5) = 2$

(a) Regression line: $y = 2 + 1.4x$

(b)

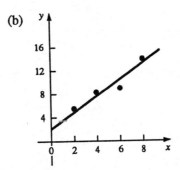

(continued)

3. *(continued)*

(c) Residuals: 0.4, 0.2, 0.8, –1.4

(d) Sum of squares error $= \sum\limits_{i=1}^{4} [y_i - (2 + 1.4x)]^2$

$$= 2.8$$

5.

x	y	x^2	xy	$1.4 + 2.2x$	$y - (1.4 + 2.2x)$
1	4	1	4	3.6	0.4
2	6	4	12	5.8	0.2
3	7	9	21	8	1
4	10	16	40	10.2	0.2
5	13	25	65	12.4	0.6
15	40	55	142		

$n = 5, \bar{x} = \dfrac{15}{5} = 3, \bar{y} = \dfrac{40}{5} = 8$

$m = \dfrac{\Sigma x_i y_i - n\bar{x}\bar{y}}{\Sigma x_i^2 - n\bar{x}^2} = \dfrac{142 - 5(3)(8)}{55 - 5(9)} = 2.2$

$b = \bar{y} - m\bar{x} = 8 - 2.2(3) = 1.4$

(a) Regression line: $y = 1.4 + 2.2x$

(b)
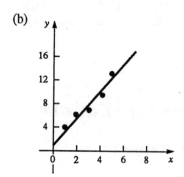

(c) Residuals: 0.4, 0.2, –1.0, 0.2, 0.6

(d) Sum of squares error $= \sum\limits_{i=1}^{4} [y_i - (1.4 + 2.2x_i)]^2$

$$= 1.6$$

7.

x	y	x^2	xy	$7 + 2.6x$	$y - (7 + 2.6x)$
–2	2	4	–4	1.8	0.2
–1	4	1	–4	4.4	0.4
0	7	0	0	7	0
1	10	1	10	9.6	0.4
2	12	4	24	12.2	0.2
0	35	10	26		

$n = 5, \bar{x} = 0, \bar{y} = \dfrac{35}{5} = 7$

$m = \dfrac{\Sigma x_i y_i - n\bar{x}\bar{y}}{\Sigma x_i^2 - n\bar{x}^2} = \dfrac{26 - 5(0)(7)}{10 - 5(0)} = 2.6$

$b = \bar{y} - m\bar{x} = 7 - 2.6(0) = 7$

(a) Regression line: $y = 7 + 2.6x$

(b)
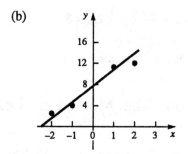

(c) Residuals: 0.2, –0.4, 0, 0.4, –0.2

(d) Sum of squares error $= \sum\limits_{i=1}^{5} [y_i - (7 + 2.6x_i)]^2$

$$= 0.4$$

9. $y = 0.74 + 2.29x$

(a) At $x = 3.5$, $y = 0.74 + 2.29(3.5) = 8.76$, estimated value.

(b) Sum of squares error $= \sum\limits_{i=1}^{5} (Y_i - y_i)^2$

$= (7 - 5.3110)^2 + (9 - 9.8841)^2$
$+ (13 - 14.4573)^2 + (13 - 14.4573)^2$
$+ (16 - 19.0305)^2 + (25 - 21.3171)^2$
$= 37.6043$
$= (1.6890)^2 + (-0.8841)^2 + (-1.4573)^2$
$+ (-3.0305)^2 + (3.6829)^2$
$= 28.5058$

11. $y = -2.78 + 3.82x$

 (a) At $x = 3.5$, $y = -2.78 + 3.82(3.5) = 10.59$, estimated value.

 (b) Sum of squares error $= \sum\limits_{i=1}^{5} (Y_i - \hat{Y}_i)^2$

 $\begin{aligned} &= (6 - 4.8660)^2 + (9 - 8.6907)^2 \\ &\quad + (19 - 23.9897)^2 + (28 - 27.8144)^2 \\ &\quad + (35 - 31.6392)^2 \\ &= (1.1340)^2 + (0.3093)^2 + (-4.9897)^2 \\ &\quad + (0.1856)^2 + (3.3608)^2 \\ &= 37.6043 \end{aligned}$

13. (a) The quadratic model has a lower sum of squares error and, therefore, fits the set of data better.

 (b) $\ln y = 1.89 + 0.593x$
 $\begin{aligned} y &= e^{1.89} e^{0.593x} = e^{1.89}(e^{0.593})^x \\ &\approx 6.619(1.809)^x \text{ or } \approx 6.619e^{0.593x} \end{aligned}$

 (c) At $x = 4.5$, $y \approx 6.619(1.809)^{4.5} \approx 95.43$, exponential model

 (d) At $x = 4.5$,
 $y \approx 4.71(4.5)^2 + 2.76(4.5) + 0.05 \approx 107.85$

15. (a) The fixed cost is $y(0) = \$2211$

 (b) The variable cost per hour $= \$0.841 =$ slope of linear regression line.

17. The beta of the mutual shares $= 0.655$, the slope of the regression line. This means that a change of 1 unit in the S&P 500 results in a change, in the same direction, of approximately 0.655 unit in the mutual shares.

19. The beta of the Neuberger Manhattan fund is 1.29, the slope of the regression line. This means that a change of 1 unit in the S&P 500 is accompanied by a change, in the same direction, of approximately 1.29 units in the Neuberger Manhattan fund.

21. $\ln y = 5.77 + 0.0785x$

 (a) $\begin{aligned} y &= e^{5.77} e^{0.0785x} = e^{5.77}(e^{0.0785})^x \\ &\approx 320.5(1.082)^x \text{ or } \approx 320.5e^{0.0785x} \end{aligned}$

 (b) $r = e^{0.0785} - 1 \approx 0.0817$, or 8.17%.

 (c)

year	x	y
1987	33	4274.8
1988	34	4623.9
1989	35	5001.5

23. (a) The quadratic model fits the data better, since the sum of squares error is less than that of the linear model.

 (b)

year	x	y (linear model)
1987	12	439.6
1988	13	460.4
1989	14	481.2

 (c) According to the quadratic model, U.S. exports are decreasing.

 (d)

year	x	y (quadratic model)
1987	12	345.9
1988	13	319.9
1989	14	286.7

25. (a)

x	y	x^2	xy
2.5	15.5	6.25	38.75
3.0	19.7	9.0	59.1
2.4	15.2	5.76	36.48
3.5	22.8	12.25	79.8
2.1	14.6	4.41	30.66
2.6	15.8	6.76	41.08
16.1	103.6	44.43	285.87

 $n = 6$, $\bar{x} = \dfrac{16.1}{6} = 2.6833$, $\bar{y} = \dfrac{103.6}{6} = 17.2667$

 $\begin{aligned} m &= \dfrac{\Sigma x_i y_i - n\bar{x}\bar{y}}{\Sigma x_i^2 - n\bar{x}^2} \\ &= \dfrac{285.87 - 6(2.6833)(17.2667)}{44.43 - 6(2.6833)^2} \approx 6.4149 \end{aligned}$

(continued)

25. *(continued)*

$$b = \bar{y} - m\bar{x} = 17.2667 - 6.4149(2.6833)$$
$$\approx 0.0535$$
Regression line: $y = 0.0535 + 6.4149x$

(b) An increase of 1 unit in the GPA leads to an increase of $6,415 in the annual starting salary.

(c) At $x = 2.9$,
$y = 0.0535 + 6.4149(2.9) \approx \$18.657$.

27. (a)   If x_2 is held constant and x_1 increases by 1 degree Fahrenheit, the crop yield will decrease by approximately 7.8 million bushels.

(b)   If x_1 is held constant and x_2 increases by 1 percentage point, the crop yield will increase by approximately 4.3 million bushels.

15–5 Lagrange Multipliers

1. $f(x,y) = x^2 - 4xy + y^2 + 200$
subject to $g(x,y) = 2x + y = 26$
Let $F(x,y,\lambda) = x^2 - 4xy + y^2 + 200 + \lambda(26 - 2x - y)$

(1) $\dfrac{\partial F}{\partial x} = 2x - 4y - 2\lambda$

(2) $\dfrac{\partial F}{\partial y} = -4x + 2y - \lambda$

(3) $\dfrac{\partial F}{\partial \lambda} = 26 - 2x - y$

Set the partial derivatives equal to zero, and solve for x and y.
From (1) and (2), $\lambda = x - 2y = -4x + 2y$
$$\Rightarrow 5x = 4y \Rightarrow y = \frac{5}{4}x$$

Substitute into (3): $26 - 2x - \dfrac{5}{4}x = 0 \Rightarrow \dfrac{13}{4}x = 26$;
$$x = 8, y = 10$$

Calculate $\begin{vmatrix} 0 & g_x & g_y \\ g_x & f_{xx} & f_{xy} \\ g_y & f_{yx} & f_{yy} \end{vmatrix} = \begin{vmatrix} 0 & 2 & 1 \\ 2 & 2 & -4 \\ 1 & -4 & 2 \end{vmatrix} = -26$

Since $|H| < 0$, $f(8,10) = 44$ is a relative minimum.

3. $f(x,y) = x^2 + 6xy + 2y^2$
subject to $g(x,y) = 4x + y = 18$
Let $F(x,y,\lambda) = x^2 + 6xy + 2y^2 + \lambda(18 - 4x - y)$

(1) $\dfrac{\partial F}{\partial x} = 2x + 6y - 4\lambda$

(2) $\dfrac{\partial F}{\partial y} = 6x + 4y - \lambda$

(3) $\dfrac{\partial F}{\partial \lambda} = 18 - 4x - y$

Set the partial derivatives equal to zero, and solve for x and y.
From (1) and (2), $\lambda = \dfrac{1}{2}x + \dfrac{3}{2}y = 6x + 4y$
$$\Rightarrow \frac{11}{2}x = -\frac{5}{2}y; \quad y = -\frac{11}{5}x$$

Substitute into (3): $18 - 4x + \dfrac{11}{5}x = 0 \Rightarrow \dfrac{9}{5}x = 18$;
$$x = 10, y = -22$$

Calculate $|H| = \begin{vmatrix} 0 & g_x & g_y \\ g_x & f_{xx} & f_{xy} \\ g_y & f_{yx} & f_{yy} \end{vmatrix} = \begin{vmatrix} 0 & 4 & 1 \\ 4 & 2 & 6 \\ 1 & 6 & 4 \end{vmatrix} = -18$

Since $|H| < 0$, $f(10,-22) = -252$ is a relative minimum.

5. $f(x,y) = x^2 + 2xy + y^2 - 3x - 5y$
subject to $g(x,y) = x + y = 18$
Let $F(x,y,\lambda) = x^2 + 2xy + y^2 - 3x - 5y + \lambda(18 - x - y)$

(1) $\dfrac{\partial F}{\partial x} = 2x + 2y - 3 - \lambda$

(2) $\dfrac{\partial F}{\partial y} = 2x + 2y - 5 - \lambda$

(3) $\dfrac{\partial F}{\partial \lambda} = 18 - x - y$

Set the partial derivatives equal to zero, and solve for x and y.
From (1) and (2), $2x + 2y - 3 = 2x + 2y - 5$
$$\Rightarrow -3 = -5$$
This conclusion is not possible. Hence there is no relative maximum or minimum.
We can also obtain this result by substituting $y = 18 - x$ into the expression for $f(x,y)$ to get
$h(x) = x^2 + 2x(18 - x) + (18 - x)^2 - 3x - 5(18 - x)$
$$= 2x + 234$$
$h'(x) = 2 \Rightarrow$ no relative maximum or minimum.
h is unbounded as x increases or decreases.

7. $f(x,y) = x^2 - 5xy + 2y^2$
   subject to $g(x,y) = 3x + y = 20$
   Let $F(x,y,\lambda) = x^2 - 5xy + 2y^2 + \lambda(20 - 3x - y)$

   (1) $\dfrac{\partial F}{\partial x} = 2x - 5y - 3\lambda$

   (2) $\dfrac{\partial F}{\partial y} = -5x + 4y - \lambda$

   (3) $\dfrac{\partial F}{\partial \lambda} = 20 - 3x - y$

   Set the partial derivatives equal to zero, and solve for x and y.
   From (1) and (2), $2x - 5y = -15x + 12y \Rightarrow x = y$
   Substitute into (3): $20 - 3x - x = 0 \Rightarrow x = 5,\ y = 5$

   Calculate $|H| = \begin{vmatrix} 0 & g_x & g_y \\ g_x & f_{xx} & f_{xy} \\ g_y & f_{yx} & f_{yy} \end{vmatrix} = \begin{vmatrix} 0 & 3 & 1 \\ 3 & 2 & -5 \\ 1 & -5 & 4 \end{vmatrix}$

   $= -68$

   Since $|H| < 0$, $f(5,5) = -50$ is a relative minimum.

9. In Exercise 2, $\lambda_0 = 2x + 6y$, evaluated at $(5,5)$, $= 40$. λ_0 measures the sensitivity of the optimal value of f to a change in C, where $g(x,y) = C = 10$ in this case. An increase of 1 unit in C increases the optimal value of f by approximately 40 units.

11. In Exercise 4, λ_0 evaluated at $x_0 = \dfrac{47}{6}$.

    $\lambda_0 = 4x + 1 = \dfrac{97}{3} \approx 32.33$

    λ_0 measures the sensitivity of the optimal value of f to a change in C, where $g(x,y) = C = 12$ in this case. An increase of 1 unit in C increases the optimal value of f by approximately 32.33 units.

13. In Exercise 7, $\lambda_0 = -5x + 4y$, evaluated at $(5,5)$, $= -5$. λ_0 measures the sensitivity of the optimal value of f to a change in C, where $g(x,y) = C = 20$ in this case. An increase of 1 unit in C decreases the optimal value of f by approximately 5 units.

15. $C(x,y) = x^2 + 4y^2 - 5xy + 2000$
    subject to $g(x,y) = x + y = 40$
    Let $F(x,y,\lambda) = x^2 + 4y^2 - 5xy + 2000$
    $\qquad\qquad\qquad + \lambda(40 - x - y)$

    (1) $\dfrac{\partial F}{\partial x} = 2x - 5y - \lambda$

    (2) $\dfrac{\partial F}{\partial y} = 8y - 5x - \lambda$

    (3) $\dfrac{\partial F}{\partial \lambda} = 40 - x - y$

    Set the partial derivatives equal to zero, and solve for x and y.

    From (1) and (2), $2x - 5y = 8y - 5x \Rightarrow y = \dfrac{7}{13}x$

    Substitute into (3):

    $40 - x - \dfrac{7}{13}x = 0 \Rightarrow x = 26,\ y = 14$

    Calculate $|H| = \begin{vmatrix} 0 & g_x & g_y \\ g_x & C_{xx} & C_{xy} \\ g_y & C_{yx} & C_{yy} \end{vmatrix} = \begin{vmatrix} 0 & 1 & 1 \\ 1 & 2 & -5 \\ 1 & -5 & 8 \end{vmatrix}$

    $= -20$

    Since $|H| < 0$, $C(26,14) = 1640$ is a relative minimum.

17. $A(u,v) = uv$ is to be maximized subject to

    $g(u,v) = v^2 + \left[\dfrac{\sqrt{2}}{2} + u\right]^2 = 1$, i.e., $x^2 + y^2 = 1$

    Let $F(u,v,\lambda) = uv + \lambda\left[1 - v^2 - \dfrac{1}{2} - \sqrt{2}u - u^2\right]$

    $\qquad\qquad = uv + \lambda\left[\dfrac{1}{2} - \sqrt{2}u - u^2 - v^2\right]$

    (1) $\dfrac{\partial F}{\partial u} = V - \lambda(\sqrt{2} + 2u)$

    (2) $\dfrac{\partial F}{\partial v} = u - \lambda(2v)$

    (3) $\dfrac{\partial F}{\partial \lambda} = \dfrac{1}{2} - \sqrt{2}u - u^2 - v^2$

    Set the partial derivatives equal to zero, and solve for u and v.

    (continued)

320   Chapter 15   Functions of Several Variables

17. *(continued)*

From (1) and (2), $\lambda = \dfrac{v}{2u + \sqrt{2}} = \dfrac{u}{2v}$

$2v^2 = 2u^2 + \sqrt{2}u$, or $v^2 = u^2 + \dfrac{\sqrt{2}}{2}u$

Substitute into (3): $\dfrac{1}{2} - \sqrt{2}u - u^2 - u^2 - \dfrac{\sqrt{2}}{2}u = 0$

$4u^2 + 3\sqrt{2}u - 1 = 0$;

$u = \dfrac{-3\sqrt{2} + \sqrt{18 + 16}}{8} = \dfrac{\sqrt{34} - 3\sqrt{2}}{8} \approx 0.1985$

$v^2 = u^2 + \dfrac{\sqrt{2}}{2}u = \dfrac{34 - 6\sqrt{68} + 18}{64} + \dfrac{\sqrt{68} - 6}{16}$

$= \dfrac{7 - \sqrt{17}}{16} \approx 0.1798$

$v = \dfrac{1}{4}\sqrt{7 - \sqrt{17}} \approx 0.4240$

Calculate $|H| = \begin{vmatrix} 0 & g_u & g_v \\ g_u & f_{uu} & f_{uv} \\ g_v & f_{vu} & f_{vv} \end{vmatrix}$

$= \begin{vmatrix} 0 & 2\left[\dfrac{\sqrt{2}}{2} + u\right] & 2v \\ 2\left[\dfrac{\sqrt{2}}{2} + u\right] & 0 & 1 \\ 2v & 1 & 0 \end{vmatrix}$

$= 8v\left[\dfrac{\sqrt{2}}{2} + u\right] > 0$

Since $|H| > 0$, $A(0.1985, 0.4240) \approx 0.08416$ is a relative maximum.

19. Maximize $V(x,y) = x^2y$
subject to $C(x,y) = (1)(4xy) + 3x^2 = 60$
Let $F(x, y, \lambda) = x^2y + \lambda(60 - 4xy - 3x^2)$

(1) $\dfrac{\partial F}{\partial x} = 2xy - \lambda(4y + 6x)$

(2) $\dfrac{\partial F}{\partial y} = x^2 - 4\lambda x$

(3) $\dfrac{\partial F}{\partial \lambda} = 60 - 4xy - 3x^2$

Set the partial derivatives equal to zero, and solve for x and y.

(continued)

19. *(continued)*

From (1) and (2), $2xy = \lambda(4y + 6x)$ and $x^2 = 4\lambda x$.

For $x \neq 0$, $\lambda = \dfrac{x}{4}$; $2xy = xy + \dfrac{3}{2}x^2$; $y = \dfrac{3}{2}x$.

Substitute into (3): $60 - 6x^2 - 3x^2 = 0$;

$x = \sqrt{\dfrac{20}{3}} = 2\sqrt{\dfrac{5}{3}} = \dfrac{2}{3}\sqrt{15}$; $y = \sqrt{15}$

Calculate $|H| = \begin{vmatrix} 0 & C_x & C_y \\ C_x & V_{xx} & V_{xy} \\ C_y & V_{yx} & V_{yy} \end{vmatrix}$,

evaluated at $\left[\dfrac{2}{3}\sqrt{15}, \sqrt{15}\right]$

We need $C_x = 4y + 6x = 8\sqrt{15}$

$C_y = 4x = \dfrac{8}{3}\sqrt{15}$

$V_{xx} = 2y = 2\sqrt{15}$

$V_{xy} = 2x = \dfrac{4}{3}\sqrt{15}$

$V_{yy} = 0$

$|H| = \begin{vmatrix} 0 & 8\sqrt{15} & \dfrac{8}{3}\sqrt{15} \\ 8\sqrt{15} & 2\sqrt{15} & \dfrac{4}{3}\sqrt{15} \\ \dfrac{8}{3}\sqrt{15} & \dfrac{4}{3}\sqrt{15} & 0 \end{vmatrix} = 640\sqrt{15}$

Since $|H| > 0$, $V\left[\dfrac{2}{3}\sqrt{15}, \sqrt{15}\right] = \dfrac{20}{3}\sqrt{15} \approx 25.82$ is a relative maximum.
Therefore, $x \approx 2.58$ ft, $y \approx 3.87$ ft
Maximum volume ≈ 25.82 cu ft

21. Minimize $f(x,y) = x^2 + y^2$,
subject to $g(x,y) = 4x + 3y = 24$
Let $F(x,y,\lambda) = x^2 + y^2 + \lambda(24 - 4x - 3y)$

(1) $\dfrac{\partial F}{\partial x} = 2x - 4\lambda$

(2) $\dfrac{\partial F}{\partial y} = 2y - 3\lambda$

(3) $\dfrac{\partial F}{\partial \lambda} = 24 - 4x - 3y$

(continued)

21. *(continued)*

Set the partial derivatives equal to zero, and solve for x and y.

From (1) and (2), $\dfrac{x}{2} = \dfrac{2y}{3} \Rightarrow y = \dfrac{3x}{4}$

Substitute into (3): $24 - 4x - \dfrac{9x}{4} = 0$

$$\Rightarrow x = \frac{96}{25}, \ y = \frac{72}{25}$$

Calculate $|H| = \begin{vmatrix} 0 & g_x & g_y \\ g_x & f_{xx} & f_{xy} \\ g_y & f_{yx} & f_{yy} \end{vmatrix} = \begin{vmatrix} 0 & 4 & 3 \\ 4 & 2 & 0 \\ 3 & 0 & 2 \end{vmatrix}$

$$= -50$$

Since $|H| < 0$, $f\left(\dfrac{96}{25}, \dfrac{72}{25}\right) = 23.04$ is a relative minimum.

The minimum distance $= \sqrt{23.04} \approx 4.8$ miles.

23. $C(x,y) = 80 + \dfrac{x^2}{8} + 100 + \dfrac{y^2}{4}$

subject to $g(x,y) = x + y = 9000$

Let $F(x,y,\lambda) = 180 + \dfrac{x^2}{8} + \dfrac{y^2}{4} + \lambda(9000 - x - y)$

(1) $\dfrac{\partial F}{\partial x} = \dfrac{x}{4} - \lambda$

(2) $\dfrac{\partial F}{\partial y} = \dfrac{y}{2} - \lambda$

(3) $\dfrac{\partial F}{\partial \lambda} = 9000 - x - y$

Set the partial derivatives equal to zero, and solve for x and y.

From (1) and (2), $\dfrac{x}{4} = \dfrac{y}{2} \Rightarrow y = \dfrac{x}{2}$

Substitute into (3): $9000 - x - \dfrac{x}{2} = 0$

$$\Rightarrow x = 6000, \ y = 3000$$

(continued)

23. *(continued)*

Calculate $|H| = \begin{vmatrix} 0 & g_x & g_y \\ g_x & C_{xx} & C_{xy} \\ g_y & C_{yx} & C_{yy} \end{vmatrix} = \begin{vmatrix} 0 & 1 & 1 \\ 1 & 1/4 & 0 \\ 1 & 0 & 1/2 \end{vmatrix}$

$$= -\frac{3}{4}$$

Since $|H| < 0$, $f(6000,3000) = 6,740,180$ is a relative minimum.

25. In Exercise 15, $\lambda_0 = 2x - 5y = 2(26) - 5(14) = -18$, evaluated at $(26,14)$
λ_0 measures the sensitivity of the optimal value of C to a change in C, where $g(x,y) = C = 40$ in this case. An increase of 1 unit in C decreases the optimal value of C by approximately 18 units.

27. In Exercise 17, $\lambda_0 = \dfrac{4u}{v} \approx \dfrac{4(0.1985)}{0.4240} = 1.8726$,

evaluated at $(0.1985, 0.4240)$.
λ_0 measures the sensitivity of the optimal value of A to a change in C, where $g(u,v) = C = 1$ in this case. An increase of 1 unit in C increases the optimal value of A by approximately 1.8726 unit. (An increase of 1 unit in C is large, relative to C, and the approximation is, consequently, not good. It would be more reasonable to increase C by 0.01. Such an increase would increase the optimal value of A by approximately 0.0187 unit.)

29. In Exercise 19,

$\lambda_0 = (2xy - 6x)/(4y) = \dfrac{1}{2}$ evaluated at $(2,6)$

$g(x,y) = C = 60$
λ_0 measures the sensitivity of the optimal value of f to a change in C, where $g(x,y) = C = 60$ in this case. An increase of 1 unit in C increases the optimal value of f by approximately $\dfrac{1}{2}$ unit.

31. In Exercise 21, $\lambda_0 = \frac{x_0}{2} \approx 0.2$. λ_0 measures the sensitivity of the optimal value of f to a change in C, where $g(x,y) = C = 24$ in this case. An increase of 1 unit in C increases the optimal value of f by approximately 1.92 unit.

33. In Exercise 23, $\lambda_0 = \frac{x_0}{4} = 1500$. λ_0 measures the sensitivity of the optimal value of C to a change in C, where $g(x,y) = C = 9000$ in this case. An increase of 1 unit in C increases the optimal value of C by approximately 1500 units.

EXTRA DIVIDENDS – The Back-Order Inventory Model

Given: Annual demand $D = 1200$ units
Ordering cost $K = \$5.00$/unit
Carrying cost $H = \$1.15$/unit
Annual stockout cost $B = \$2.40$/unit

Annual inventory cost:

$$C(Q,S) = K\frac{D}{Q} + H\frac{(Q - S)^2}{2Q} + B\frac{S^2}{2Q},$$

where Q = number of units ordered/order
S = maximum number of back orders allowed

1. $C(Q,S) = 5\left[\dfrac{1200}{Q}\right] + \dfrac{1.15(Q - S)}{2Q} + \dfrac{2.4S^2}{2Q}$

$= \dfrac{6000}{Q} + \dfrac{0.575(Q - S)^2}{Q} + \dfrac{1.2S^2}{Q}$

$= \dfrac{1}{Q}(6000 + 0.575Q^2 - 1.15QS + 1.775S^2)$

3. Minimum total annual inventory cost = $C(124, 40)$

$= \dfrac{1}{124}[5000 + 0.575(124)^2 - 1.15(124)(40)$

$\quad + 1.775(40)^2] \approx \$96.59$

5. Units ordered per order = $Q^* = 124$
Units available for sale = $124 - 40 = 84$

EXTRA DIVIDENDS – Using a Response Surface to Increase Industrial Productivity

1. $z(x_1, x_2) = 50 + 1.6x_1 - 1.4x_2 + 2x_1x_2 - 3x_1^2 - 2x_2^2$

$\dfrac{\partial z}{\partial x_1} = 1.6 + 2x_2 - 6x_1$

$\dfrac{\partial z}{\partial x_2} = -1.4 + 2x_1 - 4x_2$

Set the partial derivative equal to zero, and solve for x_1 and x_2.

$\begin{array}{l} 6x_1 - 2x_2 = 1.6 \\ 2x_1 - 4x_2 = 1.5 \end{array} \Rightarrow \begin{array}{l} 12x_1 - 4x_2 = 3.2 \\ \underline{2x_1 - 4x_2 = 1.4} \\ 10x_1 = 1.8 \end{array}$

$x_1 = 0.18$, $x_2 = -0.26$ are the values that maximize the yield, z.

3. $x_1 = \dfrac{C - 110}{10}$

$\Rightarrow C = 110 + 10x_1 = 110 + 10(0.18) = 111.8°C$

$x_2 = \dfrac{T - 55}{5}$

$\Rightarrow T = 55 + 5x_2 = 55 + 5(-0.26) = 53.7$ minutes

Review Exercises

1. $f(x,y) = 2x^2 + 7xy^3 - 6y^4 + 9$

(a) $f(1,-1) = 2 - 7 - 6 + 9 = -2$

(b) $f(-1,0) = 2 + 9 = 11$

(c) $f(2,1) = 8 + 14 - 6 + 9 = 25$

(d) $f(0,-2) = -6(16) + 9 = -87$

3. $f(x,y) = \dfrac{x^3 + 6xy^2 + 7x - 6}{x - 4}$

The domain of f is all ordered pairs (x,y) of real numbers such that $x \neq 4$.

5.

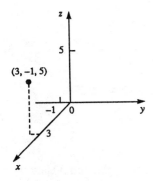

(3, –1, 5)

7. $f(x,y) = 2x^2 + 6xy - 8y^4 + 5x + 9$

(a) $f_x(x,y) = 4x + 6y + 5$

(b) $f_y(x,y) = 6x - 32y^3$

(c) $f_x(1,0) = 4 + 0 + 5 = 9$

(d) $f_y(-1,2) = -6 - 32(8) = -262$

9. $f(x,y) = x^5 e^{x+y}$

(a) $f_x(x,y) = 5x^4 e^{x+y} + x^5 e^{x+y} = x^4 e^{x+y}(5 + x)$

(b) $f_y(x,y) = x^5 e^{x+y}$

(c) $f_x(1,0) = (1)(e)(6) = 6e$

(d) $f_y(1,2) = (1)e^3 = e^3$

11. $C(x,y) = 80x + 50y + 40xy + 500$

(a) $C_x(x,y) = 80 + 40y$
    $C_x(30,40) = 80 + 1600 = 1680$

(b) $C_y(x,y) = 50 + 40x$
    $C_y(30,40) = 50 + 100 = 1250$

13. $f(x,y) = x^4 e^{x+y}$

(a) $f_x = 4x^3 e^{x+y} + x^4 e^{x+y} = x^3 e^{x+y}(4 + x)$

(b) $f_{xx} = 12x^2 e^{x+y} + 4x^3 e^{x+y} + 4x^3 e^{x+y} + x^4 e^{x+y}$
    $= x^2 e^{x+y}(12 + 8x + x^2)$

(c) $f_y = x^4 e^{x+y}$

(d) $f_{yy} = x^4 e^{x+y}$

(e) $f_{xy} = x^3 e^{x+y}(4 + x)$

(f) $f_{yx} = 4x^3 e^{x+y} + x^4 e^{x+y} = x^3 e^{x+y}(4 + x)$

(g) $f_{xx}(-1,3) = (-1)^2 e^{-1+3}(12 + 8(-1) + (-1)^2) = 5e^2$

(h) $f_{yy}(1,-2) = 1e^{1-2} = e^{-1}$

(i) $f_{xy}(1,0) = (1)(e)(5) = 5e$

15. $f(x,y) = (x^3 + y^5)^4$

(a) $f_x = 12x^2(x^3 + y^5)^3$

(b) $f_{xx} = 24x(x^3 + y^5)^3 + 108x^4(x^3 + y^5)^2$

(c) $f_y = 20y^4(x^3 + y^5)^3$

(d) $f_{yy} = 300y^8(x^3 + y^5)^2 + 80y^3(x^3 + y^5)^3$

(e) $f_{xy} = 180x^2 y^4(x^3 + y^5)^2$

(f) $f_{yx} = 180x^2 y^4(x^3 + y^5)^2$

(g) $f_{xx}(-1,3) = 24(-1)(-1 + 243)^3 + 108(1)(242)^2$
    $= -3.338148 \times 10^8$

(h) $f_{yy}(1,-2) = 300(-2)^8(1 - 32)^2 + 80(-8)(-31)^3$
    $= 92{,}871{,}040$

(i) $f_{xy}(1,0) = 0$

17. $f(x,y) = -x^2 - 2y^2 + 8xy + 32x + 12y + 60$

$$\frac{\partial f}{\partial x} = -2x + 8y + 32$$

$$\frac{\partial f}{\partial y} = -4y + 8x + 12$$

Set $\frac{\partial f}{\partial x}$ and $\frac{\partial f}{\partial y}$ equal to zero.

$$-2x + 8y = -32$$
$$\underline{2x - \ \ y = \ -3}$$
$$7y = -35 \Rightarrow y = -5, x = -4$$

Compute $AB - C^2$, where $A = \frac{\partial^2 f}{\partial x^2}$, $B = \frac{\partial^2 f}{\partial y^2}$, and $C = \frac{\partial^2 f}{\partial x \partial y}$, all evaluated at $(-4,-5)$.

$$\frac{\partial^2 f}{\partial x^2} = -2; \ \frac{\partial^2 f}{\partial y^2} = -4; \ \frac{\partial^2 f}{\partial x \partial y} = 8$$

$AB - C^2 = 8 - 64 = -56$
Since $AB - C^2 < 0$, $(-4,-5)$ is a saddle point of f.

19. $f(x,y) = -2x^3 - 8y^2 + 24xy + 12$

$$\frac{\partial f}{\partial x} = -6x^2 + 24y$$

$$\frac{\partial f}{\partial x} = -16y + 24x$$

$$\frac{\partial f}{\partial y} = 0 \Rightarrow y = \frac{3}{2}x;$$

substitute into $\frac{\partial f}{\partial x} = 0$ to get $x^2 = 6x \Rightarrow x = 0, 6$

Critical points: $(0,0)$, $(6,9)$

Compute $AB - C^2$, where $A + \frac{\partial^2 f}{\partial x^2}$, $B = \frac{\partial^2 f}{\partial y^2}$, and $C = \frac{\partial^2 f}{\partial x \partial y}$, evaluated at each critical point.

$$\frac{\partial^2 f}{\partial x^2} = -12x; \ \frac{\partial^2 f}{\partial y^2} = -16; \ \frac{\partial^2 f}{\partial x \partial y} = 24$$

At $(0,0)$, $AB - C^2 = -576 < 0 \Rightarrow (0,0)$ is a saddle point of f.
At $(6,9)$, $AB - C^2 = (-72)(-16) - 24^2 = 576$.
Since $AB - C^2 > 0$ and $f_{xx}(6,9) < 0$, $f(6,9) = 228$ is a relative maximum.

21. $R(x,y) = -10x^2 - 16y^2 + 48x + 36y$

$$\frac{\partial R}{\partial x} = -20x + 48; \quad \frac{\partial R}{\partial x} = 0 \Rightarrow x = 2.4$$

$$\frac{\partial R}{\partial y} = -32y + 36 \quad \frac{\partial R}{\partial y} = 0 \Rightarrow y = 1.125$$

Critical point: $(2.4, 1.125)$

Compute $AB - C^2$, where $A + \frac{\partial^2 R}{\partial x^2}$, $B = \frac{\partial R}{\partial y^2}$, and $C = \frac{\partial^2 R}{\partial x \partial y}$, evaluated at $(2.4, 1.125)$

$A = -20$; $B = -32$; $C = 0$
$AB - C^2 = 640$
Since $AB - C^2 > 0$ and $A < 0$,
$R(2.4, 1.125) = 77.85$ is a relative maximum.

23. $x_1 = 400 - 12p_1 + 14p_2$
$x_2 = 600 - 15p_2 + 6p_1$

(a) $R(p_1,p_2) = p_1 x_1 + p_2 x_2$
$\quad = 400p_1 - 12p_1^2 + 14p_1 p_2$
$\quad \quad + 600p_2 - 15p_2^2 + 6p_1 p_2$
$\quad = 400 \, p_1 - 12p_1^2 + 20p_1 p_2$
$\quad \quad + 600p_2 - 15p_2^2$

(b) $\frac{\partial R}{\partial p_1} = 400 - 24p_1 + 20p_2$

$$\frac{\partial R}{\partial p_2} = 20p_1 + 600 - 30p_2$$

Set $\frac{\partial R}{\partial p_1}$ and $\frac{\partial R}{\partial p_2}$ equal to zero, and solve for p_1 and p_2.

$$\begin{array}{c} -6p_1 + 5p_2 = -100 \\ \underline{2p_1 - 3p_2 = \ -60} \end{array} \rightarrow \begin{array}{c} -6p_1 + 5p_2 = -100 \\ \underline{6p_1 - 9p_2 = -180} \\ -4p_2 = -280 \\ p_2 = \quad 70 \end{array}$$

$$p_1 = -30 + \frac{3}{2} \ p_2 = 75$$

Critical point: $(75,70)$

(continued)

23. *(continued)*

Compute $AB - C^2$, where $A = \dfrac{\partial^2 R}{\partial p_1{}^2}$, $B = \dfrac{\partial^2 R}{\partial p_2{}^2}$,

and $C = \dfrac{\partial^2 R}{\partial p_1 \partial p_2}$, evaluated at (70,50).

$A = -24$; $B = -30$; $C = 20$
$AB - C^2 = 720 - 400 = 320$
Since $AB - C^2 > 0$ and $A < 0$,
$R(75,70) = 36{,}000$ is a relative maximum.

(c) $R(75,70) = \$36{,}000$

(d) $x_1 = 400 - 12(75) + 14(70) = 480$
$x_2 = 600 - 15(70) + 6(75) = 0$

25. (a)

x	y	x^2	xy
4	9	16	36
8	20	64	160
3	8	9	24
9	23	81	207
24	60	170	427

$n = 4,\ \bar{x} = \dfrac{24}{4} = 6,\ \bar{y} = \dfrac{60}{4} = 15$

$m = \dfrac{\Sigma x_i y_i - n\bar{x}\bar{y}}{\Sigma x_i{}^2 - n\bar{x}^2} = \dfrac{427 - 4(6)(15)}{170 - 4(36)} \approx 2.5769$

$b = \bar{y} - m\bar{x} = 15 - (2.5769)(6) \approx -0.4615$
Regression line: $y = -0.4615 + 2.5769x$

(b)

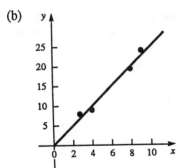

(c)

x	y	y least squares	Residual
4	9	9.8461	-0.8461
8	20	20.1537	-0.1537
3	8	7.2692	0.7308
9	23	22.7306	0.2694

(d) Sum of squares error $= \displaystyle\sum_{i=1}^{4} \text{residual}_i{}^2 \approx 1.346$

(e) At $x = 5$, $y = -0.4615 + 2.5769(5) \approx 12.42$

27. $y = 1.90x - 1.05$
The slope of the line is the beta value of the *XYZ* Mutual Fund. A change of 1 unit in the S&P 500 is accompanied by a change, in the same direction, of approximately 1.90 units in the *XYZ* fund.

29.

x	y	$\ln y$
1	130	4.8675
2	30	3.4012
3	4	1.3863
6		9.6550

$y = ab^x$, or $\ln y = \ln a + x \ln b$

$n = 3,\ \bar{x} = \dfrac{6}{3} = 2,\ \overline{\ln y} = \dfrac{9.655}{3} \approx 3.2183$

$m = \ln b = \dfrac{\Sigma x_i \ln y_i - n\bar{x}\,\overline{\ln y}}{\Sigma x_i{}^2 - n\bar{x}^3}$

$= \dfrac{15.8288 - 3(2)(3.2183)}{14 - 3(4)} \approx -1.7405$

$\ln a = \overline{\ln y} - m\bar{x} = 3.2183 - (-1.7405)(2)$
$ \approx 6.6993$
$\ln y = 6.6993 - 1.7405x$, or
$y = e^{6.6993}e^{-1.7405x} \approx 811.8(0.1754)^x$, or $811.8e^{-1.7405x}$
At $x = 1.5$, predicted value of $y = 59.65$

31. $f(x,y) = x^2 + 8xy + y^2 + 600$
subject to $x + 2y = 66$
Let $F(x,y,\lambda) = x^2 + 8xy + y^2 + 600 + \lambda(66 - x - 2y)$

(1) $\dfrac{\partial F}{\partial x} = 2x + 8y - \lambda$

(2) $\dfrac{\partial F}{\partial y} = 8x + 2y - 2\lambda$

(3) $\dfrac{\partial F}{\partial \lambda} = 66 - x - 2y$

Set the partial derivatives equal to zero, and solve for x and y.

From (1) and (2), $2x + 8y = 4x + y \Rightarrow y = \dfrac{2}{7}x$

(continued)

31. *(continued)*

Substitute into (3): $66 - x - \dfrac{4}{7}x = 0$

$$\Rightarrow x = 42, \ y = 12$$

Calculate $|H| = \begin{vmatrix} 0 & g_x & g_y \\ g_x & f_{xx} & f_{xy} \\ g_y & f_{yx} & f_{yy} \end{vmatrix} = \begin{vmatrix} 0 & 1 & 2 \\ 1 & 2 & 8 \\ 2 & 8 & 2 \end{vmatrix} = 22$

Since $|H| > 0$, $f(42,12) = 6540$ is a relative maximum.

33. In Exercise 30, $\lambda_0 = 2x_0 - 10y_0 = 56 - 240 = -184$
λ_0 measures the sensitivity of the optimal values of f to a change in C, where $g(x,y) = C = 52$ in this case. An increase of 1 unit in C decreases the optimal value of f by approximately 184 units.

35. In Exercise 32, $\lambda_0 = 2x_0 + 6y_0 = 2(14) + 6(10) = 88$
λ_0 measures the sensitivity of the optimal value of f to a change in C, where $g(x,y) = C = 24$ in this case. An increase of 1 unit in C increases the optimal value of f by approximately 88 units.

37. $V(x,y) = x^2y$
subject to $g(x,y) =$
$C(x,y) = 2(5x^2) + 4(10xy) = 10x^2 + 40xy = 480$
Let $F(x,y,\lambda) = x^2y + \lambda(480 - 10x^2 - 40xy)$

(1) $\dfrac{\partial F}{\partial x} = 2xy - 20\lambda x - 40\lambda y$

(2) $\dfrac{\partial F}{\partial y} = x^2 - 40\lambda x$

(3) $\dfrac{\partial F}{\partial \lambda} = 480 - 10x^2 - 40xy$

Set the partial derivatives equal to zero, and solve for x and y.
From (1) and (2), $xy = 10\lambda x + 20\lambda y$ and $x^2 = 40\lambda x$

For $x \neq 0$ and $\lambda \neq 0$, this becomes

$\lambda = \dfrac{x}{40}$ and $y = \dfrac{x}{4} + \dfrac{y}{2} \rightarrow y = \dfrac{x}{2}$

37. *(continued)*

Substitute into (3): $480 - 10x^2 - 20x^2 = 0$

$$30x^2 = 480$$
$$x = 4$$
$$y = \frac{x}{2} = 2$$

$\Rightarrow x = 4, \ y = 2$

Calculate $|H| = \begin{vmatrix} 0 & g_x & g_y \\ g_x & C_{xx} & C_{xy} \\ g_y & C_{yx} & C_{yy} \end{vmatrix}$

$\lambda = \dfrac{x}{40} = \dfrac{1}{10}$

$g_x = 20x + 40y = 160$
$g_y = 40x = 160$
$f_{xx} = 2y - 20\lambda = 2$
$f_{xy} = 2x - 40\lambda = 4$
$f_{yx} = 2x - 40\lambda = 4$
$f_{yy} = 0$
(evaluated at (4,2))
Evaluate $|H|$ at (4,2):

$\begin{vmatrix} 0 & 160 & 160 \\ 160 & 2 & 4 \\ 160 & 4 & 0 \end{vmatrix} = 1{,}536{,}000$

Since $|H| > 0$, $f(4,2)$ is a relative minimum.

$\lambda_0 = \dfrac{x_0}{40} = \dfrac{1}{10} = 0.10$. λ_0 measures the sensitivity of the optimal value of C to a change in C, where $g(x,y) = C = 480$ in this case. An increase of 1 unit in C increases the optimal value of C by approximately 0.10 units.